Elementary Functions with Coordinate Geometry

Elementary Functions with Coordinate Geometry

Earl W. Swokowski
MARQUETTE UNIVERSITY

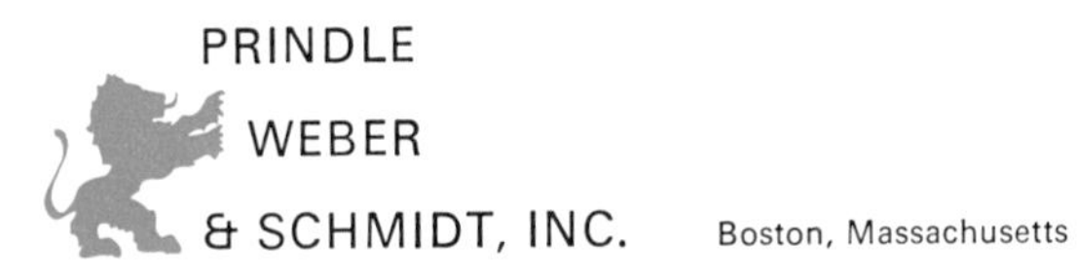

Library of Congress
Catalog Number 76-155292

SBN 87150-117-1

Printed in U.S.A.

Fifth Printing: December, 1975

Preface

This text provides the mathematical prerequisites for calculus. The topics discussed correspond closely to those recommended for precalculus students by the Committee on the Undergraduate Program of the Mathematical Association of America. Moreover, in developing these topics, the author has been guided by the manner in which they are used in the study of limits, differentiation, and integration.

The general notions of functions and their graphs are presented in Chapter One. This is followed, in Chapter Two, by a discussion of polynomials. Chapter Three contains results on continuity and extrema of functions. The approach used is graphical and no attempt is made (or should be made) to be rigorous. Instead, the objective is to have the student acquire an intuitive grasp of several concepts which are fundamental in more advanced studies. The material in Chapter Four on exponential and logarithmic functions is presented in such a way that computational aspects may be omitted without affecting the remainder of the book. The subject matter of trigonometry appears in Chapters Five and Six. Trigonometric functions of real numbers are defined first; however, the more classical descriptions in terms of angles and ratios are brought in shortly afterward. By blending these two approaches, instead of concentrating on one of them, the student should arrive at a deeper understanding of this important class of functions. Equations for straight lines and the conic sections are derived in Chapter Seven. Finally, Chapter Eight contains a brief introduction to coordinate geometry in three dimensions.

An ample supply of graded exercises is included to help the student learn the material. Exercise sets are designed so that by working either the odd-numbered or the even-numbered exercises, practice on all parts of the theory is obtained. There is a review section at the end of each chapter consisting of a list of important topics together with pertinent exercises. Also included at the end of each chapter is a section entitled *Supplementary Questions*. The word "supplementary" should be taken literally, since the questions contain ideas which are not required elsewhere in the text and hence may be omitted if so desired. These questions can be used in various ways. Thus, the instructor may wish to assign or discuss some of them in order to reinforce or extend concepts developed previously. They may also be used to inaugurate special projects for students who desire, and have the ability, to go beyond the subject matter presented in the text. As an illustration, several of the supplementary questions in Chapter Four are concerned with hyperbolic functions. This topic would be ideal for any of the uses just mentioned. Answers to the odd-numbered exercises appear at the end of the text. An answer booklet is available for the even-numbered exercises and the supplementary questions.

More than enough material is included for a one-semester course. Because of this, the text may be used in a variety of ways. Thus, for students who will proceed to a course in calculus *and* coordinate geometry, it is sufficient to concentrate on the first six chapters. However, all topics can be covered provided little emphasis is placed on numerical computations. A number of sections are oriented toward computation or manipulation and are suitable for problem sessions or quiz periods. Included in this category are the following: the material on graphs of equations and the algebra of functions in Chapter One; the sections on graphs of polynomials and synthetic division in Chapter Two; the last four sections of Chapter Four; the work on solutions of triangles in Chapters Five and Six; Sections 1, 2, and 8 of Chapter Six; and the sections on translation and rotation of axes in Chapter Seven. In addition, the subject matter which appears in Sections 2, 3, and 4 of Chapter Three is not required in subsequent chapters and may be omitted. In general, the student who masters the concepts discussed in Chapters One, Two, Four, Five, and Six will be well prepared for the calculus.

Earl W. Swokowski

Contents

chapter one Functions

chapter two Polynomials

chapter three Additional Topics on Functions

chapter four Exponential and Logarithmic Functions

chapter five The Trigonometric Functions

chapter six Analytic Trigonometry

chapter seven Coordinate Geometry in Two Dimensions

chapter eight Coordinate Geometry in Three Dimensions

Elementary Functions with Coordinate Geometry

chapter one Functions

There is much agreement that the most useful concept in mathematics is that of *function.* Indeed, it is safe to say that without the notion of function, little progress can be made in mathematics or in any of the scientific areas as we know them today. Prerequisites for the study of functions are presented in the first four sections of this chapter. The last four sections contain a discussion of functions and their graphs.

1 SETS

Throughout mathematics and other areas the concept of *set* is used extensively. A set may be thought of as a collection of objects of some type. Thus one might speak of the set of books in a library, the set of giraffes in a zoo, the set of natural numbers $1, 2, 3, \cdots$, the set of points in a plane, and so on. The objects in a given set are called the *elements* of the set. We assume that every set is *well defined* in the sense that there is some rule or property that can be used to determine whether a given object is or is not an element of the set.

Notationally, capital letters $A, B, C, R, S, \cdots$ will be used to denote sets, whereas lower-case letters $a, b, x, y, \cdots$ will represent elements of sets. If S is a set, then the symbol $a \in S$ denotes the fact that a is an element of S. Similarly, $a, b \in S$ means that a and b are elements of S. The notation $a \notin S$ signifies that a is not an element of S.

If every element of a set S is also an element of a set T, then S is called a *subset* of T and we write $S \subseteq T$, or $T \supseteq S$, which may be read "S is contained in T" or "T contains S." For example, if T is the set of letters in the English alphabet and if S is the set of vowels, then $S \subseteq T$.

It is important to note that for every set S we have $S \subseteq S$, since every element of S is an element of S. The symbol $S \nsubseteq T$ means that S is not a subset of T. In this case there is at least one element of S which is not an element of T.

We say that two sets S and T are *equal*, and write $S = T$, provided S and T contain precisely the same elements. This is equivalent to saying that $S \subseteq T$ and also $T \subseteq S$. If S and T are not equal, then we write $S \neq T$. If $S \subseteq T$ and $S \neq T$, then S is called a *proper* subset of T. In this case there exists at least one element of T which is not an element of S.

The notation $a = b$, translated "a equals b," means that a and b are symbols which represent the same element. For example, in arithmetic the symbol $2 + 3$ represents the same number as the symbol $4 + 1$ and hence we write $2 + 3 = 4 + 1$. Similarly, $a = b = c$ means that a, b, and c all represent the same element. Of course, $a \neq b$ means that a and b represent different elements. We assume that equality of elements of a set S satisfies the following three properties: (i) $a = a$ for all $a \in S$, (ii) if $a = b$, then $b = a$, and (iii) if $a = b$ and $b = c$, then $a = c$.

There are various ways of describing sets. One method, especially adapted for sets containing only a few elements, is to list all the elements within braces. For example, if S consists of the first five letters of the alphabet, we write $S = \{a, b, c, d, e\}$. When sets are given in this way, the order used in listing the elements is considered irrelevant. We could also write $S = \{a, c, b, e, d\}$, or $S = \{d, c, b, e, a\}$, and so on. This notation is also useful for describing larger sets when there is some definite pattern for the elements. As an illustration, we might specify the set $\mathbf{N}$ of *natural numbers* by writing

$$\mathbf{N} = \{1, 2, 3, 4, \cdots\},$$

where the dots may be read "and so on."

There is another convenient method for describing sets. If S consists of a set of elements from some set T, each of which has a certain property, denoted by p, then we write

$$S = \{x \in T \mid x \text{ has property } p\}$$

or, if the set T from which the elements are chosen is clear, then we merely write

$$S = \{x \mid x \text{ has property } p\}.$$

To translate this notation, for the braces read "the set of" and for the vertical bar read "such that." As a specific example, let $S = \{x \in \mathbf{N} \mid x + 2 = 7\}$. This can be read "$S$ is the set of elements x in $\mathbf{N}$ such that $x + 2 = 7$." Hence S contains only one element, the number 5. As another illustration, if $E = \{x \in \mathbf{N} \mid x \text{ is even}\}$, then E consists of the collection of all even natural numbers, that is, $S = \{2, 4, 6, 8, \cdots\}$.

Another way of describing E is to write $E = \{2n \mid n \in \mathbf{N}\}$. The set F of *odd* natural numbers can be denoted by $F = \{2n - 1 \mid n \in \mathbf{N}\}$. For example, by substituting the natural numbers 1, 2, 3, 4 for n, we obtain the elements 1, 3, 5, 7 of F.

One must distinguish between an element a of a set and the *set* consisting of the element a. Thus if $S = \{a, b, c\}$, then $a \in S$ *but not* $\{a\} \in S$. On the other hand, if we wish to discuss the *subset* of S consisting of the element a, we use the notation $\{a\}$ and write $\{a\} \subseteq S$.

The *empty set* $\varnothing$ is sometimes defined by $\varnothing = \{x \mid x \neq x\}$. The set $\varnothing$ differs from all other sets because it contains no elements. It is mainly a notational device we find convenient to use in certain instances. For example, if $S = \{x \in \mathbf{N} \mid x + 2 = 1\}$, then $S = \varnothing$, since $x + 2$ is never 1 when $x \in \mathbf{N}$. It is customary to assume that $\varnothing$ is a subset of every set S.

Let us list the subsets of the set $S = \{a, b, c\}$. There are 8 subsets in all. They are $\{a\}$, $\{b\}$, $\{c\}$, $\{a, c\}$, $\{a, b\}$, $\{b, c\}$, $\{a, b, c\}$, and $\varnothing$. In Exercise 3 the student is asked to list the 16 subsets of a set which contains 4 elements. A set of 10 elements has 1024 subsets. For obvious reasons we shall not attempt to list them!

When working with several sets, we assume that they are subsets of some larger set U, called a universal set. However, U will not always be given explicitly. If elements $x, y, z, \cdots$ are employed, without specifying any particular set, we assume they belong to some universal set U. With these remarks in mind, we state the following two definitions.

(1.1) Definition of Union

The *union* of two sets A and B, denoted by $A \cup B$, is the set $\{x \mid x \in A \text{ or } x \in B\}$.

The word "or" in this definition and generally throughout mathematics means that either $x \in A$ or $x \in B$, or possibly that x is in *both* A and B.

(1.2) Definition of Intersection

The *intersection* of two sets A and B, denoted by $A \cap B$, is the set $\{x \mid x \in A \text{ and } x \in B\}$.

Thus the intersection of two sets consists of the elements which are *common* to the sets. If $A \cap B = \varnothing$, that is, if A and B have no elements in common, then A and B are said to be *disjoint*.

EXAMPLE. If $A = \{a, b, c, d\}$, $B = \{b, c, e, f\}$ and $C = \{a, d\}$ find $A \cup B$, $A \cap B$, $A \cup C$, $A \cap C$, and $B \cap C$.

Solution: By (1.1) and (1.2) we have $A \cup B = \{a, b, c, d, e, f\}$, $A \cap B = \{b, c\}$, $A \cup C = A$, $A \cap C = C$, and $B \cap C = \varnothing$.

Sometimes sets are pictured by drawing circles, squares, or other simple closed curves in a plane, where it is understood that the points within these figures represent the elements of the sets. Thus if A and B are subsets of a set U, we might indicate this as in Fig. 1.1. Unions and

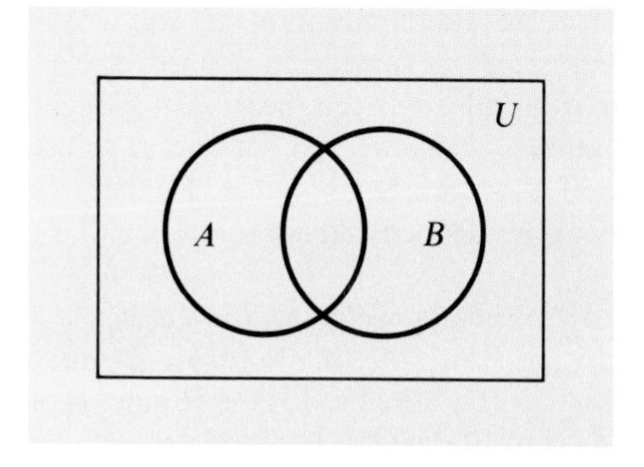

Figure 1.1

intersections can then be represented by shading appropriate parts of the figure. This is illustrated in Fig. 1.2, where we have deleted from our

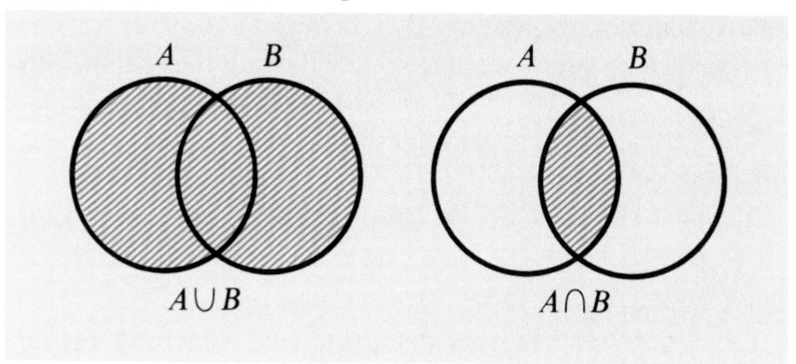

Figure 1.2

picture the universal set U. The reader should also sketch $A \cup B$ and $A \cap B$ when $A \subseteq B$, $B \subseteq A$, or $A \cap B = \varnothing$. It is important to realize that diagrams such as these are used merely to help motivate and visualize notions concerning sets and are not used to prove or serve as steps in proofs of any theorems.

Unions and intersections of more than two sets can also be considered. For example, if A, B, and C are sets, we may consider the union of $A \cup B$ and C. Thus $(A \cup B) \cup C = \{x \mid x \in A \cup B \text{ or } x \in C\}$, that is, $(A \cup B) \cup C$ is the set of all elements x which appear in at least one of the sets A, B, or C. On the other hand, first forming $B \cup C$, we could consider $A \cup (B \cup C)$. Evidently $(A \cup B) \cup C = A \cup (B \cup C)$. Similarly, one can show that $(A \cap B) \cap C = A \cap (B \cap C)$.

EXERCISES

1. Characterize each of the following as true or false, and give reasons for your answers.

   (a) $\{1, 3\} \subseteq \{3, 2, 1\}$. (b) $\{1, 2\} \subseteq \{1, 3\}$.
   (c) $\{1, 2\} = \{2, 1\}$. (d) $\{1, 2\} \subseteq \{1, 2\}$.
   (e) $\{2\} \in \{1, 2\}$. (f) $2 \in \{1, 2\}$.
   (g) $2 \subseteq \{1, 2\}$. (h) $\{2\} \subseteq 2$.

2. Determine all the subsets of the set $W = \{a, b\}$.

3. Find 16 different subsets of the set $S = \{a, b, c, d\}$.

4. Let $R = \{4, 7, 1, 5, 2\}$, $S = \{3, 5, 2\}$, $V = \{7, 2, 5\}$, and $P = \{3, 6\}$. Determine:

   (a) $R \cup S$. (b) $R \cup V$.
   (c) $R \cap S$. (d) $R \cap V$.
   (e) $R \cap P$. (f) $R \cup P$.
   (g) $(S \cup V) \cap P$. (h) $S \cup (V \cap P)$.
   (i) $(S \cup P) \cap (R \cup V)$.

5. Let S and T be sets. Under what conditions will the following be true?

   (a) $S \cup T = S$. (b) $S \cap T = S$.
   (c) $S \cap \varnothing = S$. (d) $S \cup \varnothing = S$.
   (e) $S \cap T = \varnothing$. (f) $S \cup T = \varnothing$.
   (g) $S \cap T = T \cap S$. (h) $S \cup T = T \cup S$.

6. Let sets A and B be represented by overlapping circles as in Fig. 1.1. Shade the part of U which corresponds to each of the following sets.

   (a) $\{x \in U \mid x \in A \text{ and } x \notin B\}$.
   (b) $\{x \in U \mid x \notin A \cup B\}$.
   (c) $\{x \in U \mid x \in A \cup B \text{ and } x \notin A \cap B\}$.

7. Use two circles in a plane to represent two sets A and B. Shade the regions which represent $A \cup B$ and $A \cap B$ if:

   (a) $A \subseteq B$. (b) $B \subseteq A$. (c) $A = B$. (d) $A \cap B = \varnothing$.

8. Use three overlapping circles in a plane to represent sets A, B, and C. Shade the region which represents each of the following.

   (a) $(A \cap B) \cap C$. (b) $A \cap (B \cap C)$.
   (c) $(A \cup B) \cap C$. (d) $A \cup (B \cap C)$.
   (e) $(A \cap B) \cup C$. (f) $A \cap (B \cup C)$.
   (g) $(A \cap B) \cup (A \cap C)$. (h) $(A \cup B) \cap (A \cup C)$.

2 NUMBER SYSTEMS

The set used most frequently in mathematics is the set **R** of real numbers. We refer to **R**, together with the various properties possessed by its elements, as the *real number system.* The reader is undoubtedly well acquainted with symbols such as 2, $-3/5$, $\sqrt{3}$, 0, -8.614, $0.3333\cdots$, etc. which are used to denote elements of **R**. In this section we shall list some properties of **R** and review the notation and terminology associated with real numbers.

The system **R** is *closed* relative to operations of addition (denoted by "+") and multiplication (denoted by "·"). Thus, for every $a, b \in \mathbf{R}$, there corresponds a unique element $a + b$ called the *sum* of a and b and a unique element $a \cdot b$ (sometimes written ab) called the *product* of a and b. These operations have the following properties, where all lower case letters denote arbitrary elements of **R**.

(1.3) Commutative Laws

$$a + b = b + a, \qquad ab = ba.$$

(1.4) Associative Laws

$$a + (b + c) = (a + b) + c, \qquad a(bc) = (ab)c.$$

(1.5) Distributive Laws

$$a(b + c) = ab + ac, \qquad (a + b)c = ac + bc.$$

(1.6) Identity Elements

There exist special real numbers, denoted by 0 and 1 with the following properties:

$$a + 0 = a = 0 + a, \qquad a \cdot 1 = a = 1 \cdot a.$$

(1.7) Inverse Elements

For every real number a, there is a real number denoted by $-a$ such that $a + (-a) = 0 = (-a) + a$. For every real number $a \neq 0$, there is a real number denoted by $1/a$ such that $a(1/a) = 1 = (1/a)a$.

A set which satisfies the above properties is referred to as a *field.* For this reason, (1.3)–(1.7) are sometimes called the *field properties* of the real number system.

The special real numbers 0 and 1 are referred to as *zero* and *one*, respectively. We call $-a$ the *negative* of a and $1/a$ the *reciprocal* of a. The symbol a^{-1} is often used in place of $1/a$.

Many properties of **R** can be derived from (1.3)–(1.7). For example, one can prove the *cancellation laws*, which state that if $a + c = b + c$, then $a = b$, and if $ac = bc$, where $c \neq 0$, then $a = b$. One can also show that $a \cdot 0 = 0 = 0 \cdot a$ for every $a \in \mathbf{R}$. Moreover, if $ab = 0$, then either

$a = 0$ or $b = 0$. The following rules involving negatives may be established: $-(-a) = a$, $(-a)b = -ab = a(-b)$, $(-a)(-b) = ab$ and $(-1)a = -a$.

If $a, b \in \mathbf{R}$, then the operation of *subtraction* (denoted by "$-$") is defined by $a - b = a + (-b)$. If $b \neq 0$, then *division* (denoted by "$\div$") is defined by $a \div b = a(1/b)$. The symbol a/b is often used in place of $a \div b$, and we refer to it as the *quotient of a by b* or the *fraction a over b*. The numbers a and b are called the *numerator* and *denominator*, respectively, of the fraction. It is important to note that a/b is not defined if $b = 0$, that is, *division by zero is not permissible in* $\mathbf{R}$. The following rules for quotients may be established, where all denominators are nonzero real numbers.

$$a/b = c/d \quad \text{if and only if} \quad ad = bc.$$
$$(ad)/(bd) = a/b.$$
$$(a/b) + (c/d) = (ad + bc)/(bd).$$
$$(a/b)(c/d) = (ac)/(bd).$$

If we begin with the real number 1 and successively add it to itself, we obtain the set $\mathbf{N}$ of *positive integers* (also called the *natural numbers*):

$$\{1, 1 + 1, 1 + 1 + 1, 1 + 1 + 1 + 1, \cdots\}.$$

As usual, this set is specified by writing

$$\mathbf{N} = \{1, 2, 3, 4, \cdots\}.$$

The negatives $-1, -2, -3, -4, \cdots$ of the positive integers are referred to as *negative integers*. The set $\mathbf{Z}$ of *integers* is the totality of positive and negative integers together with the real number 0 — that is,

$$\mathbf{Z} = \{\cdots, -3, -2, -1, 0, 1, 2, 3, \cdots\}.$$

A real number is called a *rational number* if it can be written in the form a/b, where a and b are integers and $b \neq 0$. Real numbers that are not rational are called *irrational*. The ratio of the circumference of a circle to its diameter is irrational. This real number, which is denoted by π, is often approximated by the decimal 3.1416 or by the rational number 22/7. We use the notation $\pi \doteq 3.1416$ to indicate that π is *approximately equal* to 3.1416.

Real numbers may be represented by *infinite decimals*. In order to obtain such a representation for rational numbers, the process of long division may be used. Thus, by long division, the decimal representation for the rational number 7434/2310 is found to be $3.2181818\cdots$, where the three dots indicate that the digits 1 and 8 repeat indefinitely. Such an expression is referred to as an *infinite repeating decimal*. Every rational number has associated with it an infinite repeating decimal and, conversely, given such a decimal it is possible to find a rational number which has, as its decimal representation, the given infinite repeating decimal. Decimal representations for irrational numbers may also be obtained; however, they are always *infinite nonrepeating*. For numerical applications, infinite decimals are approximated by finite decimals, where the number of decimal places is determined by the degree of accuracy which is desired. We sometimes write $\pi = 3.14159 \cdots$ to indicate that π can be approximated to any degree of accuracy, even though the approximation beyond the fifth decimal place is not specified.

An important subset of the real numbers is the collection of *positive real numbers*, as characterized by the following properties:

(i) If a and b are positive, then so are $a + b$ and ab.
(ii) If $a \neq 0$, then either a is positive or $-a$ is positive (but not both).

The positive integers are examples of positive real numbers, as is every rational number a/b when a and b are both positive or both negative. The nonzero real numbers which are not positive are called *negative* real numbers. The negative integers or rational numbers such as $(-2)/3$, $13/(-18)$, etc. are examples of negative real numbers. The real number 0 is considered *neither* positive nor negative.

If $a, b \in \mathbf{R}$, then we write $a > b$ or $b < a$, and say that *a is greater than b*, or *b is less than a*, if and only if $a - b$ is positive. The symbols "$>$" and "$<$" are called *inequality signs* and expressions such as $a > b$ or $b < a$ are called *inequalities*. To illustrate, $5 > 3$ since $5 - 3 = 2$, which is positive. Similarly, $-6 < -2$ since $-2 - (-6) = 4$, which is positive.

Since $a - 0 = a$, it follows that $a > 0$ if and only if a is positive. Similarly, $a < 0$ if and only if a is negative. A number of important rules can be established for inequalities. Among these are the following, where all letters represent real numbers.

(i) If $a > b$ and $b > c$, then $a > c$.
(ii) If $a > b$, then $a + c > b + c$.
(iii) If $a > b$ and $c > 0$, then $ac > bc$.
(iv) If $a > b$ and $c < 0$, then $ac < bc$.

Thus, by (iii) it is permissible to multiply both sides of an inequality by a positive real number; however, as indicated by (iv), multiplying both sides by a negative real number "reverses" the inequality sign. For example, if we multiply both sides of the inequality $5 > 3$ by -2, we obtain $-10 < -6$. Similar rules are true for "$<$." The symbol $a \geq b$, which is read "a is greater than or equal to b," means that either $a > b$ or $a = b$ (but not both). Likewise for $a \leq b$. A symbol such as $a < b < c$ means that $a < b$ and $b < c$, in which case we say that b is *between* a and c. The notation $a < b \leq c$ means $a < b$ and $b \leq c$, whereas $a \leq b < c$ means $a \leq b$ and $b < c$. As a final illustration, $a \leq b \leq c$ means $a \leq b$ and $b \leq c$.

We conclude this brief survey of real numbers with the following concept.

(1.8) Definition of Absolute Value

The *absolute value* $|a|$ of a real number a is defined as follows:

(i) If $a \geq 0$, then $|a| = a$.
(ii) If $a < 0$, then $|a| = -a$.

From this definition we see that *the absolute value of every nonzero real number is positive*, for if $a > 0$, then $|a| = a$, which is positive. On the other hand, if $a < 0$, then multiplying both sides by -1 gives us $-a > 0$ and by (ii) of (1.8) $|a| = -a > 0$; that is, $|a|$ is positive. It also follows that $|x| = |-x|$, for all $x \in \mathbf{R}$. As illustrations, since 3, $2 - \sqrt{2}$, and 0 are nonnegative we have

$$|3| = 3, \quad |2 - \sqrt{2}| = 2 - \sqrt{2}, \quad \text{and} \quad |0| = 0.$$

On the other hand, -3 and $\sqrt{2} - 2$ are negative and hence, by (ii) of (1.8),

$$|-3| = -(-3) = 3 \quad \text{and} \quad |\sqrt{2} - 2| = -(\sqrt{2} - 2) = 2 - \sqrt{2}.$$

Thus $|-3| = |3|$, $|2 - \sqrt{2}| = |\sqrt{2} - 2|$, and so on.

If we consider an inequality such as $|a| < 5$, then the letter a could represent real numbers such as $1/2$, $\sqrt{2}$, 2.7, π, 4.999, or for that matter, *any* number between 0 and 5. Moreover, since $|a| = |-a|$, a could also denote $-1/2$, $-\sqrt{2}$, -2.7, $-\pi$, -4.999, or any number between -5 and 0. Thus it appears that if $|a| < 5$, then $-5 < a < 5$. In general, it can be shown that if b is any positive real number, then

(1.9) $$|a| < b \quad \text{if and only if} \quad -b < a < b.$$

In previous discussions we have frequently used symbols to denote arbitrary elements of a set. Thus the notation $x \in \mathbf{R}$ means that x is a real number, although no *particular* real number is specified. A letter which is used to represent any element of a given set is sometimes called a *variable*. Letters near the end of the alphabet, such as $x, y, z, w, \cdots$, are often employed for variables. In some cases we wish to restrict a variable to some subset of $\mathbf{R}$. For example, given the expression $\sqrt{x}$, x must be nonnegative if a real number is to result. Similarly, given $1/x$, x may represent any number *except* 0. In general, the subset of $\mathbf{R}$ whose elements are represented by a variable is called the *domain* of the variable. The domain of a variable x is often referred to as the set of "permissible" or "allowable" values for x.

If we begin with any collection of variables and real numbers, then an *algebraic expression* is the result obtained by applying to this collection a finite number of additions, subtractions, multiplications, and divisions, together with the process of taking roots. Examples of algebraic expressions are the following:

$$x^3 - 2x + \frac{3}{\sqrt{2x}}, \qquad \frac{2xy + \sqrt[3]{3x^2}}{y - 1}.$$

When working with algebraic expressions, it is assumed that domains are chosen so that variables do not represent numbers which make the expressions meaningless. Thus it is assumed that denominators do not vanish, roots always exist, and so on. To simplify our work, we shall not always explicitly state the domains of variables. If meaningless expressions occur when certain numbers are substituted for the variables, then these numbers are *not* in the domains of the variables.

If x is a variable, then expressions such as

$$x + 3 = 0, \quad x^2 - 5 = 4x, \quad (x^2 - 9)\sqrt[3]{x + 1} = 0$$

are called *equations* in the variable x. Notice that when certain numbers are substituted for x in these equations, true statements are obtained, whereas other numbers produce false statements. Thus the equation $x + 3 = 0$ leads to a false statement for every value of x except -3. If 2 is substituted for x in the second equation we obtain $4 - 5 = 8$, or $-1 = 8$, a false statement. On the other hand, if we let $x = 5$, then we obtain $(5)^2 - 5 = 4 \cdot 5$, or $20 = 20$, which is true.

In general, if p and q are algebraic expressions in a variable x, then an expression of the form $p = q$ is called an *algebraic equation* in x. If a true statement is obtained when x is replaced by some real number a from the domain of x, then a is called a *solution* or a *root* of the equation. The *solution set* of an equation is the set of all solutions. To *solve* an equation means to find its solution set. If the solution set is the entire domain of x, then the equation is called an *identity*. For example, $\sqrt{1 - x^2} = \sqrt{(1 - x)(1 + x)}$ is an identity since it is true for *every* number in the domain of x. On the other hand, if there are numbers in the domain of x which are *not* solutions, then the equation is called a *conditional equation*. Analogous definitions can be given for *inequalities* which involve a variable x. Thus, the solution set of the inequality $x + 3 < 0$ consists of all real numbers less than -3. To *solve* an inequality $p < q$ means to find its solution set.

Although real numbers are adequate for many mathematical and scientific problems, there is a serious defect in the system when it comes to solving some equations. Indeed, since the square of a real number is never negative, the solution set of an equation of the form $x^2 = -4$ is empty if the domain of x is restricted to **R**. In certain important applications it is necessary to have available a number system which *contains* the real numbers and which has the additional property that equations such as $x^2 = -4$ *do* have solutions. Fortunately, it is possible to construct such a system: the *system of complex numbers.*

Complex numbers may be represented by symbols of the form $a + bi$, where a and b are real numbers and i is a symbol (not a real number) with the property $i^2 = -1$. Two complex numbers $a + bi$ and $c + di$ are said to be *equal*, and we write $a + bi = c + di$ if and only if $a = c$ and $b = d$. Operations of addition, subtraction, multiplication and division are defined *just as though* all letters denote real numbers with the additional stipulation that i^2 is the same as -1. For example, the formulas for addition and multiplication of two complex numbers $a + bi$ and $c + di$ are

$$(1.10)\qquad \begin{aligned}(a + bi) + (c + di) &= (a + c) + (b + d)i\\ (a + bi)(c + di) &= (ac - bd) + (ad + bc)i.\end{aligned}$$

The set of complex numbers, together with the operations of addition and multiplication as defined by (1.10), satisfy the field properties (1.3)–(1.7). Moreover, we may regard **R** as a subset of the set of complex numbers by identifying the real number a with the complex number $a + 0i$. Since we will be concerned primarily with real numbers, we shall often use the term "number" in place of "real number." The term "complex number" will never be abbreviated in this way.

It is easy to see that an equation such as $x^2 = -4$ has a complex solution $2i$ since $(2i)^2 = 2^2 i^2 = 4(-1) = -4$. Similarly $-2i$ is a solution.

We sometimes use the symbol $\sqrt{-1}$ in place of i and write $\sqrt{-13} = \sqrt{13}i$, $2 + \sqrt{-5} = 2 + \sqrt{5}i$, and so on. It can be shown that a *quadratic equation* of the form $ax^2 + bx + c = 0$, where $a, b, c \in \mathbf{R}$ and $a \neq 0$ has roots given by the *quadratic formula* $(-b \pm \sqrt{b^2 - 4ac})/2a$. If $b^2 - 4ac < 0$, then these roots are complex numbers. For example, applying the quadratic formula to the equation $x^2 - 4x + 13 = 0$ we obtain $(4 \pm \sqrt{16 - 52})/2$ or $(4 \pm \sqrt{-36})/2$, which may be written as $(4 \pm 6i)/2$. Thus the equation has the two complex solutions $2 + 3i$ and $2 - 3i$.

EXERCISES

1. In each of the following, replace the comma between the given pair of real numbers with the appropriate symbol $<$, $>$, or $=$.
   (a) $-2, -5$ (b) $-2, 5.$
   (c) $|-5|, |2|.$ (d) $2/3, 0.66.$
   (e) $|3 - 6|, |6 - 3|.$ (f) $\pi, 22/7.$
2. Same as Exercise 1 for each of the following.
   (a) $-3, 0.$ (b) $-8, -3.$
   (c) $|-8|, |-3|.$ (d) $|\frac{2}{3} - \frac{3}{4}|, \frac{1}{15}.$
   (e) $\sqrt{2}, 1.4.$ (f) $4053/1110, 3.6513.$

3. Rewrite each of the following numbers without using symbols for absolute values.
   (a) $|2 - 5|$.
   (b) $|-5| + |-2|$.
   (c) $|5| + |-2|$
   (d) $|-5| - |-2|$.
   (e) $|\sqrt{3} - 2|$.
   (f) $\dfrac{-2}{|-2|}$.
4. Same as Exercise 3 for each of the following.
   (a) $-|4 - 5|$.
   (b) $|3 - \pi|$.
   (c) $|-5| - |-8|$.
   (d) $|-2|^2$.
   (e) $|2 - \sqrt{4}|$.
   (f) $|-0.614|$.
5. State the rules for "$<$" which correspond to (i)–(iv) on p. 8.
6. Prove rules (i)–(iv) on p. 8.

In Exercises 7–12 use (i)–(iv) on p. 8 or Exercise 5 to change the given inequality to the form $a > k$ where $k \in \mathbf{R}$.

7. $3a > 8$.
8. $-3a < 8$.
9. $-2a + 1 < 5$.
10. $6 > 9 - 3a$.
11. $5a + 1 > 2a - 6$.
12. $4a - 3 > 2a + 5$.

In Exercises 13–18 change the given inequality to the form $k < a < l$ for some $k, l \in \mathbf{R}$.

13. $|a + 4| < 9$.
14. $|-a| < 4$.
15. $|2a - 1| < 5$.
16. $|3a + 5| < 5$.
17. $-2 < 3a + 1 < 7$.
18. $1 < 4 + 5a < 3$.
19. Show that if $a, b \in \mathbf{R}$ then $a > b$ if and only if $-a < -b$.
20. If a and b are positive, prove that $a > b$ if and only if $a^2 > b^2$. Is this true if a or b is negative?

In Exercises 21–28, find the solution sets of the given equations.

21. $(3x + 1)(2x + 5) = (1 - 6x)(4 - x)$.
22. $\dfrac{x + 1}{x} = 1 + \dfrac{1}{x}$.
23. $2x^2 - 7x - 15 = 0$.
24. $\dfrac{1}{x^2} + \dfrac{2}{x} = 3$.
25. $\sqrt{3x + 2} - 5 = 0$.
26. $|3x - 2| = 5$.
27. $x^2 - 8x + 25 = 0$.
28. $2x^2 + 4x + 3 = 0$.

Find the solution sets of the following inequalities:

29. $5 - 3x < 7 + 2x$.
30. $\frac{1}{2}x - \frac{3}{4} < 2 - \frac{1}{3}x$.
31. $|3x - 5| < 1$.
32. $|7 - 2x| > 1$.
33. $|2x - 1| > 3$.
34. $4/(x^2 + 9) > 0$.
35. $(x - 5)(x + 2) < 0$.
36. $x^2 > 4x$.

3 COORDINATE SYSTEMS

It is possible to associate the set **R** of real numbers with the points on a straight line l in such a way that for each real number a there corresponds one and only one point and, conversely, to each point P on l there corresponds precisely one real number. Such an association between two sets is referred to as a one-to-one correspondence. We begin by choosing an arbitrary point O on l, called the *origin*, and associate with it the real number 0. We then select any other point P_1 and associate with it the real number 1. The line segment OP_1 from O to P_1 is said to have *unit length*. Proceeding from P_1, we lay off successive segments of unit length along l, obtaining points $P_2, P_3, \cdots$, which are associated with the real numbers $2, 3, \cdots$, respectively. This process corresponds to the algebraic operation of adding 1 to itself many times. Similarly, working on the opposite side of O from P_1, we locate points $P_{-1}, P_{-2}, \cdots$ corresponding to the negative integers $-1, -2, \cdots$. In this manner we set up a correspondence between the set of integers and certain equispaced points on l (see Fig. 1.3).

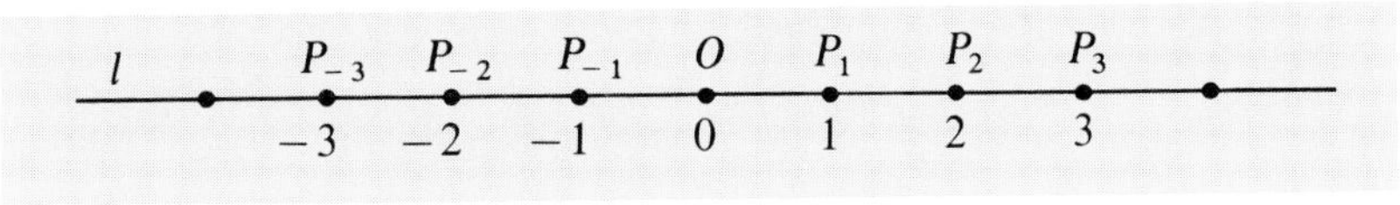

l P_{-3} P_{-2} P_{-1} O P_1 P_2 P_3

-3 -2 -1 0 1 2 3

Figure 1.3

For any positive integer n, a line segment can be subdivided into n equal parts. If the segment OP_1 is subdivided in this way, the endpoint of the first such subdivision is associated with the rational number $1/n$. By counting off m such segments, we can associate a point on l with any positive rational number $m/n = m(1/n)$. Thus the point corresponding to $13/5$ is $13/5$ units from the origin, or $3/5$ of the way from P_2 to P_3. The negative rational numbers are handled in like manner.

Certain points on l associated with irrational numbers can be determined. For example, the point corresponding to $\sqrt{2}$ can be found by striking off a circular arc as indicated in Fig. 1.4. Not all points corre-

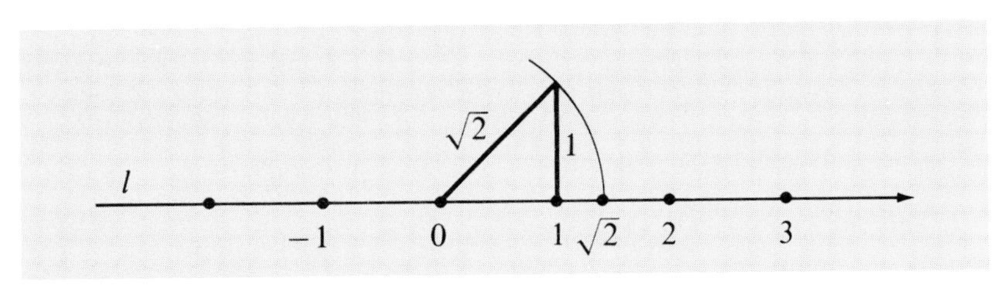

Figure 1.4

sponding to irrational numbers can be constructed in this way. The number $\pi = 3.14159\cdots$ is irrational, but no such construction is possible. However, its position on l can be approximated to within any degree of accuracy. Thus we could locate successively the points corresponding to 3, 3.1, 3.14, 3.141, 3.1415, $\cdots$, and so on. It can be shown that to every irrational number there corresponds a unique point on l and, conversely, every point that is not associated with a rational number corresponds to an irrational number.

The number x that is associated with a point X on l is called the *coordinate* of X. Such an assignment of coordinates to points on l is called a *coordinate system* for l, and l is called a *coordinate line*. A coordinate line l is often pictured as a horizontal line, the positive real numbers being taken as coordinates of points to the right of O and the negative real numbers as coordinates for points to the left of O. A direction can be assigned to l by taking the *positive direction* along l as that in which coordinates increase and the *negative direction* as that in which they decrease. The positive direction is noted by placing an arrowhead on l as shown in Fig. 1.4. If coordinates are assigned in this manner, it is evident that if the real numbers a and b are coordinates of points A and B, respectively, then $a < b$ if and only if B lies to the right of A.

We now assign numerical values to segments on l.

(1.11) Definition of Length

Let a and b be the coordinates of two points A and B respectively on a coordinate line l, and let AB denote the line segment from A to B. The *length* $d(A, B)$ of AB is defined by $d(A, B) = |b - a|$.

The number $d(A, B)$ is called the *distance between* A and B. Note that since $|b - a| = |a - b|$, we have $d(A, B) = d(B, A)$. The distance between the origin O and any point A is $d(O, A) = |0 - a| = |-a| = |a|$.

EXAMPLE 1. Let A, B, C, and D have coordinates -5, -3, 1, and 6, respectively (see Fig. 1.5). Find $d(A, B)$, $d(C, B)$, $d(O, A)$, and $d(C, D)$.

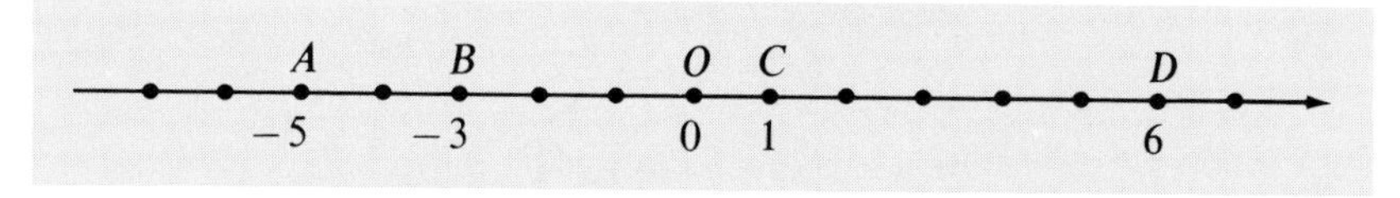

Figure 1.5

Solution: By (1.11) we have

$$d(A, B) = |-3 - (-5)| = |-3 + 5| = |2| = 2,$$

$$d(C, B) = |-3 - 1| = |-4| = 4,$$

$$d(O, A) = |-5 - 0| = |-5| = 5,$$

$$d(C, D) = |6 - 1| = |5| = 5.$$

In some cases we wish to work with distances along l which take into account the direction of l, as in the next definition.

(1.12) Definition of Directed Distance

Let a and b be the coordinates of two points A and B respectively on a coordinate line l. The *directed distance* $\overline{AB}$ from A to B is defined by $\overline{AB} = b - a$.

Since $\overline{BA} = a - b$, we have $\overline{AB} = -\overline{BA}$. Evidently, the point B is to the right of A if and only if $\overline{AB} > 0$ and is to the left of A if and only if $\overline{AB} < 0$. In this way the directed distance indicates the *direction* from A to B with respect to the direction assigned to l.

EXAMPLE 2. If A, B, C, and D are the points in Example 1, find the directed distances $\overline{AB}$, $\overline{BA}$, $\overline{CB}$, and $\overline{OA}$.

Solution: By (1.12) we have $\overline{AB} = -3 - (-5) = 2$, $\overline{BA} = -5 - (-3) = -2$, $\overline{CB} = -3 - 1 = -4$, and $\overline{OA} = -5 - 0 = -5$.

Coordinate systems can also be introduced in planes. To accomplish this the notion of *ordered pair* is used. As mentioned earlier, if S is a set and $a, b \in S$, then, as far as subsets are concerned, there is no difference between $\{a, b\}$ and $\{b, a\}$. However, an ordered pair is not merely a subset. It consists of two elements a and b in which one of the elements is designated the "first" element and the other the "second" element. The symbol (a, b) is used to denote the ordered pair consisting of the elements a and b with first element a and second element b. We consider two ordered pairs (a, b) and (c, d) equal, and write $(a, b) = (c, d)$, if and only if $a = c$ and $b = d$. This implies, in particular, that $(a, b) \neq (b, a)$ when $a \neq b$.

In order to introduce what is called a *rectangular*, or *Cartesian*, *coordinate system* in a given plane, we consider two mutually perpendicular coordinate lines in the plane which intersect in the origin O on each line. We take one of the lines horizontally, with positive direction to the right, and the other line vertically, with positive direction upward. For clarity, we place an arrowhead on each line to indicate the positive direction (see Fig. 1.6). The two lines are called *coordinate axes* and the point O is called the *origin*. More specifically, the horizontal line is referred to as the *x-axis* and the vertical line as the *y-axis*, and we label them x and y respectively. In spite of the fact that the symbols x and y are used to denote lines as well as numbers, there should be no misunderstanding as to exactly what these letters represent when they appear alongside of coordinate lines, as in Fig. 1.6. In certain applications different labels, such as d, t, etc., are used for these coordinate lines. Unless it is specified otherwise, the same unit of length is chosen on each axis. We refer to the given plane as a *coordinate plane* or, with the above notation for coordinate axes, as the *xy-plane*. The coordinate axes divide the plane into four parts, called the first, second, third, and fourth quadrants and labeled I, II, III, and IV, respectively, as in Fig. 1.6.

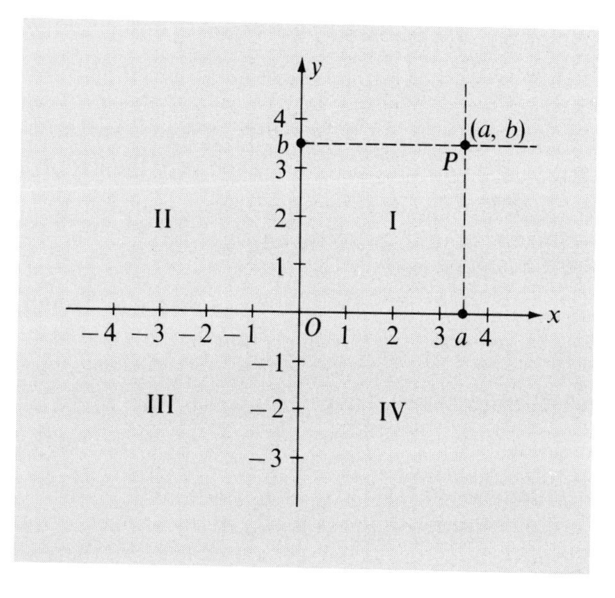

Figure 1.6

We may now assign to each point in the xy-plane a unique ordered pair of real numbers. If P is a point on the plane, we construct lines through P, perpendicular to the x- and y-axes and intersecting these axes at points with coordinates a and b respectively. Then P is assigned the pair (a, b) (see Fig. 1.6.). The number a is called the *x-coordinate*, or *abscissa*, of P, and the number b is called the *y-coordinate*, or *ordinate*, of P. For convenience we shall say that *P has coordinates* (a, b).

Conversely, every ordered pair (a, b) of real numbers determines a point P in the xy-plane having these coordinates. We merely construct lines perpendicular to the x-axis and y-axis at the points having coordinates a and b respectively. The intersection of these two lines is the desired point.

We have established a one-to-one correspondence between points in the xy-plane and the set of all ordered pairs of real numbers. It will sometimes be convenient, in our work, to refer to the *point* (a, b), meaning the point with abscissa a and ordinate b. Again, for ease in certain statements, we shall use the notation $P(a, b)$ to denote the point P having coordinates (a, b).

To *plot* a point $P(a, b)$ means to locate, on a coordinate plane, the point P with coordinates (a, b). This point is represented by a dot in the appropriate position. Needless to say, the dot is not the point, but merely a device for visualizing this mathematical object.

The coordinates a and b of a point $P(a, b)$ are *directed* distances. For example, if a is positive, then we know that P lies to the right of the y-axis, whereas if a is negative, then P is to the left of the y-axis. Note that abscissas are positive for points in quadrants I or IV and negative for points in quadrants II or III. On the other hand, ordinates are positive for points in quadrants I or II and negative for points in quadrants III or IV. In Fig. 1.7 we have plotted some typical points on a coordinate plane.

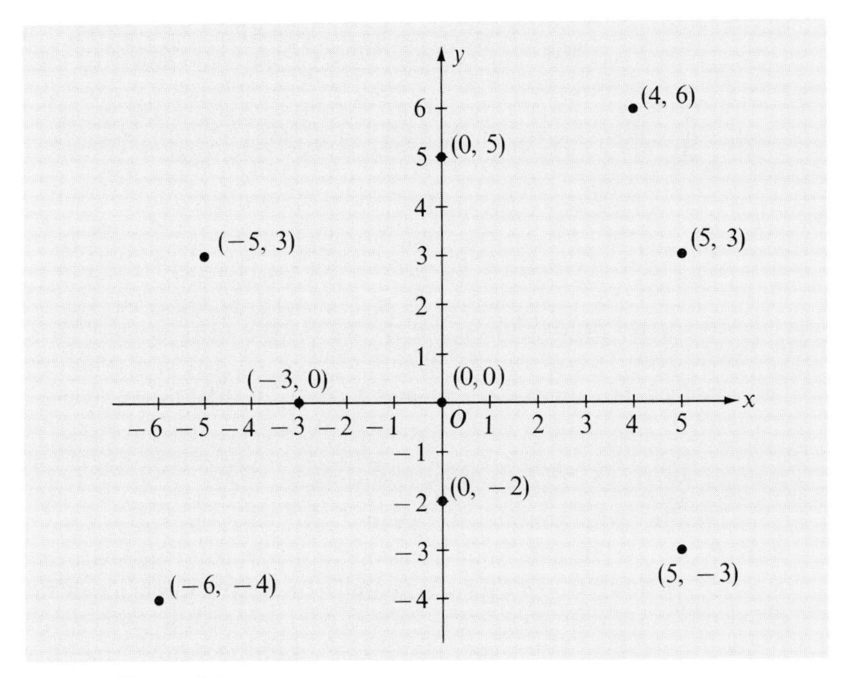

Figure 1.7

We shall now derive a formula for finding the distance between any two points on a coordinate plane. The distance between two points P and Q will be denoted by $d(P, Q)$. If $P = Q$, then we have $d(P, Q) = 0$, whereas if $P \neq Q$, then the distance is considered positive. Thus the number $d(P, Q)$ is *not* a directed distance and we may write $d(P, Q) = d(Q, P)$.

Let us consider any two points $P_1(x_1, y_1)$ and $P_2(x_2, y_2)$. If the points lie on the same horizontal line, then $y_1 = y_2$, and we may denote the points by $P_1(x_1, y_1)$ and $P_2(x_2, y_1)$. If lines are constructed through P_1 and P_2 parallel to the y-axis, then they intersect the x-axis at $A_1(x_1, 0)$ and $A_2(x_2, 0)$, as shown in (i) of Fig. 1.8, and we have $d(P_1, P_2) =$

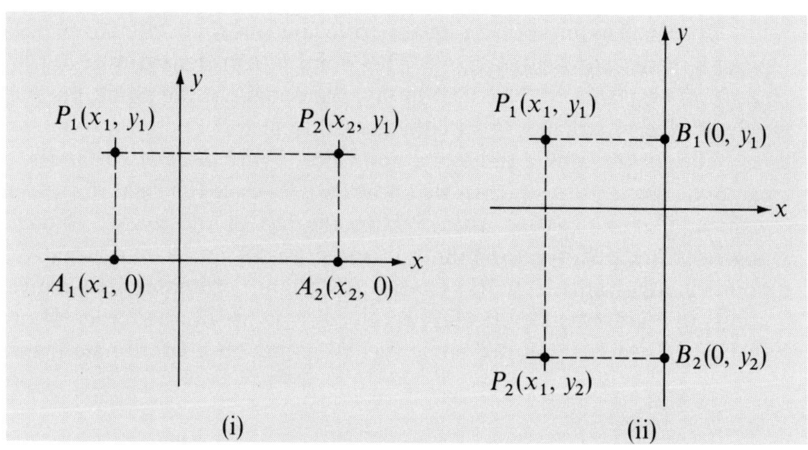

Figure 1.8

$d(A_1, A_2)$. However, by (1.11), $d(A_1, A_2) = |x_2 - x_1|$ and hence

(1.13) $$d(P_1, P_2) = |x_2 - x_1|.$$

Since $|x_2 - x_1| = |x_1 - x_2|$, this formula is valid whether P_1 lies to the left of P_2 or to the right of P_2. Moreover, the formula is independent of the quadrants in which the points lie.

In similar fashion, if P_1 and P_2 lie on the same vertical line, then $x_1 = x_2$ and we may express the points as $P_1(x_1, y_1)$ and $P_2(x_1, y_2)$. If we consider the points $B_1(0, y_1)$ and $B_2(0, y_2)$ on the y-axis, as shown in (ii) of Fig. 1.8, then we have

(1.14) $$d(P_1, P_2) = d(B_1, B_2) = |y_2 - y_1|.$$

Finally, let us consider the general case, where the points $P_1(x_1, y_1)$ and $P_2(x_2, y_2)$ do not lie on the same horizontal or vertical line. We construct a line through $P_1(x_1, y_1)$ parallel to the x-axis and a line through $P(x_2, y_2)$ parallel to the y-axis. These lines intersect at some point P_3. Since P_3 has the same y-coordinate as P_1 and the same x-coordinate as P_2, we can denote this point by $P_3(x_2, y_1)$ (see Fig. 1.9). From the previous

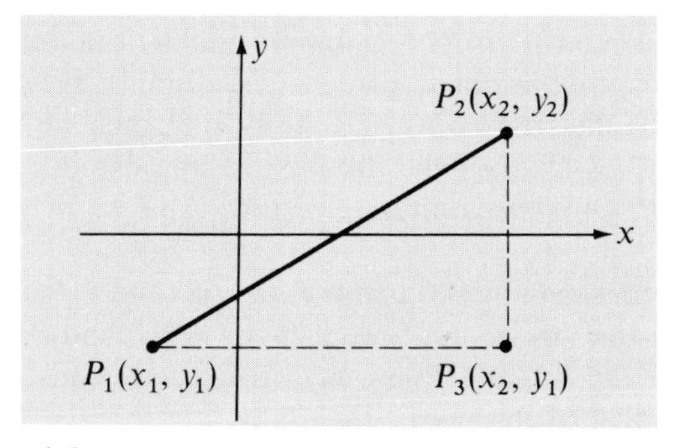

Figure 1.9

discussion $d(P_1, P_3) = |x_2 - x_1|$ and $d(P_3, P_2) = |y_2 - y_1|$. Since P_1, P_2, and P_3 form a right triangle with hypotenuse from P_1 to P_2, we have, by the Pythagorean Theorem,

$$[d(P_1, P_2)]^2 = [d(P_1, P_3)]^2 + [d(P_3, P_2)]^2.$$

Applying (1.13) and (1.14) gives us

$$[d(P_1, P_2)]^2 = |x_2 - x_1|^2 + |y_2 - y_1|^2.$$

Using the fact that $d(P_1, P_2)$ is nonnegative and that $|a|^2 = a^2$ for every $a \in \mathbf{R}$, we obtain the formula given in (1.15).

(1.15) Distance Formula

For all points $P_1(x_1, y_1)$ and $P_2(x_2, y_2)$ in a coordinate plane,

$$d(P_1, P_2) = \sqrt{(x_2 - x_1)^2 + (y_2 - y_1)^2}.$$

Although in our proof of the distance formula we referred to Fig. 1.9, our argument is independent of the quadrants in which the points P_1 and P_2 lie. When (1.15) is used, it is immaterial which point is labeled P_1 and which is labeled P_2. The formula is also valid if the points P_1 and P_2 lie on the same horizontal line, for in this case we have $y_1 = y_2$ and the formula gives us $d(P_1, P_2) = \sqrt{(x_2 - x_1)^2 + 0^2}$ and the latter equality is the same as $d(P_1, P_2) = |x_2 - x_1|$, which gives us (1.13). Similarly, if P_1 and P_2 lie on a vertical line, (1.15) reduces to (1.14).

EXAMPLE 1. Prove that the triangle with vertices $A(-1, -3)$, $B(6, 1)$, and $C(2, -5)$ is a right triangle.

Solution: By the distance formula we have

$$d(A, B) = \sqrt{(-1 - 6)^2 + (-3 - 1)^2} = \sqrt{49 + 16} = \sqrt{65},$$
$$d(B, C) = \sqrt{(6 - 2)^2 + (1 + 5)^2} = \sqrt{16 + 36} = \sqrt{52},$$
$$d(A, C) = \sqrt{(-1 - 2)^2 + (-3 + 5)^2} = \sqrt{9 + 4} = \sqrt{13}.$$

Hence $[d(A, B)]^2 = [d(B, C)]^2 + [d(A, C)]^2$, which proves that the triangle is a right triangle with hypotenuse joining A to B.

EXERCISES

1. Let A, B, and C be points on a coordinate line with coordinates -2, 5, and 1, respectively. Find: (a) $d(A, B)$, (b) $d(B, C)$, (c) $d(C, B)$, (d) $d(A, C)$.
2. Same as Exercise 1 if A, B, and C have coordinates 3, -6, and 8, respectively.

3. If A, B, and C are as in Exercise 1, find: (a) $\overline{AB}$, (b) $\overline{BC}$, (c) $\overline{CB}$, (d) $\overline{AC}$.
4. Same as Exercise 3 if A, B, and C are as in Exercise 2.
5. Plot the following points on a rectangular coordinate system: $A(-2, 3)$; $B(2, -3)$; $C(-2, -3)$; $D(0, -5)$; $E(-3/2, 0)$; $F(\pi, -\sqrt{2})$.
6. Plot $A(0, 0)$; $B(1, 1)$; $C(4, 4)$; $D(-3, -3)$. Describe the set of points $S = \{(x, x) \mid x \in \mathbf{R}\}$.
7. Describe the set of all points $P(x, y)$ in a coordinate plane such that: (a) $xy = 0$. (b) $xy > 0$. (c) $x/y < 0$.
8. List the coordinates of five points with abscissa -3. Describe the set of all such points.

In Exercises 9–12, find the distance $d(A, B)$ between the given points A and B.

9. $A(6, -2)$, $B(2, 1)$.
10. $A(-4, -1)$, $B(2, 3)$.
11. $A(0, -7)$, $B(-1, -2)$.
12. $A(4, 5)$, $B(4, -4)$.

In Exercises 13 and 14 prove that the triangle with the indicated vertices is a right triangle and find its area.

13. $A(2, 1)$, $B(-1, 4)$, $C(-2, -3)$.
14. $A(-7, -1)$, $B(-4, 5)$, $C(-1, -4)$.
15. Prove that the following points are vertices of a square: $A(-2, 5)$, $B(8, 2)$, $C(-5, -5)$, $D(5, -8)$.
16. Prove the following points are vertices of a parallelogram: $A(0, 0)$, $B(-5, 1)$, $C(8, 3)$, $D(3, 4)$.
17. Given $A(-2, 1)$ and $B(4, -3)$, prove that $P(1, -1)$ is on the perpendicular bisector of AB.
18. If A and B are as in Exercise 17, find a formula which expresses the fact that $P(x, y)$ is on the perpendicular bisector of AB.
19. If J, K, and L are any three points on a coordinate line l, prove that $\overline{JK} + \overline{KL} = \overline{JL}$.
20. Show that if J, K, and L are points on a coordinate line, then it is not always true that $d(J, K) + d(K, L) = d(J, L)$.

4 RELATIONS AND THEIR GRAPHS

If S and T are sets, let us consider the collection of ordered pairs (s, t) that can be obtained by letting s range through all elements of S and letting t range through all elements of T. This collection is called the *product set* of S and T and will be denoted by $S \times T$. Thus

$$S \times T = \{(s, t) \mid s \in S, t \in T\}.$$

EXAMPLE 1. Given $A = \{a, b, c\}$ and $B = \{c, d\}$, find $A \times B$, $B \times A$, and $B \times B$.

Solution: From the definition of product set we have

$$A \times B = \{(a, c), (b, c), (c, c), (a, d), (b, d), (c, d)\},$$
$$B \times A = \{(c, a), (c, b), (c, c), (d, a), (d, b), (d, c)\},$$
$$B \times B = \{(c, c), (c, d), (d, c), (d, d)\}.$$

As illustrated by the third part of Example 1, the sets S and T need not be different. We shall often work with subsets of product sets. Such subsets are given special names as in the next definition.

(1.16) Definition of Relation

A *relation* between two sets S and T is a subset of $S \times T$.

If W is a relation between S and T (that is, if $W \subseteq S \times T$) and if an ordered pair (s, t) is in W, we say that *s is related to t*. As an illustration, let S denote the set of all points in a plane, T the set of all lines, and let W denote the subset of $S \times T$ defined by

$$W = \{(s, t) \mid s \text{ lies on } t\}.$$

Thus the point s is related to the line t if and only if s lies on t.

If $S = T$ in (1.16), it is customary to use the phrase "relation *on* S" in place of "relation *between* S and S." For example, if W is the subset of $\mathbf{R} \times \mathbf{R}$ defined by

$$W = \{(a, b) \mid a < b\},$$

then W is a relation on $\mathbf{R}$ and a real number a is related to a real number b if and only if a is less than b. Thus 2 is related to 5, -3 is related to 2, $\sqrt{2}$ is related to 1.5, and so on. We might refer to W as the "less than" relation on $\mathbf{R}$.

In the discussion to follow, our primary interest will be in relations on $\mathbf{R}$ — that is, in subsets of $\mathbf{R} \times \mathbf{R}$. Since such a relation W is a set of ordered pairs of real numbers, we may speak of the point $P(x, y)$ in a coordinate plane which corresponds to the ordered pair (x, y) in W. This leads to the next definition.

(1.17) Definition of Graph of a Relation

If W is a relation on $\mathbf{R}$, then the *graph* of W is the set of all points in a coordinate plane which correspond to the ordered pairs in W.

In order to simplify statements we shall not always include the phrase "relation on $\mathbf{R}$" but merely specify a relation by defining a certain subset of $\mathbf{R} \times \mathbf{R}$. This is illustrated in the next example.

EXAMPLE 2. Find the graph of $W = \{(x, y) \mid |x| \leq 2, |y| \leq 1\}$.

Solution: From (1.9) x and y satisfy the indicated inequalities if and only if $-2 \leq x \leq 2$ and also $-1 \leq y \leq 1$. Therefore the graph of W consists of all points within and on the boundary of the rectangular region shown in Fig. 1.10.

The graph of the relation W in Example 2 consists of a *region* in the plane. Graphs of relations often consist of lines, circles, or other curves. In order to find the graph of such a relation W we often plot a number of points $P(x, y)$, where $(x, y) \in W$, until some pattern emerges and then draw a smooth curve (or curves) through appropriate points. This is sometimes referred to as "sketching the graph of the relation."

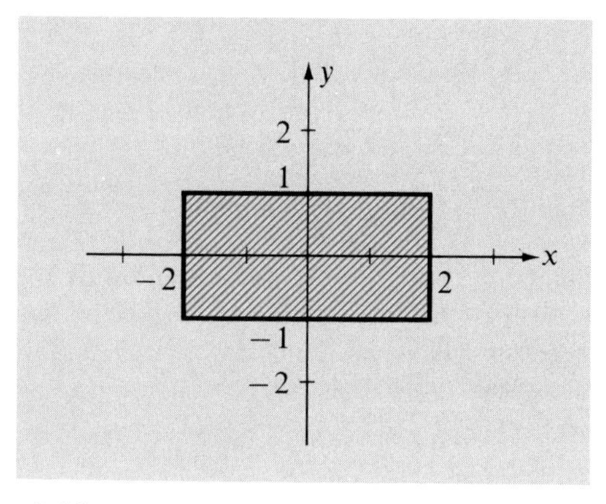

Figure 1.10

EXAMPLE 3. Sketch the graph of $W = \{(x, y) \mid y = 2x - 1\}$.

Solution: We begin by finding points with coordinates of the form (x, y), where $(x, y) \in W$. It is convenient to list these coordinates in tabular form as shown below.

x	-2	-1	0	1	2	3
y	-5	-3	-1	1	3	5

Plotting, it appears that the points with these coordinates all lie on a straight line, and we sketch the graph accordingly (see Fig. 1.11).

Since the relation of Example 3 is defined by a rather simple algebraic expression, we substituted only integral values for x. For more complicated expressions, points which have rational or even irrational abscissas are sometimes plotted. It is impossible to sketch the entire graph in Example 3,

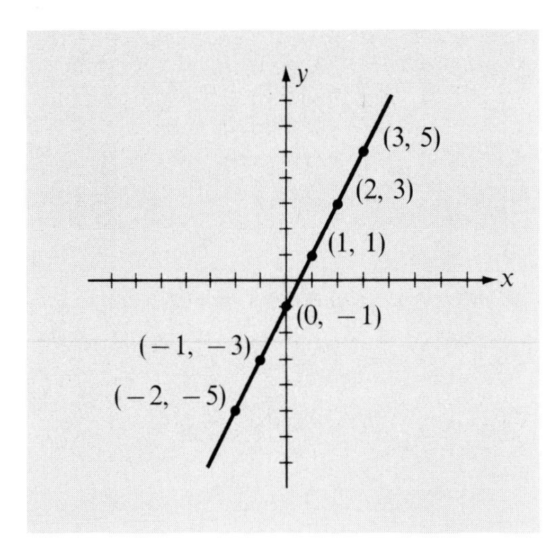

Figure 1.11

since x may be assigned values which are numerically as large as desired. Nevertheless, we often call a drawing of the type given in Fig. 1.11 "the graph of the relation" or a "sketch of the graph" where it is understood that the drawing is only a device for visualizing the actual graph and the line does not terminate as shown in the figure. In general, when sketching a graph one should illustrate enough of the graph so that the remaining parts are evident. In some cases only a few points need be plotted, whereas in others it may be necessary to plot very many points before the shape of the graph is found.

The relation in Example 3 is determined by the equation $y = 2x - 1$, in the sense that for any $x \in \mathbf{R}$, the equation can be used to find a number y such that $(x, y) \in W$. Given an equation involving x and y, we say that an ordered pair (a, b) is a *solution* of the equation if, when a is substituted for x and b for y, equality is obtained. For example, $(2, 3)$ is a solution of $y = 2x - 1$, since substitution of 2 for x and 3 for y leads to $3 = 4 - 1$. The *solution set* of such an equation is the set of all solutions and hence is a relation on $\mathbf{R}$. Two equations in x and y are said to be *equivalent* if they have the same solution sets.

The *graph of an equation* in x and y is defined as the graph of its solution set and the phrase "sketch the graph of an equation" means to sketch the graph of its solution set. Notice that the solutions of the equation $y = 2x - 1$ are the pairs (a, b) such that $b = 2a - 1$ and hence the solution set is identical with the relation W given in Example 3. Consequently, the graph of the equation $y = 2x - 1$ is the same as the graph of W.

EXAMPLE 4. Sketch the graph of the equation $y = 2x - x^2$.

Solution: The graph in question is the same as the graph of the relation $W = \{(x, y) \mid y = 2x - x^2\}$ (Why?). Tabulating coordinates of several points on the graph, we obtain the table shown below.

x	-2	-1	0	1	2	3	4
y	-8	-3	0	1	0	-3	-8

A portion of the graph is sketched in Fig. 1.12.

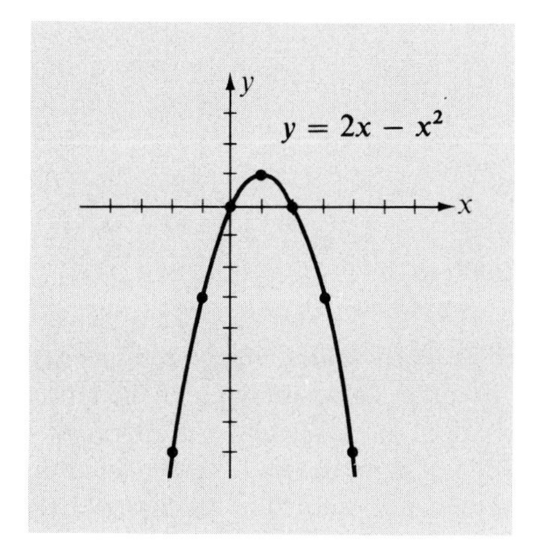

Figure 1.12

EXAMPLE 5. Sketch the graph of the equation $y^2 = x$.

Solution: The solution set of the equation $y^2 = x$ is the *union* of the solution sets of the two equations $y = \sqrt{x}$ and $y = -\sqrt{x}$; that is, for each positive value of x there correspond *two* values of y which give solutions, namely $\sqrt{x}$ and $-\sqrt{x}$. We tabulate coordinates of points on the graph as shown below. See Fig. 1.13 for a portion of the graph.

x	0	1	2	3	4	9
y	0	± 1	$\pm\sqrt{2}$	$\pm\sqrt{3}$	± 2	± 3

If S is a set of points or a geometric figure in a coordinate plane, it is sometimes possible to find an equation for S, in the sense that the graph of the equation is S. We shall demonstrate how this can be accomplished if S is a circle.

If $C(h, k)$ is a point in a coordinate plane, then a circle in the plane with center C and radius r may be defined as the collection of all points in the plane that are r units from C. If $P(x, y)$ denotes an arbitrary point

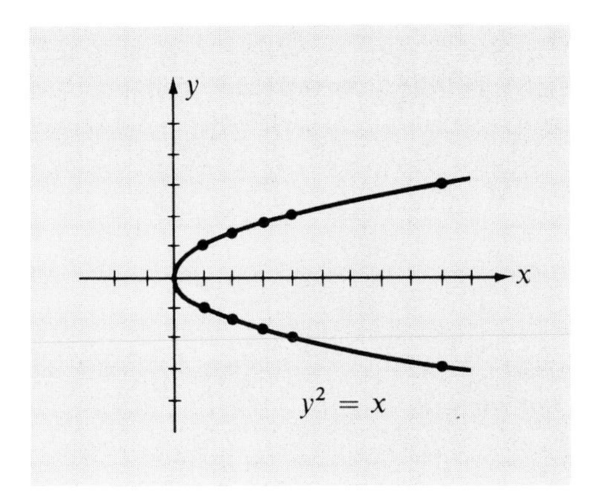

Figure 1.13

in the plane, then as shown in Fig. 1.14, P is on the circle if and only if $d(C, P) = r$, that is, if and only if

$$\sqrt{(x - h)^2 + (y - k)^2} = r.$$

The graph of the solution set of the latter equation gives us the circle.

An equivalent equation is

$$(x - h)^2 + (y - k)^2 = r^2, \qquad r > 0. \tag{1.18}$$

We shall call (1.18) the *standard equation* of a circle of radius r and center (h, k). If $h = 0$ and $k = 0$, then (1.18) reduces to

$$x^2 + y^2 = r^2, \tag{1.19}$$

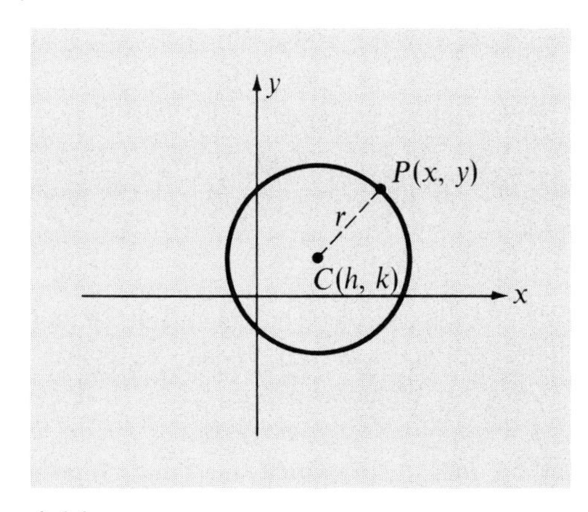

Figure 1.14

which is an equation of a circle of radius r with center at the origin (see Fig. 1.15). If $r = 1$, we sometimes refer to the graph of (1.19) as a *unit circle* with center at the origin. Note that if a point $P(x, y)$ is on this unit circle, then $x^2 + y^2 = 1$.

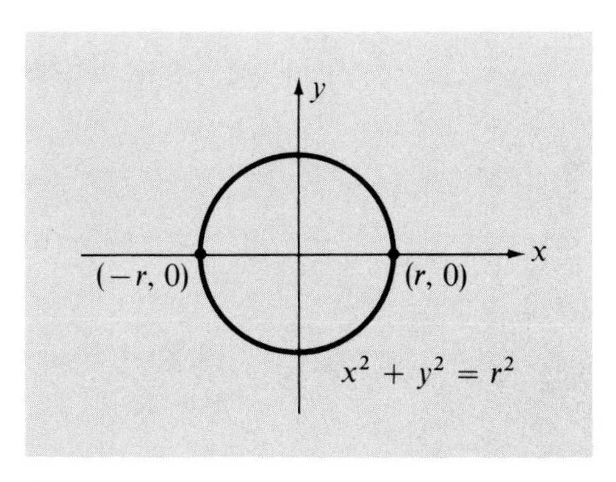

Figure 1.15

EXAMPLE 6. Find an equation of the circle which has center $C(-2, 3)$ and which passes through the point $D(4, 5)$.

Solution: Since D is on the circle, the radius r is $d(C, D)$. By the distance formula (1.15) we have

$$r = \sqrt{(-2 - 4)^2 + (3 - 5)^2} = \sqrt{36 + 4} = \sqrt{40}.$$

Hence by (1.18) with $h = -2$ and $k = 3$,

$$(x + 2)^2 + (y - 3)^2 = 40$$

is an equation of the circle. Squaring the indicated terms and simplifying, we have an equivalent equation:

$$x^2 + y^2 + 4x - 6y - 27 = 0.$$

EXERCISES

1. If $S = \{1, 2, 3\}$ and $T = \{x, y\}$, list the elements of each of the following.
   (a) $S \times T$. (b) $T \times S$. (c) $S \times S$. (d) $T \times T$.
2. (a) If S contains three elements and T contains two elements, how many elements are in the product set $S \times T$?
   (b) How many elements are in $S \times T$ if S contains four elements and T contains three elements?
   (c) Give a reason for calling $S \times T$ a *product* set.

In Exercises 3–8 sketch the graph of the given relation W on **R**.

3. $W = \{(x, y) \mid y = -3/2\}$.
4. $W = \{(x, -x) \mid x > 0\}$.
5. $W = \{(x, y) \mid y = x\}$.
6. $W = \{(x, y) \mid y > x\}$.
7. $W = \{(x, y) \mid |x| \geq 1, |y| \leq 1\}$.
8. $W = \{(x, y) \mid |x| = 1\}$.

In Exercises 9–14 sketch the graph of the equation after plotting a sufficient number of points.

9. $y^2 = 4 + x$.
10. $y^2 = 4 - x$.
11. $2x + 11 = 3 - x$.
12. $y(x - 4) - 3x = x(y - 3) + 10$.
13. $y^2 = x^3$.
14. $x^2 = y^3$.

In Exercises 15–20 find the equation of a circle satisfying the stated condition.

15. Center $C(-2, 4)$, radius 3.
16. Center $C(-3, -1)$, radius $\sqrt{5}$.
17. Center $C(3, -5)$, radius 2.
18. Center $C(-1/2, -5/2)$, passing through $P(1, 1)$.
19. Center $C(4, -7)$, passing through the origin.
20. Center $C(-6, 8)$, tangent to the x-axis.

In Exercises 21–26 sketch the graph of the circle which has the given equation.

21. $x^2 + y^2 = 4$.
22. $x^2 = 16 - y^2$.
23. $(x + 5)^2 + (y - 2)^2 = 9$.
24. $(x - 4)^2 + y^2 = 16$.
25. $(x + 1)^2 + (y + 2)^2 = 2$.
26. $4x^2 + 4y^2 = 1$.

5 FUNCTIONS

The word "function" is not a very descriptive title for the concept it represents. The term "correspondence" would be more fitting. The type of correspondence we have in mind occurs *outside* the subject of mathematics as well as within this discipline. The idea is this: we are given two sets X and Y, not necessarily distinct. A *correspondence* from X to Y can be thought of as a rule that associates with each element of X *one and only one* element of Y. As an illustration, let X denote the set of books in a library and Y the set of integers. If with each book we associate the number of pages in the book, we obtain a correspondence from X to Y, since for each $x \in X$ there is assigned precisely one $y \in Y$. Note that there may be elements of Y which are not associated with elements of X. The negative integers in Y are in this category, since a book cannot have a negative number of pages! Also, it is unlikely that any element of X has for its associate in Y the number 1,000,000,000.

Sometimes functions are represented pictorially by diagrams of the type shown in Fig. 1.16. The arrow indicates that the element y of Y is associated with the element x of X. We might imagine a whole family of arrows of this type, each arrow connecting an element of X to some specific element of Y. We have pictured X and Y as disjoint sets. In

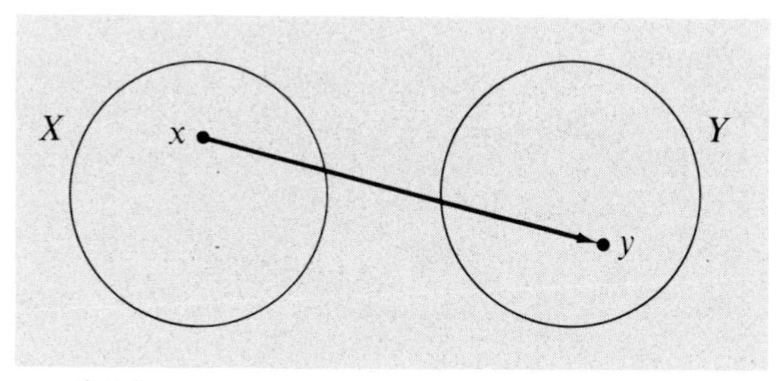

Figure 1.16

general, X and Y may have elements in common. As a matter of fact, we often take $X = Y$.

As another illustration, let X denote the set of all people who were born in the United States in the year 1970 and let Y denote the set of all states in the United States. If with each $x \in X$ we associate the state in which x was born, a correspondence from X to Y is established. In particular, with each person born in Chicago in 1970, we would associate the state of Illinois. This example brings out an important fact about correspondences. With each element of X there is associated precisely one element of Y; however, an element of Y may be associated with different elements of X.

In most of our work X and Y will be sets of numbers. To illustrate, we let X and Y both denote the set **R** of real numbers. With each real number $x \in X$ let us associate its square, x^2. Thus with 3 we associate 9, with -5 we associate 25, with $\sqrt{2}$ the number 2, and so on. This gives us a correspondence from **R** to **R**.

All the examples of correspondences we have given are *functions*, as defined in (1.20) below. At the end of this section we shall state another definition of function which is often used. It is customary (but not a definite requirement) to use letters near the middle of the alphabet, such as f, g, h, etc., to denote functions. Sometimes capital letters, F, G, H, etc., are also used.

(1.20) Definition of Function

A *function* f from a set X to a set Y is a correspondence that associates with each element x of X a unique element y of Y. The element y is called the *image* of x under f and is denoted by $f(x)$. The set X is called the *domain* of the function. The *range* of the function is the set of all images of elements of X.

In (1.20) we have introduced the notation $f(x)$ for the element which is associated with x. This is usually read "f of x" or "f at x." Sometimes $f(x)$ is called the *value* of f at x. In terms of the pictorial representation given earlier, we might now sketch a diagram as in Fig. 1.17, where we show several of the arrows which indicate the correspondence between elements of X and Y. It is important to note that with each element x of X there is associated precisely one image $f(x)$ in Y; however, different elements of X, such as w and z in Fig. 1.17, may have the same image in Y.

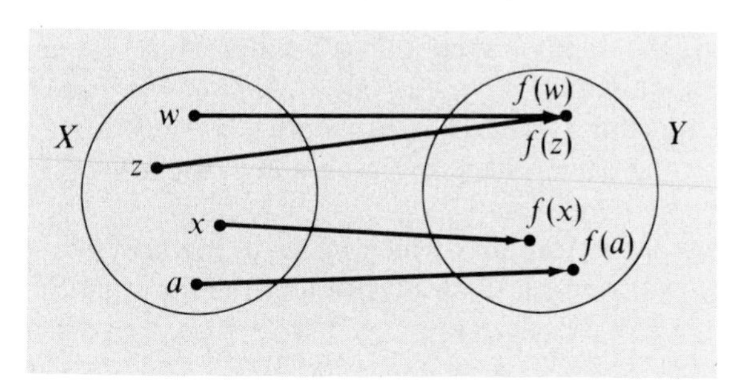

Figure 1.17

Beginning students are sometimes confused by the symbols f and $f(x)$. It should be remembered that f is used to represent the *function*. It is neither an element of X nor an element of Y. On the other hand, $f(x)$ *is an element* Y, namely the element which f associates with x.

EXAMPLE 1. Let f be the function with domain $\mathbf{R}$ such that $f(x) = x^2$ for all $x \in \mathbf{R}$. Find $f(-6)$, $f(\sqrt{3})$, and $f(a)$, where $a \in \mathbf{R}$. What is the range of f?

Solution: The function f associates with each real number x its square, x^2. Values of f — that is, images of elements of $\mathbf{R}$ — may be found by substituting for x in the equation $f(x) = x^2$. Thus

$$f(-6) = (-6)^2 = 36, \quad f(\sqrt{3}) = (\sqrt{3})^2 = 3, \quad \text{and} \quad f(a) = a^2.$$

If S denotes the range of f, then by definition

$$S = \{f(a) \mid a \in \mathbf{R}\} = \{a^2 \mid a \in \mathbf{R}\}.$$

Evidently, the range of f is the set of all nonnegative real numbers.

Occasionally one of the notations

$$X \xrightarrow{f} Y, \quad f\colon X \to Y, \quad \text{or} \quad f\colon x \to f(x)$$

is used to signify that f is a function from X to Y. It is not unusual in this event to say *f sends X into Y* or *f sends x into $f(x)$*. Thus, in Example 1, f sends x into x^2 and we may write $f\colon x \to x^2$.

Two functions f and g from X to Y are said to be *equal*, and we write $f = g$, provided $f(x) = g(x)$ for all $x \in X$. If g is the function from $\mathbf{R}$ to $\mathbf{R}$ such that $g(x) = \frac{1}{2}(2x^2 - 6) + 3$ for all $x \in \mathbf{R}$ and if $f(x) = x^2$, then we have $g = f$ (Why?).

In order to describe a function f from X to Y, it is necessary to specify the image $f(x)$ of *each* element x of X. This can be done in a number of ways. As in our first illustrations of correspondence, we could give a verbal statement which determines each image. Another common method is to use an equation, as was done in Example 1. When the latter method is employed, the symbol used for the variable is immaterial — that is, expressions such as $f(x) = x^2$, $f(s) = s^2$, $f(t) = t^2$, and so on, all define the same function f. This is true because if a is any number in the domain of f, then the same image a^2 is obtained regardless of which of these expressions is employed.

Many of the formulas which occur in mathematics and the sciences determine functions. As an illustration, consider the formula $A = \pi r^2$, for the area A of a circle of radius r. For each positive real number assigned to r there corresponds a unique value of A. This gives us the function f having as its domain the set X of positive real numbers and where $f(r) = \pi r^2$ for every $r \in X$. We may, therefore, write $A = f(r)$. In a case such as this the symbol r, which represents an arbitrary number from the domain of f, is called an *independent variable.* The symbol A, which represents a number from the range of f, is called a *dependent variable,* since its value *depends* on the number assigned to r. In a situation where two variables r and A are related in this manner, it is customary to use the phrase "*A is a function of r.*"

To cite another example, if an automobile travels at a uniform rate of 50 miles per hour, then the distance d (miles) traversed in time t (hours) is given by $d = 50t$. Thus the distance d is a function of the time t.

As we have already pointed out, different elements in the domain X of a function f may have the same image in Y. For example, if $f(x) = x^2$ then $f(-2) = f(2)$, even though $-2 \neq 2$. On the other hand, if a function f from X to Y has the property that whenever $x_1 \neq x_2$ in X, then $f(x_1) \neq f(x_2)$ in Y, we call the function a *one-to-one function* (or *correspondence*) from X to Y. If the range of f is Y and f is one-to-one, then sets X and Y are said to be in *one-to-one correspondence.* In this case each element of Y is the image of precisely one element of X. The association between real numbers and points on a coordinate line described in Section 3 is an example of a one-to-one correspondence.

If f is a function from X to X and if $f(x) = x$ for all $x \in X$, then f is called the *identity function* on X. In this case every element x is sent into itself.

We call a function f from X to Y a *constant function* if there is some (fixed) element c in Y such that $f(x) = c$ for every $x \in X$. In this event the image of every element of X is the same element of Y. If we represent a constant function by means of a diagram such as in Fig. 1.17, then every arrow from X would terminate at the same point in Y.

In the remainder of our work we shall restrict ourselves primarily to functions whose domains and ranges are subsets of $\mathbf{R}$, the set of real numbers.

Example 2. Let X denote the set of nonnegative real numbers and let f be the function from X to $\mathbf{R}$ defined by $f(x) = \sqrt{x} + 1$, for all $x \in X$. Find $f(1), f(5/2), f(\sqrt[3]{2})$, and $f(b + c)$, where $b, c \in X$. What is the range of f?

Solution: Finding images under f is simply a matter of substituting the appropriate number for x in the expression for $f(x)$. Thus

$$f(1) = \sqrt{1} + 1 = 2,$$
$$f(5/2) = \sqrt{5/2} + 1 = \tfrac{1}{2}\sqrt{10} + 1,$$
$$f(\sqrt[3]{2}) = \sqrt{\sqrt[3]{2}} + 1 = \sqrt[6]{2} + 1,$$
$$f(b + c) = \sqrt{b + c} + 1.$$

The range of f is $\{f(a) \mid a \geq 0\} = \{\sqrt{a} + 1 \mid a \geq 0\}$. Thus the range of f is the set of all real numbers y such that $y \geq 1$.

In certain areas of mathematics, such as calculus, it is important to carry out manipulations such as those given in the next example.

Example 3. Let f be the function with domain $\mathbf{R}$ defined by $f(x) = x^2 + 3x - 2$ for all $x \in \mathbf{R}$. Let $a, t \in \mathbf{R}, t \neq 0$. Find

$$\frac{f(a + t) - f(a)}{t}.$$

Solution: We have

$$f(a + t) = (a + t)^2 + 3(a + t) - 2$$

and

$$f(a) = a^2 + 3a - 2.$$

Hence

$$\begin{aligned}\frac{f(a + t) - f(a)}{t} &= \frac{(a^2 + 2at + t^2) + (3a + 3t) - 2 - (a^2 + 3a - 2)}{t}\\ &= \frac{2at + t^2 + 3t}{t}\\ &= 2a + t + 3.\end{aligned}$$

One may use the concept of relation to obtain an alternate approach to the definition of function. We shall conclude this section by discussing this approach.

A function f from X to Y determines a relation W between X and Y, that is, a certain set of ordered pairs of the form (x, y), where $x \in X$, $y \in Y$. Specifically, W is defined by

$$W = \{(x, f(x)) \mid x \in X\},$$

or equivalently by

$$W = \{(x, y) \mid x \in X, y = f(x)\}.$$

In either case, note that W is found by considering the totality of pairs such that in each pair the first element is in X and the second element is the image, under f, of the first element. Thus, in Example 2, W consists of all pairs of the form $(x, \sqrt{x} + 1)$, where x is a nonnegative real number. It is important to note that for each $x \in X$ there is one and only one pair $(x, y) \in W$ having x as its first element.

Conversely, if we *begin* with a relation W between X and Y such that each element of X appears *exactly once* as a first element of a pair, then W determines a function from X to Y. To see this, note that for any $x \in X$ there is a *unique* pair (x, y) in W. By letting y correspond to x, we obtain a function from X to Y.

It follows from the preceding discussion that statement (1.21) below could serve equally well as a definition of function. We prefer, however, to think of it as an alternate approach to this concept and we shall refer to it only sparingly in this book.

(1.21) Alternate Definition of a Function

A *function* from a set X to a set Y is a relation W between X and Y such that for each $x \in X$, there is exactly one ordered pair (x, y) in W having x as its first element.

Using (1.21), the function f of Example 1 would be described as the relation $W = \{(x, x^2) \mid x \in \mathbf{R}\}$. Similarly, the relation

$$W = \{(x, x^2 + 3x - 2) \mid x \in \mathbf{R}\},$$

determines the function given in Example 3.

EXERCISES

1. If f is the function with domain $\mathbf{R}$ such that $f(x) = x^2 - 3x + 5$, find:

   (a) $f(1)$. (b) $f(3)$. (c) $f(0)$.
   (d) $f(\sqrt{2} - 1)$. (e) $f(a)$. (f) $f(-a)$.
   (g) $-f(a)$. (h) $f(a + h)$. (i) $f(a) + f(h)$.
   (j) $\dfrac{f(a + h) - f(a)}{h}$. (k) $\dfrac{f(x + h) - f(x)}{h}$.

2. Same as Exercise 1 if $f(x) = 3x^2 - x + 2$.
3. Same as Exercise 1 for the function f whose domain is

$$X = \{x \mid x > -1\} \quad \text{and} \quad f(x) = 1/(x + 1).$$

4. Same as Exercise 1 if $f(x) = 1/(x + 1)^2$.

5. If $g(x) = 1/(x^2 + 1)$, find:
   (a) $g(1/a)$. (b) $1/g(a)$. (c) $g(a^2)$.
   (d) $(g(a))^2$. (e) $g(\sqrt{x})$. (f) $\sqrt{g(x)}$.
6. Same as Exercise 5 if $g(x) = 1/x$.
7. Let f be the function from $X = \{x \mid x \geq \frac{1}{2}\}$ to **R** defined by $f(x) = \sqrt{2x - 1}$, $x \in X$. What number is sent into 4 by f? If $a > 0$, what number is sent into a? Find the range of f.
8. Same as Exercise 7 if f is defined by $f(x) = x^3 - 1$.

In Exercises 9–12 find the largest subset of **R** that can serve as domain of the function f defined by the indicated expression.

9. $f(x) = \sqrt{x - 1}$.
10. $f(x) = \dfrac{1}{3x - 2}$.
11. $f(x) = \dfrac{1}{x^2 + 1}$.
12. $f(x) = \sqrt{1 - x^2}$.

In Exercises 13–16 determine whether the function f with domain **R** defined by the given expression is one-to-one.

13. $f(x) = 2x - 5$.
14. $f(x) = 3 - 7x^2$.
15. $f(x) = 2x^2 + 3x - 4$.
16. $f(x) = |x|$.

A function f is termed *even* if $f(-a) = f(a)$ for all a in the domain X of f, whereas f is *odd* if $f(-a) = -f(a)$ for all $a \in X$. In Exercises 17–22 determine which functions defined by the indicated expressions are even, odd, or neither even nor odd.

17. $f(x) = x^4 + 2x^2 - 5$.
18. $f(x) = x^2 - 3x + 2$.
19. $f(x) = 3x^3 - 5x$.
20. $f(x) = \sqrt[3]{x^3 + 1}$.
21. $f(x) = \dfrac{x}{\sqrt{1 + x^4}}$.
22. $f(x) = 5$.

In Exercises 23–26 determine which of the given relations on **R** determine functions in the sense of (1.21).

23. $\{(x, y) \mid y = 3x - 5\}$.
24. $\{(x, y) \mid y = x^2 + 1\}$.
25. $\{(x, y) \mid x^2 + y^2 = 1\}$.
26. $\{(x, y) \mid y^2 = x^2\}$.
27. $\{(x, y) \mid x = 4\}$.
28. $\{(x, y) \mid y = 1\}$.
29. $\{(x, y) \mid x + y = 0\}$.
30. $\{(x, y) \mid xy = 0\}$.

6 GRAPHS OF FUNCTIONS

Algebraic expressions were defined on p. 10. A function f is called *algebraic* if there is an algebraic expression p in a variable x such that $f(x) = p$ for all x in the domain of f. In this and the next two chapters, we shall work primarily with such functions. As illustrations, the functions f and g defined as follows are algebraic:

$$f(x) = \sqrt{x} + 3, \qquad g(x) = \frac{3x^3 + 2}{x - 1}.$$

We will not usually specify the domain or the range of an algebraic function. Rather it will be *assumed* that the domain is the totality of real numbers for which the expression p is meaningful. With this understanding, the domain of f defined above is the set of nonnegative real numbers and the domain of g is the set of all real numbers different from 1. If x is in the domain of f we sometimes say that f is *defined* at x, or $f(x)$ *exists*. The terminology f is *undefined* at x, or $f(x)$ does *not* exist, means that x is not in the domain of f.

Henceforth, when we use the expression "f is a function," it is assumed that the domain is some (perhaps unspecified) subset of **R**, and similarly for the range. In any particular examples the reader should be able to find the domain and range.

(1.22) Definition of Graph of a Function

The *graph* of a function f is the graph of the relation

$$\{(x, f(x)) \mid x \text{ is in the domain of } f\}.$$

The graph of f can also be described as the set of all points $P(x, y)$ in a coordinate plane such that $y = f(x)$, where x is in the domain of f. Thus the graph of f is the same as the graph of the equation $y = f(x)$. Note that if $P(x, y)$ is on the graph of f, then the ordinate y of P is the functional value $f(x)$. If we consider f to be a relation W, as in (1.21), then the graph of f is the graph of W.

EXAMPLE 1. Sketch the graph of the function f defined by $f(x) = 2x - 1$.

Solution: By our previous remarks, it is assumed that the domain of f is **R**. By (1.22) the graph of f is the graph of $\{(x, 2x - 1) \mid x \in \mathbf{R}\}$ and hence is identical with the graph of the relation considered in Example 3 of Section 4. The graph is sketched in Fig. 1.11.

A function f, such as that given in Example 1, which is defined by an expression of the form $f(x) = ax + b$, where $a, b \in \mathbf{R}$, is called a *linear function*. It will be shown in Chapter Seven that the graph of a linear function is always a straight line.

It is often useful to find the points at which the graph of a function intersects the x-axis. The abscissas of these points are called the *x-intercepts* of the graph and are found by locating all points with zero ordinates, that is, all points $(x, f(x))$ such that $f(x) = 0$. In Example 1 the x-intercept is $\frac{1}{2}$. A number a such that $f(a) = 0$ is also called a *zero* of the function f.

If the number 0 is in the domain of f, then $f(0)$ is called the *y-intercept* of the graph of f. It is the ordinate of the point at which the graph intersects the y-axis. The graph of a function can have at most one y-intercept (Why?). The y-intercept in Example 1 is -1.

EXAMPLE 2. Sketch the graph of the function f defined by $f(x) = x^2 - 3$.

Solution: We list coordinates $(x, f(x))$ of some points on the graph of f in tabular form, as shown below.

x	-3	-2	-1	0	1	2	3	4
$f(x)$	6	1	-2	-3	-2	1	6	13

The x-intercepts are the solutions of the equation $f(x) = 0$, that is, of $x^2 - 3 = 0$. These are $\pm\sqrt{3}$. The y-intercept is $f(0) = -3$. Plotting

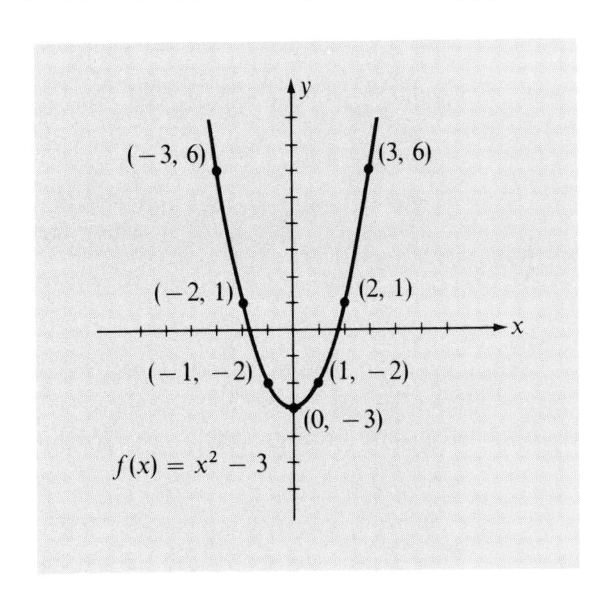

Figure 1.18

points, we find that the graph has the form shown in Fig. 1.18, where again we cannot show the entire graph (Why not?).

EXAMPLE 3. Sketch the graph of f if $f(x) = |x|$ for all $x \in \mathbf{R}$.

Solution: If $x > 0$, then $f(x) = x$ and we obtain the set of points (x, x) on the graph of f. Negative values of x give rise to the table shown below.

x	-1	-2	-3	-4
$f(x)$	1	2	3	4

More generally, we obtain all points of the form $(-a, a)$, $a > 0$. Part of the graph is shown in Fig. 1.19.

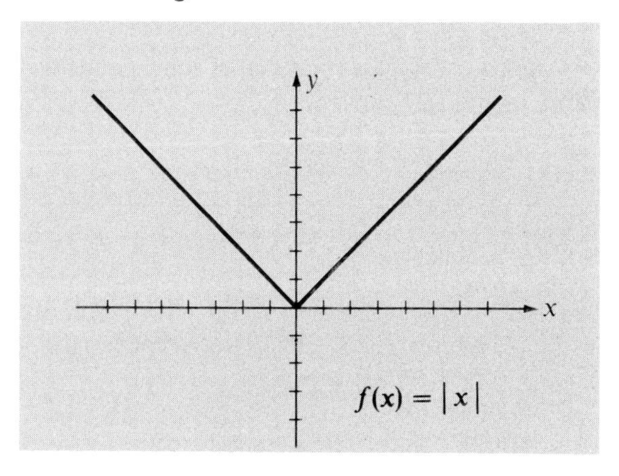

Figure 1.19

Notice that no portion of the graph in Example 3 appears *below* the x-axis. A region of the coordinate plane in which there is no graph is called an *excluded region.*

EXAMPLE 4. Sketch the graph of f if $f(x) = \sqrt{x - 1}$.

Solution: The domain of f cannot include values of x such that $x - 1 < 0$, since $f(x)$ is not real in this case. Therefore the set of points (x, y) with $x < 1$ is an excluded region for the graph. We have the following table.

x	1	2	3	4	5	6
$f(x)$	0	1	$\sqrt{2}$	$\sqrt{3}$	2	$\sqrt{5}$

The sketch is shown in Fig. 1.20. The x-intercept is 1 and there is no y-intercept (Why?).

EXAMPLE 5. Sketch the graph of f if $f(x) = 1/x$.

Solution: The domain of f is the set of all nonzero real numbers. Before constructing a table, let us make some general observations. First, when x is positive, so is $f(x)$, and hence quadrant IV is an excluded region.

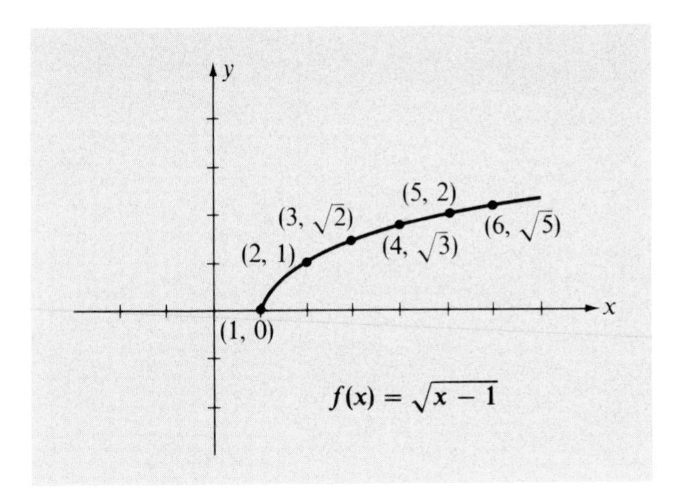

Figure 1.20

Quadrant II is also an excluded region, since when $x < 0$, then also $f(x) < 0$. If x is close to zero, the ordinate $1/x$ is very large numerically. As x increases through positive values, $1/x$ decreases and is close to zero when x is large. Similarly, if we let x take on numerically large negative values, the ordinate $1/x$ is close to zero. From these remarks and the following table, we obtain the sketch given in Fig. 1.21. Note that this graph has *neither* an x-intercept nor a y-intercept.

x	$\frac{1}{10}$	$\frac{1}{2}$	1	2	5	10	$-\frac{1}{2}$	-1	-2	-4
$f(x)$	10	2	1	$\frac{1}{2}$	$\frac{1}{5}$	$\frac{1}{10}$	-2	-1	$-\frac{1}{2}$	$-\frac{1}{4}$

EXAMPLE 6. Sketch the graph of the constant function defined by $f(x) = k$, where $k \in \mathbf{R}$.

Solution: The graph of f consists of all points with coordinates (x, k), where $x \in \mathbf{R}$. This includes $(-1, k)$, $(0, k)$, $(2, k)$, and so on. Since all the ordinates equal k, the graph is a line parallel to the x-axis with y-intercept k. A sketch for the case $k > 0$ is shown in Fig. 1.22.

Functions do not always have to be defined by algebraic expressions as in the previous examples, nor do the graphs always consist of one unbroken line or curve. This is illustrated in the next example.

EXAMPLE 7. If x is any real number, then there exist consecutive integers n and $n + 1$ such that $n \leq x < n + 1$. Let us define a function f from $\mathbf{R}$ to $\mathbf{R}$ as follows: if $x \in \mathbf{R}$ and $n \leq x < n + 1$, then $f(x) = n$. Sketch the graph of f.

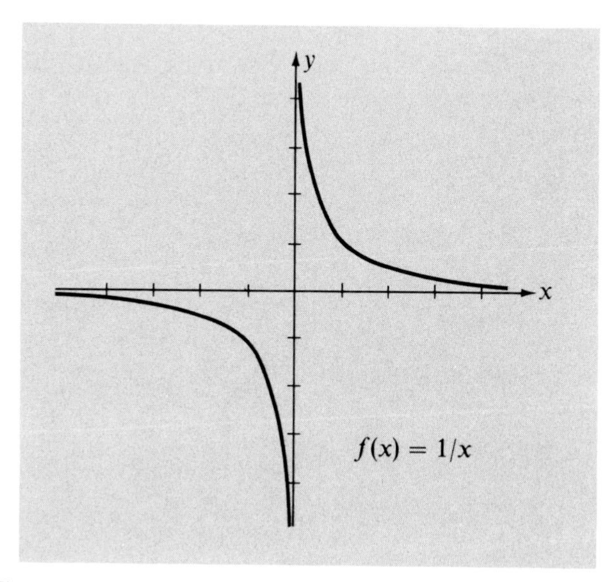

Figure 1.21

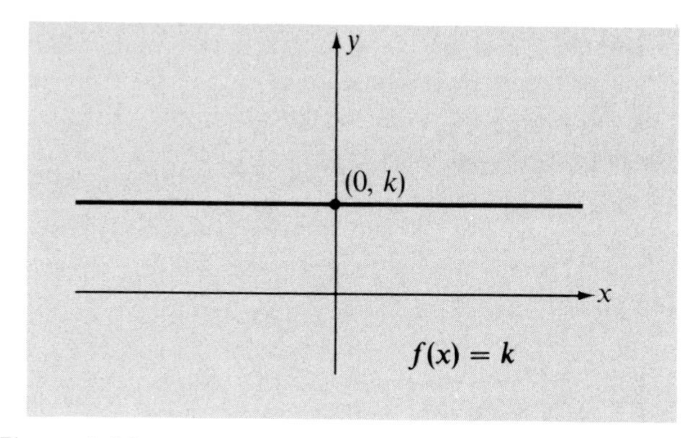

Figure 1.22

Solution: Instead of making the usual table, we list abscissas and ordinates as follows:

Values of x	$f(x)$
$-2 \le x < -1$	-2
$-1 \le x < 0$	-1
$0 \le x < 1$	0
$1 \le x < 2$	1
$2 \le x < 3$	2

We see that f behaves in the same manner as a constant function when x is between integer values of x and hence the corresponding part of the graph is a segment of a horizontal line.

Part of the graph of f is given in Fig. 1.23. This function is sometimes called a *step function*. The "steps" continue indefinitely in both directions. The graph has one y-intercept and an infinite number of x-intercepts.

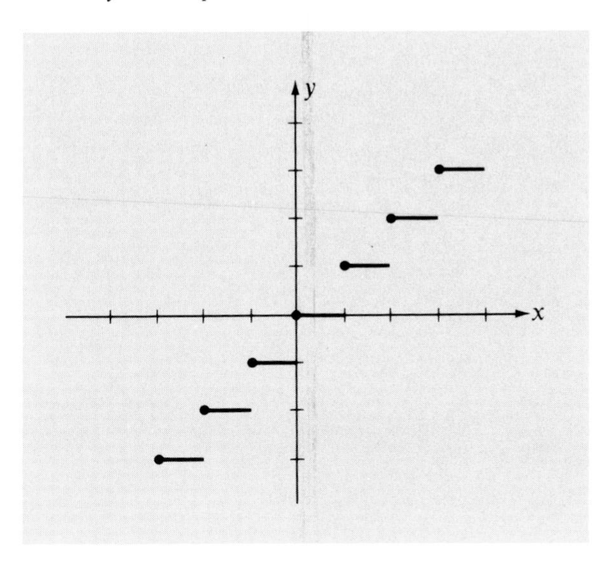

Figure 1.23

As a final remark, let us re-emphasize the fact that since each x in the domain of a function f produces precisely *one* $f(x)$, there can be only *one* point on the graph corresponding to each x. Consequently, for graphs of functions it is impossible to obtain a sketch such as that shown in Fig. 1.13 or Fig. 1.15, where a vertical line pierces the graph in more than one point.

EXERCISES

What subset of **R** is assumed to be the domain of f if $f(x)$ is as given by Exercises 1–4?

1. $f(x) = \dfrac{3x - 5}{x^2 - 4}$.
2. $f(x) = \sqrt{x^2 + 1}$.
3. $f(x) = \dfrac{1}{\sqrt{2 + x}\sqrt{3 - x}}$.
4. $f(x) = \dfrac{x}{x(x - 1)}$.

In the following exercises sketch the graph of the function f defined as indicated.

5. $f(x) = 3x + 5$.
6. $f(x) = 3x - 5$.
7. $f(x) = -3x + 5$.
8. $f(x) = -3x - 5$.
9. $f(x) = 4$.
10. $f(x) = -4$.
11. $f(x) = -x^2$.
12. $f(x) = 1 - x^2$.

13. $f(x) = 2 - x^2$.

14. $f(x) = x^2 - 1$.

15. $f(x) = \dfrac{1}{x - 1}$.

16. $f(x) = \dfrac{1}{x + 1}$.

17. $f(x) = \dfrac{1}{(x - 1)^2}$.

18. $f(x) = \dfrac{-1}{x + 1}$.

19. $f(x) = \sqrt{25 - x^2}$.

20. $f(x) = \sqrt{x^2 - 25}$.

21. $f(x) = |1 + x|$.

22. $f(x) = 1 + |x|$.

23. $f(x) = x\,|x|$.

24. $f(x) = \dfrac{x}{|x|}$.

25. $f(x) = \sqrt{1 - x}$.

26. $f(x) = 1 - \sqrt{x}$.

27. $f(x) = \begin{cases} 2 & \text{if } x < 0 \\ -2 & \text{if } x \geq 0. \end{cases}$

28. $f(x) = \begin{cases} 0 & \text{if } x \in \mathbf{Z} \\ 1 & \text{if } x \notin \mathbf{Z}. \end{cases}$

29. $f(x) = \begin{cases} 2 & \text{if } x < -2 \\ -x & \text{if } -2 \leq x \leq 2 \\ -2 & \text{if } x > 2. \end{cases}$

30. $f(x) = \begin{cases} \sqrt{-x} & \text{if } x < 0 \\ 3 & \text{if } x = 0 \\ 2x & \text{if } x > 0. \end{cases}$

7 THE ALGEBRA OF FUNCTIONS

In mathematics and its applications it is common to encounter functions that are defined in terms of sums, differences, products and quotients of various expressions. The function h defined by $h(x) = x^2 + \sqrt{5x + 1}$ is an example. We may regard $h(x)$ as a sum of functional values of the "simpler" functions f and g defined by $f(x) = x^2$ and $g(x) = \sqrt{5x + 1}$. According to the definition given below, the function h is the *sum* of the functions f and g.

In the following discussion f and g will denote arbitrary functions and D will denote the intersection of their domains. Thus if $x \in D$, then both $f(x)$ and $g(x)$ exist. The *sum* of f and g is the function s defined by

$$s(x) = f(x) + g(x), \qquad x \in D.$$

In order to indicate that values of s are found by *adding* values of f and g it is convenient to denote the function s by the symbol $f + g$. Since f and g are functions and not numbers, the "+" used between f and g is not to be considered as addition of real numbers. As we have said above, it is used to indicate that the image of x under $f + g$ is $f(x) + g(x)$. Notationally this is written

$$(f + g)(x) = f(x) + g(x), \qquad x \in D,$$

or, in terms of the arrow notation described on p. 29, we have

$$(f + g)\colon x \to f(x) + g(x).$$

Similarly, the *difference* $f - g$ of f and g is defined by

$$(f - g)(x) = f(x) - g(x), \qquad x \in D.$$

In like manner, the *product* of f and g is a function denoted by fg and defined by

$$(fg)(x) = f(x)g(x), \qquad x \in D,$$

whereas the *quotient* of f by g is the function denoted by f/g and defined by

$$(f/g)(x) = f(x)/g(x), \qquad x \in D \quad \text{and} \quad g(x) \neq 0.$$

EXAMPLE 1. Let f and g be defined by $f(x) = \sqrt{4 - x^2}$ and $g(x) = 3x + 1$, respectively. Find the sum, difference, and product of f and g, and the quotient of f by g.

Solution: By our agreement on algebraic functions, the domain of f is $X = \{x \mid -2 \leq x \leq 2\}$ and the domain of g is **R**. Therefore the intersection of their domains is X. The required functions are given by

$$(f + g)(x) = \sqrt{4 - x^2} + 3x + 1, \qquad -2 \leq x \leq 2$$

$$(f - g)(x) = \sqrt{4 - x^2} - (3x + 1), \qquad -2 \leq x \leq 2$$

$$(fg)(x) = \sqrt{4 - x^2}\,(3x + 1), \qquad -2 \leq x \leq 2$$

$$(f/g)(x) = \frac{\sqrt{4 - x^2}}{3x + 1}, \qquad -2 \leq x \leq 2, \qquad x \neq -\tfrac{1}{3}.$$

The definitions of sum and product may be extended to more than two functions in the obvious way.

If a function s is a sum of two functions f and g, then the graph of s may be found from the graphs of f and g by a technique called *addition of ordinates.* Thus consider $s(x) = f(x) + g(x)$, $x \in D$. We begin by sketching the graphs of the equations $y = f(x)$ and $y = g(x)$ on the same coordinate axes, as illustrated by the dashes in Fig. 1.24. If $x_1 \in D$, then

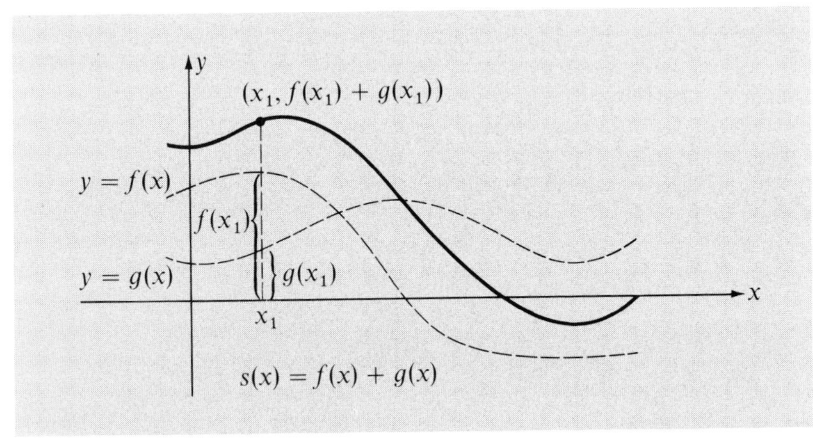

Figure 1.24

$s(x_1) = f(x_1) + g(x_1)$; that is, the ordinate of the point on the graph of s with abscissa x_1 is the *sum* of the corresponding ordinates of points on the graphs of f and g. After drawing a vertical line through the point with coordinates $(x_1, 0)$, the ordinates $f(x_1)$ and $g(x_1)$ may be added geometrically by means of a compass or ruler, as is illustrated in Fig. 1.24. Of course, if either $f(x_1)$ or $g(x_1)$ is negative, then a *subtraction* of ordinates may be employed.

By using this technique for all x we obtain the graph of s, as illustrated by the heavy curve in Fig. 1.24.

EXAMPLE 2. Use the method of addition of ordinates to sketch the graph of the function f defined by

$$f(x) = x + \frac{1}{x}.$$

Solution: The graphs of the equation $y = x$ and $y = 1/x$ are illustrated by the dashes in Fig. 1.25. We leave it as an exercise for the student to verify that the indicated graph of f may be obtained by adding ordinates.

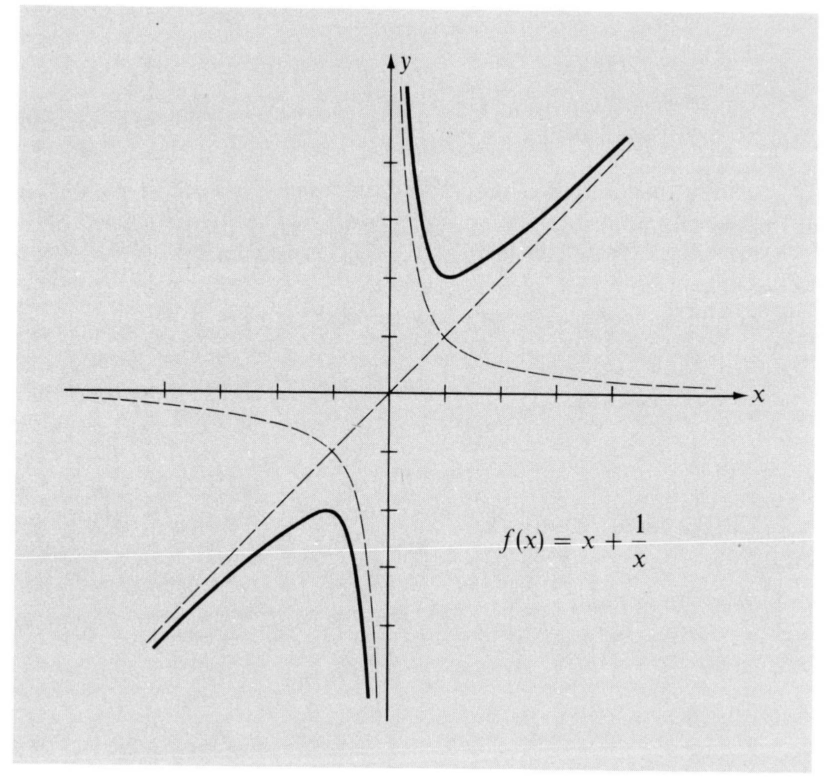

Figure 1.25

We conclude this section by describing another method of using two functions f and g to obtain a third function. Suppose X, Y, and Z are sets and let f be a function from X to Y and g a function from Y to Z. In terms of the arrow notation introduced on p. 29, we have

$$X \xrightarrow{f} Y \xrightarrow{g} Z$$

— that is, f sends X into Y and g sends Y into Z. A function from X to Z may be defined in a natural way. Given $x \in X$, the number $f(x)$ is in Y. Since the domain of g is Y we may then find the image of $f(x)$ under g. Of course, this element of Z is written as $g(f(x))$. By associating $g(f(x))$ with x, we obtain a function from X to Z called the *composite function* of g by f. This is illustrated pictorially in Fig. 1.26, where the dashes

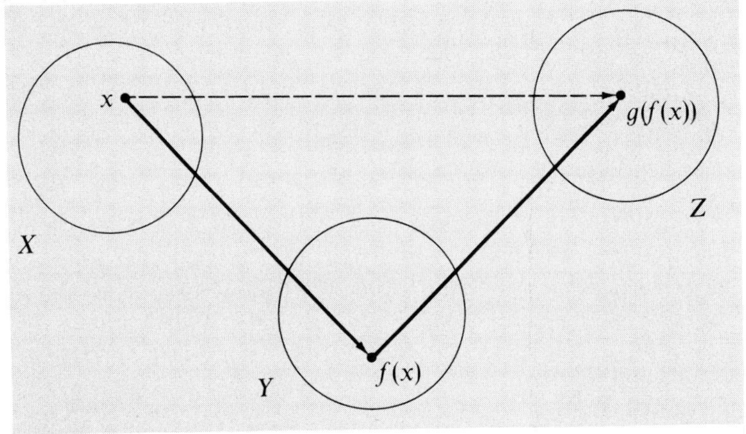

Figure 1.26

indicate the correspondence we have defined from X to Z. Since we have described a method for using two functions f and g to obtain a third function, we shall use an operational symbol $\circ$ and denote the latter function by $g \circ f$. The following definition summarizes our remarks.

(1.23) Definition of Composite Function

If f is a function from X to Y and g is a function from Y to Z, then the *composite function* $g \circ f$ is the function from X to Z defined by

$$(g \circ f)(x) = g(f(x)),$$

for all $x \in X$.

Actually, it is not essential that the domain of g be all of Y. The requirement is that the domain of g *contain* the range of f, since it is only necessary to find $g(f(a))$ for each $a \in X$. In certain cases one may wish to restrict x to some *subset* of X so that $f(x)$ is in the domain of g. This is illustrated in the next example.

EXAMPLE 3. If the functions f and g are defined by $f(x) = x - 2$ and $g(x) = 5x + \sqrt{x}$ find $(g \circ f)(x)$.

Solution: Formal substitutions give us the following:

$$\begin{aligned} (g \circ f)(x) &= g(f(x)) && \text{(definition of } g \circ f) \\ &= g(x - 2) && \text{(definition of } f) \\ &= 5(x - 2) + \sqrt{x - 2} && \text{(definition of } g) \\ &= 5x - 10 + \sqrt{x - 2} && \text{(simplifying).} \end{aligned}$$

The domain X of f is the set of all real numbers. However, we see from the last equality that $(g \circ f)(x)$ is a real number only if $x - 2 \geq 0$. Thus when working with the composite function $g \circ f$ we must restrict x to the subset $\{x \mid x \geq 2\}$ of X.

If $X = Y = Z$, then it is possible to find $f(g(x))$, where, as usual, this notation means to first obtain the image of x under g and then apply f to $g(x)$. In this way we obtain a function from Z to X called the *composite function* of f by g and denoted by $f \circ g$. Thus, by definition,

$$(f \circ g)(x) = f(g(x)),$$

for all $x \in Z$.

EXAMPLE 4. Let the functions f and g be given by $f(x) = x^2 - 1$ and $g(x) = 3x + 5$. Find $(f \circ g)(x)$ and $(g \circ f)(x)$.

Solution: We have the following identities:

$$\begin{aligned} (f \circ g)(x) &= f(g(x)) && \text{(definition of } f \circ g) \\ &= f(3x + 5) && \text{(definition of } g) \\ &= (3x + 5)^2 - 1 && \text{(definition of } f) \\ &= 9x^2 + 30x + 24 && \text{(simplifying).} \end{aligned}$$

Also,

$$\begin{aligned} (g \circ f)(x) &= g(f(x)) && \text{(definition of } g \circ f) \\ &= g(x^2 - 1) && \text{(definition of } f) \\ &= 3(x^2 - 1) + 5 && \text{(definition of } g) \\ &= 3x^2 + 2 && \text{(simplifying).} \end{aligned}$$

We see by Example 4 that $f(g(x))$ and $g(f(x))$ are not always the same — that is, $f \circ g \neq g \circ f$.

EXERCISES

If the functions f and g are defined as in Exercises 1–4, find the sum, difference, and product of f and g, and the quotient of f by g.

1. $f(x) = 3x^2, \quad g(x) = \dfrac{1}{2x - 3}.$
2. $f(x) = \sqrt{x + 3}, \quad g(x) = \sqrt{x + 3}.$
3. $f(x) = x + \dfrac{1}{x}, \quad g(x) = x - \dfrac{1}{x}.$
4. $f(x) = x^3 + 3x, \quad g(x) = 3x^2 + 1.$

In Exercises 5 and 6 use the method of addition of ordinates to sketch the graph of the function f.

5. $f(x) = x^2 + x + 1.$
6. $f(x) = \sqrt{x} + x - 2.$

In Exercises 7–14 find $(f \circ g)(x)$ and $(g \circ f)(x)$ if f and g are as defined by the given expressions.

7. $f(x) = 2x^2 + 5, \quad g(x) = 4 - 7x.$
8. $f(x) = 1/(3x + 1), \quad g(x) = 2/x^2.$
9. $f(x) = x^3, \quad g(x) = x + 1.$
10. $f(x) = \sqrt{x^2 + 4}, \quad g(x) = 7x^2 + 1.$
11. $f(x) = 3x^2 + 2, \quad g(x) = 1/(3x^2 + 2).$
12. $f(x) = 7, \quad g(x) = 4.$
13. $f(x) = \sqrt{2x + 1}, \quad g(x) = x^2 + 3.$
14. $f(x) = 6x - 12, \quad g(x) = \frac{1}{6}x + 2.$

8 INVERSE FUNCTIONS

In Example 4 of the previous section we considered two functions f and g such that $f(g(x)) \neq g(f(x))$. In certain cases it may happen that equality *does* occur. Of major importance is the case in which not only are $f(g(x))$ and $g(f(x))$ identical, but each of these expressions also equals x. Needless to say, f and g have to be very special functions in order for this to happen. In the following discussion we indicate the manner in which they will be restricted.

Suppose f is a *one-to-one function* with domain X and range Y. Therefore, as mentioned on p. 30, each element of Y is the image of precisely one element of X. Another way of saying this is that *each element of Y can be written in one and only one way in the form $f(x)$*, for some $x \in X$. We may then define a function g from Y to X by writing

(1.24) $$g(f(x)) = x,$$

for all $x \in X$. What this amounts to is *reversing* the original correspondence from X to Y. If f is represented geometrically by drawing arrows as in Fig. 1.16, then g can be represented by simply *reversing* these arrows. It follows that g is a one-to-one function with domain Y and range X.

Since g is a function from Y to X, then for any $y \in Y$ we have $g(y) \in X$ (see Fig. 1.27). Since $g(y)$ is in the domain of f, we can find its image,

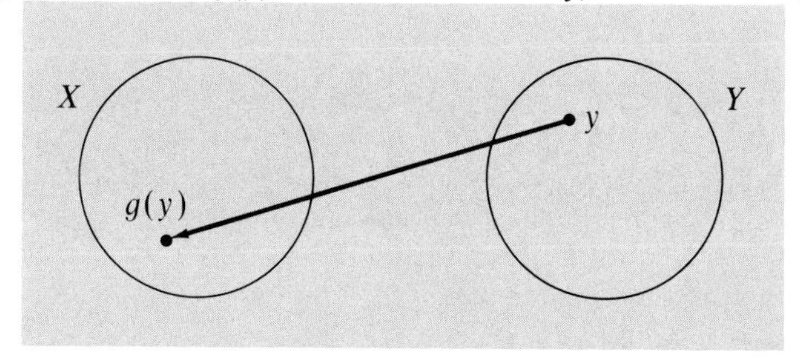

Figure 1.27

$f(g(y))$. However, $y \in Y$ implies that

(1.25) $$y = f(x), \text{ for a unique } x \in X.$$

Consequently

$$\begin{aligned} f(g(y)) &= f(g(f(x))) && (1.25) \\ &= f(x) && (1.24) \\ &= y && (1.25) \end{aligned}$$

for all $y \in Y$. Since the notation used for the variable is immaterial (see p. 30), we can write

$$f(g(x)) = x, \qquad \text{for all} \quad x \in Y.$$

The functions f and g are called *inverse functions* of one another, according to the following definition.

(1.26) Definition of Inverse Function

Let f be a one-to-one function with domain X and range Y. Then a function g with domain Y and range X is called the *inverse function* of f if

$$f(g(x)) = x, \qquad \text{for all} \quad x \in Y,$$

and

$$g(f(x)) = x, \qquad \text{for all} \quad x \in X.$$

There can be only one inverse function of f. Moreover, if g is the inverse function of f, then, by (1.26), $(g \circ f)(x) = x$. Since every $x \in X$ is sent into itself by $g \circ f$, $g \circ f$ is the *identity function* on X (see p. 30). Similarly, since $(f \circ g)(x) = x$, for all $x \in Y$, $f \circ g$ is the identity function on

Y. It is for this reason that g is called the *inverse* of f. Sometimes the symbol f^{-1} is used to denote the inverse function of f. In this event we have

$$(f^{-1} \circ f)(x) = f^{-1}(f(x)) = x, \qquad \text{for all} \quad x \in X,$$

and

$$(f \circ f^{-1})(x) = f(f^{-1}(x)) = x, \qquad \text{for all} \quad x \in Y.$$

The symbol -1 used in this notation should not be confused with the exponent -1 used for real numbers even though it behaves in a similar manner. It is employed here merely as a notation for representing inverse functions.

Inverse functions are very important in the study of trigonometry. In Chapter Four we shall discuss two other important classes of inverse functions. The following examples give less important illustrations of this concept.

EXAMPLE 1. If f is the function defined by $f(x) = 3x - 5$, find the inverse function of f.

Solution: It is not difficult to show that f is a one-to-one function with domain and range **R**. Hence the inverse function g exists. By (1.26) we must have $f(g(x)) = x$, for all $x \in \mathbf{R}$. Applying the definition of f, this equality gives us

$$3g(x) - 5 = x.$$

Solving for $g(x)$, we obtain

$$g(x) = \frac{x + 5}{3}. \tag{1.27}$$

To verify that the function g defined by (1.27) is actually the inverse function of f, we must check to see whether the two conditions stated in (1.26) are fulfilled. Thus

$$\begin{aligned} f(g(x)) &= f\left(\frac{x+5}{3}\right) && \text{(1.27)} \\ &= 3\left(\frac{x+5}{3}\right) - 5 && \text{(definition of } f) \\ &= x && \text{(simplifying).} \end{aligned}$$

Also,

$$\begin{aligned} g(f(x)) &= g(3x - 5) && \text{(definition of } f) \\ &= \frac{(3x - 5) + 5}{3} && \text{(1.27)} \\ &= x && \text{(simplifying).} \end{aligned}$$

Hence (1.27) defines the inverse function g of f.

EXAMPLE 2. Find the inverse function of f if the domain of f is the set of nonnegative real numbers and $f(x) = x^2 - 3$, for all x in the domain of f.

Solution: We have restricted the domain so that f is one-to-one. If we allowed negative values, then different numbers such as 1 and -1 would have the same image under f. The range of f is $\{y \mid y \geq -3\}$. If g is the inverse function of f, then, by (1.26), $f(g(x)) = x$, for all x. Using the definition of f, this gives us

$$[g(x)]^2 - 3 = x.$$

Solving for $g(x)$, we obtain the two possibilities $g(x) = \pm\sqrt{x + 3}$, $x \geq -3$. If we let

$$g(x) = \sqrt{x + 3}, \tag{1.28}$$

then

$$\begin{aligned} f(g(x)) &= f(\sqrt{x + 3}) && (1.28) \\ &= (\sqrt{x + 3})^2 - 3 && (\text{definition of } f) \\ &= (x + 3) - 3 && (\text{definition of square root}) \\ &= x && (\text{simplifying}). \end{aligned}$$

Also,

$$\begin{aligned} g(f(x)) &= g(x^2 - 3) && (\text{definition of } f) \\ &= \sqrt{(x^2 - 3) + 3} && (1.28) \\ &= x && (\text{simplifying, since } x \geq 0). \end{aligned}$$

This proves that (1.28) defines the inverse function g of f.

If we check $g(x) = -\sqrt{x + 3}$, we still obtain $f(g(x)) = x$. In this case, however, $g(f(x)) = -x$ and consequently the expression does *not* define the inverse function of f.

An alternate method may often be used for finding the inverse of a one-to-one function f with domain X and range Y. If we begin with the equation $y = f(x)$, then given $x \in X$, its image $y \in Y$ may be found by means of this equation. In order to find the inverse function f^{-1}, we wish to *reverse* this procedure, in the sense that given $y \in Y$, we can find $x \in X$ by means of some equation. Since x and y are related by means of $y = f(x)$, it follows that if we can *solve the latter equation for x in terms of y*, we will arrive at the inverse function f^{-1}.

EXAMPLE 3. Rework Examples 1 and 2 using the technique described in the preceding paragraph.

Solution: In Example 1, the function f is defined by $f(x) = 3x - 5$. We let $y = 3x - 5$ and then solve the equation for x in terms of y, obtaining $x = (y + 5)/3$. The latter equation enables us, when given y, to find x.

Letting $g(y) = (y + 5)/3$, then as illustrated in Fig. 1.27, we have a function from Y to X which reverses the correspondence defined by f. From our remarks about the symbol used for the independent variable (see p. 30), we may replace y by x in the expression which defines g, obtaining $g(x) = (x + 5)/3$. This agrees with (1.27).

In like manner, in Example 2, where $f(x) = x^2 - 3$, we begin by considering the equation $y = x^2 - 3$. Solving for x gives us $x = \pm\sqrt{y + 3}$. As in the preceding paragraph, let

$$g(y) = \sqrt{y + 3} \quad \text{or} \quad g(y) = -\sqrt{y + 3},$$

or equivalently

$$g(x) = \sqrt{x + 3} \quad \text{or} \quad g(x) = -\sqrt{x + 3}.$$

It is now necessary to see which of these leads to the inverse function of f. This can be done as in the solution of Example 2.

There is an interesting relationship between the graphs of a function f and its inverse function f^{-1}. We note first that if a is sent into b by f, then b is sent into a by f^{-1}; that is, if $b = f(a)$, $a = f^{-1}(b)$. These equations imply that if the point with coordinates (a, b) is on the graph of f then the point (b, a) is on the graph of f^{-1}. As an illustration, in Example 2

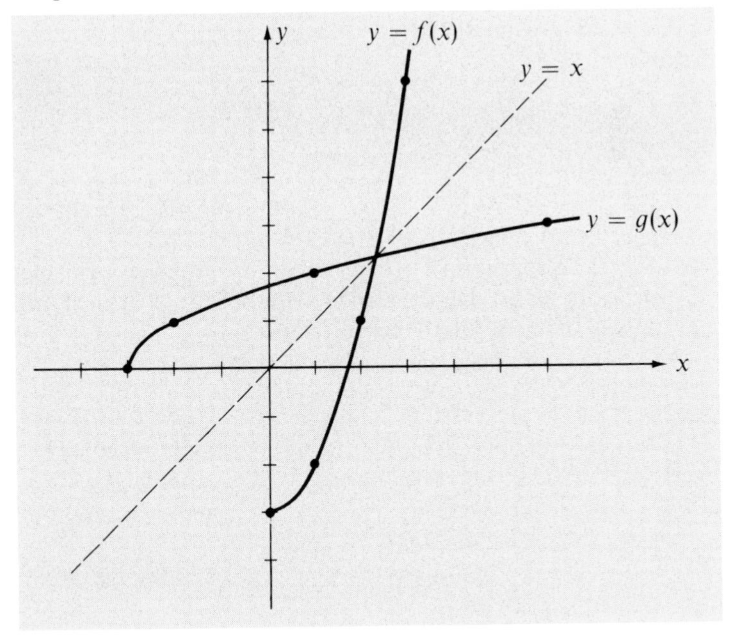

Figure 1.28

we found that the functions f and g defined by $f(x) = x^2 - 3$ and $g(x) = \sqrt{x + 3}$ are inverse functions of one another when x is suitably restricted. Some points on the graph of f are $(0, -3)$, $(1, -2)$, $(2, 1)$, and $(3, 6)$. Corresponding points on the graph of g are $(-3, 0)$, $(-2, 1)$, $(1, 2)$, and $(6, 3)$. The graphs of f and g are sketched on the same coordinate axes in Fig. 1.28. If the page is "folded" along the line having equation $y = x$ (as indicated by the dashes in the figure), then the graphs of f and g coincide. We sometimes say that the two graphs are *reflections* of one another through the line with equation $y = x$. This is typical of the graphs of all functions f which have inverse functions f^{-1}.

EXERCISES

In Exercises 1–4 prove that the functions f and g which are defined as indicated are inverse functions of one another.

1. $f(x) = 9x + 2, \quad g(x) = \frac{1}{9}x - \frac{2}{9}.$
2. $f(x) = x^3 + 1, \quad g(x) = \sqrt[3]{x - 1}.$
3. $f(x) = \sqrt{2x + 1}, \quad g(x) = \frac{1}{2}x^2 - \frac{1}{2}, \quad x \geq 0.$
4. $f(x) = \dfrac{1}{x^2 - 1}, \quad g(x) = \sqrt{\dfrac{1 + x}{x}}, \quad x > 1.$

In Exercises 5–10 find the inverse function of f if f is as defined by the given expression.

5. $f(x) = 8 + 11x.$
6. $f(x) = \dfrac{1}{8 + 11x}, \quad x > 0.$
7. $f(x) = 6 - x^2, \quad 0 \leq x \leq 2.$
8. $f(x) = 2x^3 - 5.$
9. $f(x) = \sqrt{7x - 2}, \quad x \geq \frac{2}{7}.$
10. $f(x) = \sqrt{1 - 4x^2}, \quad 0 \leq x \leq \frac{1}{2}.$
11. Sketch the graphs of f and g given in Exercise 1 on the same coordinate plane. Do the same for the functions defined in Exercise 3.
12. If f is a one-to-one function from X to Y and if g is defined by (1.24) prove that g is a one-to-one function from Y to X.

9 REVIEW EXERCISES

Oral

Define or explain the following concepts.

1. The union and intersection of two sets.
2. The commutative and associative laws of addition and multiplication.
3. The distributive laws.
4. Rational and irrational numbers.
5. The absolute value of a real number.
6. Coordinate line.
7. Rectangular coordinate system in a plane.
8. A relation between two sets.
9. The graph of an equation.
10. Function.
11. The domain and range of a function.
12. The graph of a function.
13. The composite function of two functions.
14. The inverse of a function.

Written

1. If $R = \{1, 5, 6\}$, $S = \{2, 3, 4\}$, and $T = \{6, 2\}$ find:
   (a) $R \cap (S \cup T)$. (b) $T \cap (R \cup S)$.
   (c) $R \cup (S \cap T)$. (d) $(R \cup S) \cap (R \cup T)$.

2. Express each of the following statements in terms of inequalities:
   (a) x is nonnegative. (b) x is between 4 and -3.
   (c) x is not greater than 2.

3. Rewrite each of the following numbers without using the absolute value symbol:
   (a) $|\pi - 22/7|$. (b) $|2 - \sqrt{4}|$. (c) $-|-3|$.

4. Find the solution sets of the following equations:
   (a) $\dfrac{5 - 3x}{2x - 7} = \dfrac{6x + 5}{1 - 4x}$. (b) $\dfrac{3}{x^2} - \dfrac{5}{x} + 2 = 0$.
   (c) $|5x - 2| = |2x + 13|$. (d) $x - \sqrt{2x - 5} = 4$.

5. Solve the following inequalities:
   (a) $5/(7 - 2x) < 0$. (b) $|x - 10| < 0.3$.
   (c) $|3x + 2| > 1$. (d) $x^2 < 4x$.

6. Describe the set of all points $P(x, y)$ in a coordinate plane such that:
   (a) $x = 5$. (b) $y = -2$.
   (c) $xy < 0$. (d) $x = y$.

7. Prove that $A(-3, 4)$, $B(2, -1)$, and $C(9, 6)$ are vertices of a right triangle and find its area.

8. Sketch the graphs of the following equations:
   (a) $x^2 + y = 4$. (b) $x^2 + y^2 = 4$.
   (c) $x^3 + y = 4$. (d) $x^2 + (y + 3)^2 = 9$.

9. (a) Find an equation for the circle with center $C(4, -5)$ and passing through $P(-1, 2)$.
   (b) Find an equation for the circle of radius 1 with center in the second quadrant which is tangent to both the x- and y-axes.

10. In each of the following state which subset of **R** is assumed to be the domain of f and sketch the graph of f.
    (a) $f(x) = \sqrt{5 - 2x}$. (b) $f(x) = -\sqrt{16 - x^2}$.
    (c) $f(x) = 1000$. (d) $f(x) = 4/x^2$.

11. Find $(f + g)(x)$, $(f - g)(x)$, $(fg)(x)$, $(f/g)(x)$, $(f \circ g)(x)$, and $(g \circ f)(x)$ if:
    (a) $f(x) = x^2 + 3x + 1$, $g(x) = 2x - 1$.
    (b) $f(x) = x^2 + 4$, $g(x) = \sqrt{2x + 5}$.
    (c) $f(x) = 5x + 2$, $g(x) = 1/x^2$.

12. Find the inverse function of f if f is defined by
    (a) $f(x) = 5 - 7x$. (b) $f(x) = 4x^2 + 3$, $x \geq 0$.

Supplementary Questions

1. Define the union $\cup A_i$ and the intersection $\cap A_i$ of an infinite number of sets $A_1, A_2, \cdots, A_k, \cdots$. Give examples of these concepts where each A_i is a set of real numbers.
2. What is the product set of a nonempty set and the empty set? If A, B, and C are sets, what is the difference between $(A \times B) \times C$ and $A \times (B \times C)$? Extend the notion of product set to three sets by defining $A \times B \times C$. Generalize to *any* finite number of sets.
3. Show that the operation of subtraction on **R** is neither commutative nor associative. Is 0 an identity element relative to subtraction?
4. If n is a positive integer, describe a geometric construction which may be used to locate the point on a coordinate line which corresponds to $1/n$.
5. If A and B are points on a coordinate line with coordinates a and b, respectively, prove that the midpoint of the line segment AB has coordinate $(a + b)/2$. If $P_1(x_1, y_1)$ and $P_2(x_2, y_2)$ are points in a

coordinate plane, what are the coordinates of the midpoint of the line segment P_1P_2?

6. Describe the construction of an *oblique* coordinate system in a plane by beginning with two coordinate lines (labeled the x- and y-axes) which intersect at an angle other than a right angle. Sketch the graph of the equation $y = x^2$ on such a coordinate plane.
7. Solve the inequality $|x - 4| \leq |x + 2|$ by sketching the graphs of $y = |x - 4|$ and $y = |x + 2|$ on the same coordinate plane.
8. If $X = \{a, b, c\}$ and $Y = \{r, s, t\}$, how many functions are there from X to Y? How many are one-to-one? What if $X = \{a, b, c\}$ and $Y = \{r, s\}$?
9. Prove that the linear function f defined by $f(x) = ax + b$ has an inverse function if $a \neq 0$. Does a constant function have an inverse? Does the identity function have an inverse?
10. Describe how one can determine graphically whether a function is even or odd. Prove that (a) the product of two odd functions is even, (b) the product of two even functions is even, and (c) the product of an even function and an odd function is odd.

chapter two Polynomials

Polynomial functions occur frequently in both elementary and advanced phases of mathematics. In this chapter we shall study such functions and their algebraic counterpart, polynomials. An important part of our work will be an analysis of the zeros of polynomial functions.

1 THE ALGEBRA OF POLYNOMIALS

A function p which is defined by an expression of the form $p(x) = ax^n$, where $a \in \mathbf{R}$ and n is a nonnegative integer, may be thought of as a product of the constant function g given by $g(x) = a$ and the function h defined by $h(x) = x^n$. The number a is called the coefficient of x^n. A finite sum of such functions is called a *polynomial function*. Thus f is a polynomial function if $f(x)$ can be expressed in the form

$$f(x) = a_n x^n + a_{n-1} x^{n-1} + \cdots + a_1 x + a_0, \qquad (2.1)$$

for all $x \in \mathbf{R}$, where the coefficients a_i are real numbers and the exponents are nonnegative integers. The expression on the right in (2.1) is called a *polynomial in* x (with real coefficients) and each expression $a_k x^k$ is called a *term* of the polynomial. In the discussion to follow we shall often use the phrase "the polynomial $f(x)$" when referring to expressions of this type. Of course, to each such polynomial there corresponds a polynomial function f, and vice versa.

The coefficient a_n of the highest power of x in (2.1) is known as the *leading coefficient* of $f(x)$ and, if $a_n \neq 0$, we say that $f(x)$ has *degree* n. If a coefficient a_i is zero we often abbreviate the expression (2.1) by deleting the term $a_i x^i$. If *all* the coefficients of a polynomial are zero it is called the *zero polynomial* and is denoted by 0. It is customary not to assign a degree to the zero polynomial.

If some of the coefficients are negative, then for convenience we often use minus signs between appropriate terms. For example, instead of $3x^2 + (-5)x + (-7)$ we write $3x^2 - 5x - 7$ for this polynomial of degree 2. Polynomials in other variables may also be considered. For example $\frac{2}{5}z^2 - 3z^7 + 8 - \sqrt{5}z^4$ is a polynomial in z of degree 7. Given such a polynomial, we ordinarily arrange the terms in order of decreasing powers of the variable, namely $-3z^7 - \sqrt{5}z^4 + \frac{2}{5}z^2 + 8$.

According to the definition of degree, if $f(x) = c$, where c is a nonzero real number, then $f(x)$ is a polynomial of degree 0. Such polynomials (together with the zero polynomial) are called *constant polynomials*, because the corresponding polynomial function f is a constant function. A polynomial of the form $a_1x + a_0$, $a_1 \neq 0$, is of degree 1 and is called a *linear* polynomial. Polynomials of degrees 2, 3, 4, and 5 are called *quadratic*, *cubic*, *quartic*, and *quintic* polynomials respectively. Similar names are assigned to other polynomials. Expressions such as $(1/x) + 3x$, $(x - 5)/(x^2 + 2)$, and $3x^2 + \sqrt{x} - 2$ are not polynomials, since they cannot be written as sums of the form (2.1).

If we allow the coefficients a_i in (2.1) to be any complex numbers, then $f(x)$ is called a polynomial in x with complex coefficients. Much of the theory in this chapter will hold for polynomials of this type as well as for polynomials in which all coefficients are real. We shall, therefore, use the letter F to denote either the system of real numbers or the system of complex numbers, and $F[x]$ will denote the set of all polynomials in x with coefficients in F.

Although in our work x denotes a variable which may be assigned real or complex values, there are parts of mathematics where it has other connotations. Consequently it is often convenient to regard x merely as a symbol with no specific meaning attached to it. By so doing, polynomials may be applied to a variety of systems and the meaning of x is determined by the particular system being studied. When x is used in this way, it is often referred to as an *indeterminate* rather than a variable. To develop a theory of polynomials from this point of view, one might consider all *formal expressions* of the type given in (2.1), where the coefficients are chosen from some system. By defining operations on these expressions in a suitable way, one can obtain a theory which is applicable to the various situations where polynomials are needed. We do not intend to use this approach. Rather, we shall continue to regard x as a variable and the coefficients as real (or complex) numbers. Actually, the development of polynomials by means of indeterminates is very similar to the discussion given below.

By inserting terms with zero coefficients if necessary, we can assume that in a discussion of a finite set of polynomials the same powers of x appear. As an illustration, given the polynomials $f(x) = 4x^3 - 3x + 2$ and $g(x) = x^2 + 5$, we can rewrite them as $f(x) = 4x^3 + 0x^2 - 3x + 2$ and $g(x) = 0x^3 + x^2 + 0x + 5$.

Now let $f(x)$ and $g(x)$ be any two polynomials in $F[x]$. By the preceding remarks, we can write

(2.2)
$$f(x) = a_nx^n + a_{n-1}x^{n-1} + \cdots + a_1x + a_0,$$
$$g(x) = b_nx^n + b_{n-1}x^{n-1} + \cdots + b_1x + b_0,$$

where $a_i, b_i \in F$ and where n is some nonnegative integer. The polynomials $f(x)$ and $g(x)$ are *equal* and we write $f(x) = g(x)$ if and only if coefficients of like powers of x are the same — that is, $a_0 = b_0$, $a_1 = b_1, \cdots, a_n = b_n$. To *add* $f(x)$ and $g(x)$, we add corresponding coefficients. Thus

(2.3)
$$f(x) + g(x) = (a_n + b_n)x^n + \cdots + (a_1 + b_1)x + (a_0 + b_0).$$

Many of the field properties (1.3)–(1.7) which hold in **R** are true in $F[x]$. The commutative law $f(x) + g(x) = g(x) + f(x)$ follows easily, since, for each k, $a_k + b_k = b_k + a_k$. Similarly, the associative law is a consequence of the fact that the associative law is valid for the coefficients. The zero polynomial

$$0 = 0x^n + 0x^{n-1} + \cdots + 0x + 0$$

is the additive identity, since, by (2.3), $f(x) + 0 = f(x)$. The additive inverse $-f(x)$ of $f(x)$ is given by

(2.4)
$$-f(x) = (-a_n)x^n + \cdots + (-a_1)x + (-a_0),$$

since the sum of $f(x)$ and $-f(x)$ is the zero polynomial. The rule for subtracting the polynomials (2.2) is given by

(2.5)
$$f(x) - g(x) = (a_n - b_n)x^n + \cdots + (a_1 - b_1)x + (a_0 - b_0).$$

Multiplication in $F[x]$ is carried out in the usual way, using properties of real numbers. For this operation there is no advantage in assuming that the highest exponent of x is the same for each polynomial. If $f(x)$ has degree n and $g(x)$ has degree m, we may write

(2.6)
$$f(x) = a_nx^n + a_{n-1}x^{n-1} + \cdots + a_1x + a_0, \qquad a_n \neq 0,$$
$$g(x) = b_mx^m + b_{m-1}x^{m-1} + \cdots + b_1x + b_0, \qquad b_m \neq 0.$$

Then

(2.7)
$$f(x)g(x) = a_nb_mx^{n+m} + (a_nb_{m-1} + a_{n-1}b_m)x^{n+m-1} + \cdots + (a_1b_0 + a_0b_1)x + a_0b_0.$$

Note that for each expression a_ib_j occurring in a term of (2.7) the sum $i + j$ of the subscripts is equal to the exponent of x for that term. In general, *the coefficient of* x^k *in* (2.7) is

(2.8)
$$a_kb_0 + a_{k-1}b_1 + a_{k-2}b_2 + \cdots + a_1b_{k-1} + a_0b_k.$$

Thus the coefficient of x^2 is $a_2b_0 + a_1b_1 + a_0b_2$, the coefficient of x^3 is $a_3b_0 + a_2b_1 + a_1b_2 + a_0b_3$, and so on.

As a special case of (2.7), if $f(x) = a_0$ is a constant polynomial, then

$$a_0 g(x) = a_0 b_m x^m + a_0 b_{m-1} x^{m-1} + \cdots + a_0 b_1 x + a_0 b_0.$$

That is, to find the product $a_0 g(x)$ we multiply each coefficient of $g(x)$ by a_0. Of course, this is simply a generalization of the distributive law (1.5).

Since $a_n \neq 0$ and $b_m \neq 0$ in (2.6), $a_n b_m \neq 0$ and it follows from (2.7) that $f(x)g(x)$ has degree $n + m$. This provides the following useful result.

(2.9) Theorem

The degree of the product of two nonzero polynomials equals the sum of the degrees of the two polynomials.

As an illustration of (2.9), if $f(x) = 5x^3 - x + 1$ and $g(x) = 2x^2 + 3$, then the term of highest degree in $f(x)g(x)$ is $10x^5$. Therefore the degree of the product is 5, which is the sum of the degrees of the two given polynomials.

Owing to the complexity of (2.7), it is rather tedious to check the field properties (1.3)–(1.7) which deal with multiplication. Suffice it to say that multiplication in $F[x]$ can be shown to be commutative and associative and also that the distributive laws are valid. Evidently, the zero-degree polynomial 1 is the identity element relative to multiplication. The only field property which is *not* true in $F[x]$ is the part of (1.7) about multiplicative inverses, for consider a polynomial $f(x)$ of degree 1 or higher. Now *if* there exists a polynomial $g(x)$ such that $f(x)g(x) = 1$, then applying (2.9) the degree of $f(x)g(x)$ is at least 1, whereas the degree of the constant polynomial 1 is 0. Since equal polynomials must have the same degree, we have reached a contradiction. Consequently no such $g(x)$ can exist.

(2.10) Theorem

If $f(x)$ and $g(x)$ are polynomials such that $f(x)g(x) = 0$, then either $f(x) = 0$ or $g(x) = 0$.

Proof: We shall give an indirect proof. Thus we let $f(x)g(x) = 0$ and suppose that *both* $f(x)$ and $g(x)$ are not the zero polynomial. Then both polynomials have degrees, say n and m, and, by (2.9), $f(x)g(x)$ has degree $n + m$. On the other hand, the zero polynomial 0 has *no* degree! Therefore 0 cannot equal $f(x)g(x)$. This contradiction establishes (2.10).

(2.11) Cancellation Law for Polynomials

Suppose $f(x)$, $g(x)$, and $h(x)$ are polynomials, with $h(x) \neq 0$. If $f(x)h(x) = g(x)h(x)$, then $f(x) = g(x)$.

Proof: From the hypothesis we obtain

$$[f(x) - g(x)]h(x) = 0.$$

Since $h(x) \neq 0$, we have, by (2.10), $f(x) - g(x) = 0$ and the theorem is proved.

EXAMPLE 1. If $f(x) = x^5 - 4x^3 - 5x^2 + 3$ and $g(x) = x^3 - 5x^2 - 2x + 7$, find the polynomials $f(x) + g(x)$ and $f(x) - g(x)$.

Solution: To find $f(x) + g(x)$ we add coefficients of like powers of x. This gives us

$$f(x) + g(x) = x^5 - 3x^3 - 10x^2 - 2x + 10.$$

In order to keep track of the coefficients it is sometimes convenient to use the following scheme:

$$\begin{array}{rcrrrrr} f(x) & = & x^5 & - 4x^3 & - 5x^2 & & + 3 \\ g(x) & = & & x^3 & - 5x^2 & - 2x & + 7 \\ \hline f(x) + g(x) & = & x^5 & - 3x^3 & - 10x^2 & - 2x & + 10 \end{array}$$

Similarly, by subtracting corresponding coefficients we obtain $f(x) - g(x) = x^5 - 5x^3 + 2x - 4$.

EXAMPLE 2. If $f(x) = 2x^3 + 3x - 1$ and $g(x) = x^2 - x + 4$, find the product $f(x)g(x)$.

Solution: By labeling the coefficients appropriately we could use formula (2.7) to find $f(x)g(x)$; however, this process is unnecessarily lengthy. Since polynomials obey laws similar to those for real numbers we may proceed as follows:

$$\begin{aligned} f(x)g(x) &= (2x^3 + 3x - 1)(x^2 - x + 4) \\ &= (2x^3 + 3x - 1)x^2 + (2x^3 + 3x - 1)(-x) \\ &\quad + (2x^3 + 3x - 1)4 && \text{(distributive law)} \\ &= 2x^5 + 3x^3 - x^2 - 2x^4 - 3x^2 + x \\ &\quad + 8x^3 + 12x - 4 && \text{(Why?)} \\ &= 2x^5 - 2x^4 + (3 + 8)x^3 + (-1 - 3)x^2 \\ &\quad + (1 + 12)x - 4 && \text{(distributive law)} \\ &= 2x^5 - 2x^4 + 11x^3 - 4x^2 + 13x - 4 \\ & && \text{(simplifying).} \end{aligned}$$

Sometimes, this work is arranged as follows:

$$\begin{array}{rrrrrrcl} & & 2x^3 & & + 3x & - 1 & & \\ & & & x^2 & - x & + 4 & & \\ \hline 2x^5 & & + 3x^3 & - x^2 & & & = & (2x^3 + 3x - 1)x^2 \\ & - 2x^4 & & - 3x^2 & + x & & = & (2x^3 + 3x - 1)(-x) \\ & & 8x^3 & & + 12x & - 4 & = & (2x^3 + 3x - 1)4 \\ \hline 2x^5 & - 2x^4 & + 11x^3 & - 4x^2 & + 13x & - 4 & = & \text{sum of the above.} \end{array}$$

EXERCISES

In Exercises 1–6 perform the indicated operations and find the degree of the resulting polynomial.

1. $(x^3 - 2x^2 + 3x + 5) + (x^4 - 3x^3 + 5x - 2)$.
2. $(x^5 - 3x + 1) + (x^2 + 3x - x^4 - 1)$.
3. $(x^3 + 7 - 5x + 2x^2) - (x^3 - 3x^2 + 7 + 5x)$.
4. $(a^2 + 1 + 3a^4) - (1 - a^3 + 4a^4)$.
5. $(2y^3 - y + 5)(3y^2 + 2y - 4)$.
6. $(x^2 + 1)(x^3 + 1)(x^4 + 1)$.
7. Illustrate (2.9) with the polynomials $2x^5 - x + 1$ and $x^2 - 3x^4$.
8. If $f(x)$ and $g(x)$ are polynomials in x of degree 5, does it follow that $f(x) + g(x)$ has degree 5? Explain.
9. Give examples of polynomials $f(x)$ and $g(x)$ of degree 3 such that the degree of $f(x) + g(x)$ is (a) 3, (b) 2, (c) 1, (d) 0.
10. If $f(x)$, $g(x)$, and $h(x)$ are polynomials such that $f(x)g(x)h(x) = 0$, prove that at least one of the polynomials is 0.
11. Use (2.9) to prove that a linear polynomial in $F[x]$ is prime — that is, it cannot be expressed as a product of two polynomials of positive degree.
12. Use (2.9) to prove that the degree of the product of three nonzero polynomials equals the sum of the degrees of the polynomials.
13. Let $\mathbf{Z}[x]$ denote the subset of $F[x]$ consisting of polynomials with coefficients in the set $\mathbf{Z}$ of integers. Prove that $\mathbf{Z}[x]$ is closed relative to addition and multiplication. Which of the field properties (1.3)–(1.7) are true in $\mathbf{Z}[x]$?
14. Determine which of (2.9), (2.10), and (2.11) are true for the set $\mathbf{Z}[x]$ defined in Exercise 13.

2 GRAPHS OF POLYNOMIAL FUNCTIONS

If $f(x)$ is a polynomial with coefficients in F and if $f(c) = 0$ for some $c \in F$, then c is called a *zero* of $f(x)$ (or of the polynomial function f). For example, the polynomial $f(x) = x^2 - 9$ has zeros 3 and -3. Since we now wish to discuss graphs, we shall in the remainder of this section restrict ourselves to polynomials with real coefficients. In this event the zeros of f are the abscissas of points at which the graph of f crosses the x-axis (Why?).

In Chapter One we discussed graphs of certain polynomial functions. In particular, if $f(x)$ is a polynomial of degree 0 (or the zero polynomial), then $f(x) = k$ for some real number k. As shown in Example 6 on p. 37, the graph of f is a straight line parallel to the x-axis, with y-intercept k (see Fig. 1.20).

If $f(x)$ is a polynomial of degree 1, then $f(x) = ax + b$, where $a, b \in \mathbf{R}$ and $a \neq 0$. As pointed out on p. 34, f is called a *linear function* and the graph of f is a straight line. A linear function has precisely one zero, $-b/a$, for if $f(c) = 0$ then $ac + b = 0$ and hence $c = -b/a$.

The general polynomial $f(x)$ of degree 2 with real coefficients may be written as

$$f(x) = ax^2 + bx + c, \tag{2.12}$$

where $a, b, c \in \mathbf{R}$ and $a \neq 0$. In this case f is called a *quadratic function.* The graph of f or, equivalently, of the equation $y = ax^2 + bx + c$ is called a *parabola*, and it can be shown that it always has the same general shape as shown in Figs. 1.12 and 1.18. In Fig. 1.18 we say that the parabola *opens upward*; this will always be the case if $a > 0$. In Fig. 1.12 we have $a = -1 < 0$ and the parabola *opens downward.* The high (or low) point on a parabola of this type is called the *vertex.* If the parabola opens upward, then $f(x)$ has its smallest value when x is the abscissa of the vertex. If the parabola opens downward, then $f(x)$ takes on its largest value at this number.

In Chapter Seven, where parabolas are discussed in greater detail, we shall state the definition of a parabola and derive equations for their graphs.

To find the x-intercepts of the graph of the quadratic function defined by (2.12) we solve the quadratic equation $ax^2 + bx + c = 0$. By the quadratic formula (see p. 11), if $b^2 - 4ac > 0$, then there are two real and unequal solutions and the graph has two x-intercepts. If $b^2 - 4ac = 0$, then there is one (double) root and the graph is tangent to the x-axis. Finally, if $b^2 - 4ac < 0$, then there are no x-intercepts. Assuming that $a > 0$, we have sketched these cases in Fig. 2.1.

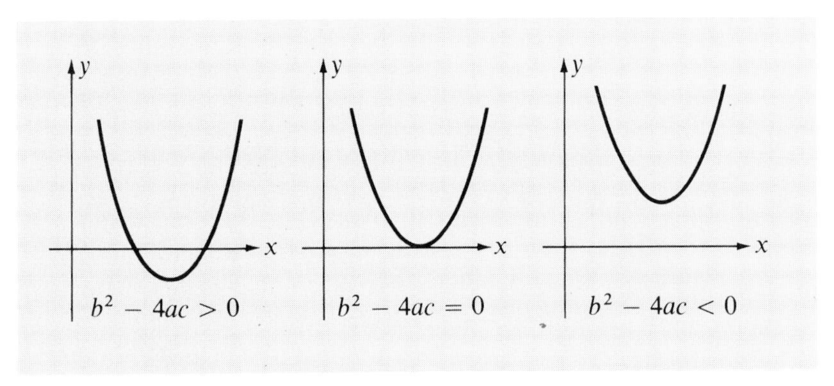

Figure 2.1

As the degree increases, the graphs of polynomial functions become more complicated, however, they always have a smooth appearance with a number of "hills" and "valleys" as illustrated in Fig. 2.2. This number may be large when the degree is large, but this is not necessarily always true. For example, the graph of the polynomial function f defined by

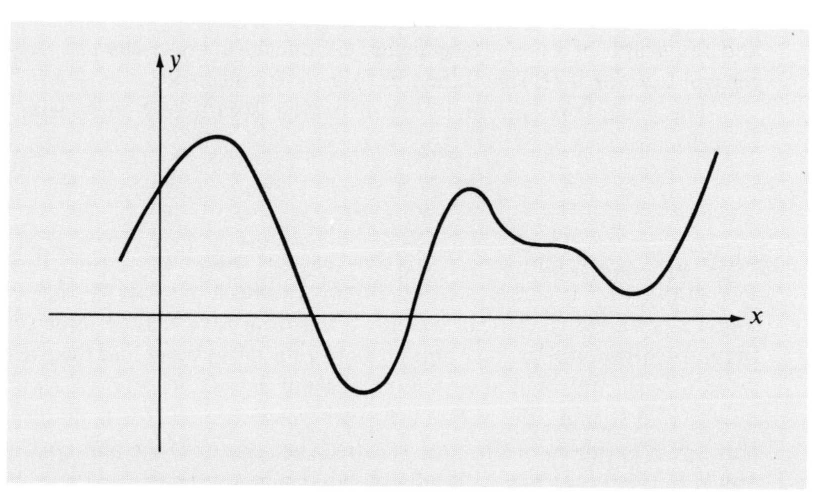

Figure 2.2

$f(x) = 1 - x^6$, shown in Fig. 2.3, has one very flat "hill," as may be verified by plotting points.

EXAMPLE 1. If $f(x) = -x^3 - x^2 + 2x$, sketch the graph of f.

Solution: It is convenient to express $f(x)$ as a product in the following way:

$$f(x) = -x(x^2 + x - 2) = -x(x + 2)(x - 1).$$

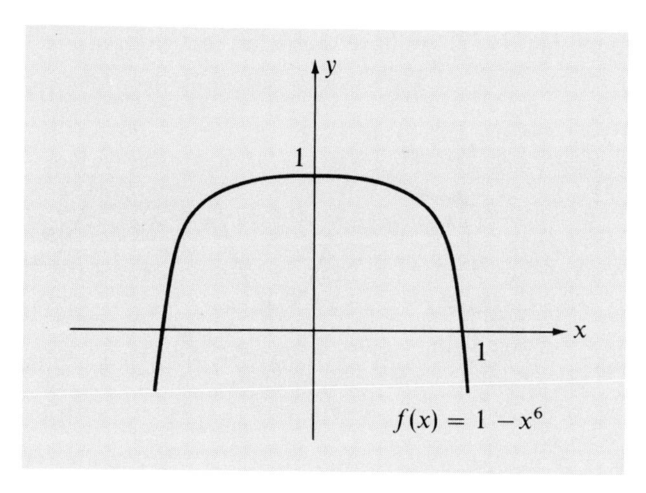

Figure 2.3

It follows that f has zeros -2, 0, and 1, and these are also the x-intercepts of the graph. The corresponding points on the graph divide the x-axis into four parts, and the sign of $f(x)$ in each part may be determined by investigating the signs of the three factors $-x$, $x + 2$, and $x - 1$. The procedure may be systematized as follows:

	Sign of factors			Sign of product
	$-x$	$x + 2$	$x - 1$	$-x(x + 2)(x - 1)$
$x < -2$	$+$	$-$	$-$	$+$
$-2 < x < 0$	$+$	$+$	$-$	$-$
$0 < x < 1$	$-$	$+$	$-$	$+$
$x > 1$	$-$	$+$	$+$	$-$

Since $f(x)$ is the product of the three factors, we see that $f(x)$ is positive (and hence the graph lies above the x-axis) if $x < -2$ or $0 < x < 1$. On the other hand, $f(x)$ is negative (and the graph lies below the x-axis) if $-2 < x < 0$ or $x > 1$. Coordinates of several other points on the graph are given in the following table:

x	-3	-1	$1/2$	$3/2$
$f(x)$	12	-2	$5/8$	$-21/8$

The graph is sketched in Fig. 2.4. A more complete analysis of graphs of this and other polynomial functions requires methods of the calculus.

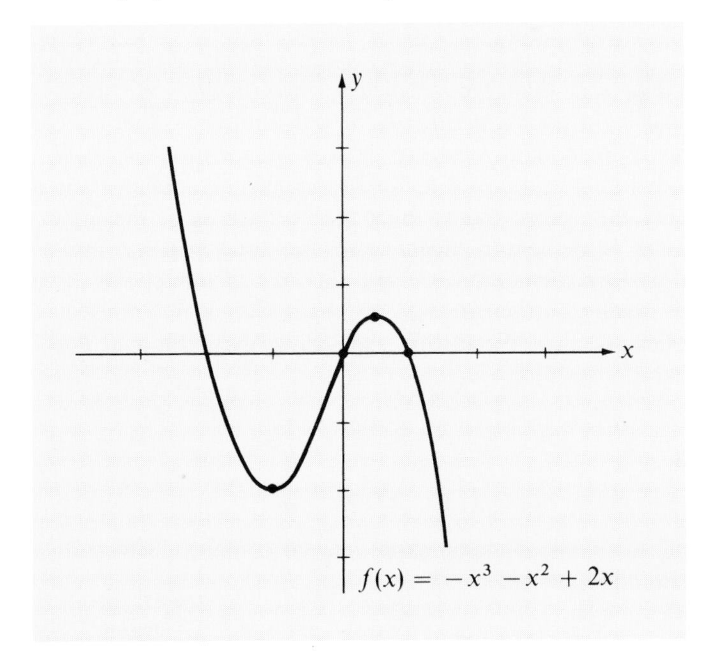

Figure 2.4

EXERCISES

In Exercises 1–12 sketch the graph of the polynomial function f which is defined as indicated.

1. $f(x) = 3x - 5$.
2. $f(x) = \frac{1}{2}x + 2$.
3. $f(x) = -6x + 1$.
4. $f(x) = -4$.
5. $f(x) = x^2 - 8x + 15$.
6. $f(x) = x^2 - 6x + 9$.
7. $f(x) = x^3 - 6x^2 + 7$.
8. $f(x) = x^3 - 3x^2 - x + 3$.
9. (a) $f(x) = x^3$. (b) $f(x) = x^5$. (c) $f(x) = x^7$.
10. (a) $f(x) = x^2$. (b) $f(x) = x^4$. (c) $f(x) = x^6$.
11. (a) $f(x) = x^3 + 5$. (b) $f(x) = x^3 - 5$.
12. (a) $f(x) = x^2 + 4$. (b) $f(x) = x^2 - 4$.

3 PROPERTIES OF DIVISION

If a polynomial is written as a product of other polynomials, then each of the latter polynomials is called a *factor* of the original polynomial. Since $x^2 - 1 = (x + 1)(x - 1)$, we see that $x + 1$ and $x - 1$ are factors of $x^2 - 1$. If real coefficients are used, then any polynomial has as a factor *every* nonzero real number c. As an illustration, given $3x^2 - 5x + 2$ and any nonzero real number c, we can write this polynomial in the form

$$c\left(\frac{3}{c}x^2 - \frac{5}{c}x + \frac{2}{c}\right).$$

A factor c of this type is called a *trivial factor*. We shall be interested primarily in *nontrivial* factors of polynomials — that is, factors which contain polynomials of degree greater than zero.

Before carrying out factorizations of polynomials, it is necessary to specify the system from which the coefficients of the factors are to be chosen. In elementary algebra one often uses the rule that if a polynomial with *integer* coefficients is given, then the factors should be polynomials with integer coefficients. If we begin with a polynomial with *rational* coefficients, then the factors should also have rational coefficients. As illustrations we have

$$x^2 + x - 6 = (x + 3)(x - 2)$$

and

$$4x^2 - \tfrac{9}{16} = (2x - \tfrac{3}{4})(2x + \tfrac{3}{4}).$$

In more advanced mathematics, real or complex coefficients are allowed. As examples we have

$$x^2 - 2 = (x + \sqrt{2})(x - \sqrt{2})$$

and

$$x^2 + 1 = (x + i)(x - i).$$

An integer $a > 1$ is *prime* if it cannot be written as a product of two positive integers greater than 1. In similar fashion, if S denotes a set of numbers, then a polynomial with coefficients in S is said to be *prime*, or *irreducible* over S, if it cannot be expressed as a product of two polynomials of positive degree with coefficients in S. Note that a polynomial may be irreducible over one set S but not over another. For example $x^2 - 2$ is irreducible over the rational numbers — however, it can be expressed as a product of two polynomials of degree 1 if *real* coefficients are allowed. Similarly, $x^2 + 1$ is irreducible over the real numbers but not over the complex numbers. Every polynomial $ax + b$ of degree 1 in $F[x]$ is irreducible over F, for if $ax + b = f(x)g(x)$ where $f(x)$, $g(x) \in F[x]$, then by (2.9), the sum of the degrees of $f(x)$ and $g(x)$ is 1, which implies that one of the latter polynomials has degree 0.

EXAMPLE 1. Express each of the following as a product of irreducible polynomials:

(a) $6x^2 - 7x - 3$. (b) $3x^3 + 2x^2 - 12x - 8$.

Solution: (a) If we write $6x^2 - 7x - 3 = (ax + b)(cx + d)$, the product of a and c is 6, whereas the product of b and d is -3. Trying various possibilities, we arrive at the factorization $6x^2 - 7x - 3 = (2x - 3)(3x + 1)$.

(b) We employ the distributive law as follows:

$$\begin{aligned} 3x^3 + 2x^2 - 12x - 8 &= x^2(3x + 2) - 4(3x + 2) \\ &= (x^2 - 4)(3x + 2) \\ &= (x + 2)(x - 2)(3x + 2). \end{aligned}$$

If a polynomial $g(x)$ is a factor of a polynomial $f(x)$ we often say that $f(x)$ is *divisible* by $g(x)$. For example, $x^2 - 1$ is divisible by $x - 1$ and by $x + 1$. A division process may also be introduced when $g(x)$ is *not* a factor of $f(x)$. The problem here is similar to that in the set **Z** of integers. For example, the number 24 has positive factors 1, 2, 3, 4, 6, 8, 12, and 24, so we say that 24 is *divisible* by these numbers. A different situation exists if we divide 24 by nonfactors such as 5, 7, 15, 32, etc. In these cases a process called *long division* is used which yields a *quotient* and *remainder*. The reader undoubtedly remembers the process from elementary arithmetic, where the work involved in dividing 4126 by 23 might be arranged as follows:

```
      179
23 | 4126
     23
     182
     161
      216
      207
        9
```

The number 179 is called the *quotient* and 9 the *remainder* in the division of 4126 by 23. To complete the terminology, 23 is called the *divisor* and 4126 the *dividend.* The remainder should always be less than the divisor, for otherwise the quotient can be increased. Sometimes the above result is written

$$\tfrac{4126}{23} = 179 + \tfrac{9}{23}.$$

Upon multiplication by 23, this becomes $4126 = 23 \cdot 179 + 9$. The latter form is very useful for theoretical purposes and can be generalized to arbitrary integers. Specifically, it can be shown that if a and b are integers with $b > 0$, then there exist unique integers q and r such that

$$a = bq + r, \tag{2.13}$$

where $0 \le r < b$. The integer q is called the *quotient* and r the *remainder* in the division of a by b.

A similar discussion can be given for polynomials. For example, the polynomial $x^4 - 16$ is divisible by $x - 2$, $x + 2$, $x^2 - 4$, $x^2 + 4$, etc. On the other hand, $x^2 + 3x + 1$ is not a factor of $x^4 - 16$. However, by another process, called *long division*, we write

$$\begin{array}{r|l}
 & x^2 - 3x + 8 \\
\hline
x^2 + 3x + 1 & x^4 \qquad\qquad\qquad\qquad - 16 \\
 & x^4 + 3x^3 + x^2 \\
\hline
 & -3x^3 - x^2 \\
 & -3x^3 - 9x^2 - 3x \\
\hline
 & 8x^2 + 3x - 16 \\
 & 8x^2 + 24x + 8 \\
\hline
 & -21x - 24
\end{array}$$

obtaining the quotient $x^2 - 3x + 8$ and the remainder $-21x - 24$. In this division we proceed as indicated until we arrive at a polynomial (the remainder) which is either 0 or has smaller degree than the divisor. We shall assume familiarity with this process and not attempt to justify it here. As with integers, the result is often written

$$\frac{x^4 - 16}{x^2 + 3x + 1} = (x^2 - 3x + 8) + \left(\frac{-21x - 24}{x^2 + 3x + 1}\right)$$

or, multiplying by $x^2 + 3x + 1$,

$$x^4 - 16 = (x^2 + 3x + 1)(x^2 - 3x + 8) + (-21x - 24),$$

which has the same general form as (2.13).

The previous example illustrates the following theorem, which we state without proof.

(2.14) Division Algorithm for Polynomials

If $f(x)$ and $g(x)$ are polynomials in $F[x]$ and if $g(x) \neq 0$, then there exist unique polynomials $q(x)$ and $r(x)$ such that

$$f(x) = g(x)q(x) + r(x),$$

where either $r(x) = 0$ or the degree of $r(x)$ is less than the degree of $g(x)$. The polynomial $q(x)$ is called the *quotient* and $r(x)$ the *remainder* in the division of $f(x)$ by $g(x)$.

An interesting special case of (2.14) occurs when $f(x)$ is divided by a polynomial of the form $x - c$, where $c \in F$. By (2.14), if the remainder is not 0, then it must have smaller degree than the divisor $x - c$. This implies, however, that the remainder has degree less than 1—that is, degree 0. This, in turn, means that the remainder is a nonzero element of F. Consequently we may write

$$f(x) = (x - c)q(x) + d, \tag{2.15}$$

where $d \in F$ (possibly $d = 0$).

If c is substituted for x in (2.15), we obtain

$$f(c) = (c - c)q(c) + d,$$

which reduces to $f(c) = d$. Substituting for d in (2.15) yields

$$f(x) = (x - c)q(x) + f(c). \tag{2.16}$$

We have proved the following theorem.

(2.17) Remainder Theorem

If a polynomial $f(x)$ is divided by $x - c$, then the remainder is $f(c)$.

EXAMPLE 2. Verify the Remainder Theorem if $f(x) = x^3 - 3x^2 + x + 5$ and $c = 2$.

Solution: We note first that $f(2) = 2^3 - 3 \cdot 2^2 + 2 + 5 = 3$. Therefore, according to (2.17), the remainder, when $f(x)$ is divided by $x - 2$, should be 3. By long division we have

$$\begin{array}{r|l}
 & x^2 - x - 1 \\
\hline
x - 2 & x^3 - 3x^2 + x + 5 \\
 & x^3 - 2x^2 \\
\hline
 & \quad - x^2 + x \\
 & \quad - x^2 + 2x \\
\hline
 & \qquad\quad - x + 5 \\
 & \qquad\quad - x + 2 \\
\hline
 & \qquad\qquad\quad 3
\end{array}$$

which is what we wished to show.

The following important result is a consequence of (2.17).

(2.18) Factor Theorem

A polynomial $f(x)$ has a factor $x - c$ if and only if $f(c) = 0$.

Proof: From (2.16) we have $f(x) = (x - c)q(x) + f(c)$. If $f(c) = 0$, then $f(x) = (x - c)q(x)$; that is, $x - c$ is a factor of $f(x)$. Conversely, if $x - c$ is a factor, then the remainder upon division by $x - c$ must be 0 and, using the Remainder Theorem, we have $f(c) = 0$.

The Factor Theorem is quite useful in exhibiting factors of polynomials in certain problems, as is illustrated below.

EXAMPLE 3. Show that $x - 2$ is a factor of the polynomial $f(x) = x^3 - 4x^2 + 3x + 2$.

Solution: We have $f(2) = 8 - 16 + 6 + 2 = 0$ and hence, by the Factor Theorem, $x - 2$ is a factor of $f(x)$. Of course, another method of solution would be to divide $f(x)$ by $x - 2$ and show that the remainder is 0. The quotient in this division would be another factor of $f(x)$.

EXERCISES

Express each of the following as products of polynomials which are irreducible over **R**.

1. $2x^2 - 9x - 5$.
2. $12x^2 + 32x + 5$.
3. $x^8 - 1$.
4. $27y^3 - 64$.
5. $2x^3 - x^2 - 2x + 1$.
6. $x^2 + x + 1$.

In Exercises 7–12 find the quotient $q(x)$ and the remainder $r(x)$ if $f(x)$ is divided by $g(x)$.

7. $f(x) = 2x^4 - x^3 + 7x + 3, \quad g(x) = x^2 + 2x - 5$.
8. $f(x) = x^3 - x^2 + 4, \quad g(x) = x^2 - 4x$.
9. $f(x) = 3x^3 + 6x, \quad g(x) = 2x^2 - 8$.
10. $f(x) = x^4 + x^3 - x^2 + 4x - 3, \quad g(x) = 2x^3 + x^2 - 4x + 1$.
11. $f(x) = 2x^3 - x^2 + 1, \quad g(x) = 3x^4 - 7$.
12. $f(x) = 6x + 9, \quad g(x) = 9x^2 + 6$.

In Exercises 13–16 use the Remainder Theorem to find $f(c)$. Check by substituting c for x.

13. $f(x) = x^4 - 3x^3 + x^2 - 1, \quad c = 2$.
14. $f(x) = 2x^3 - 4x + 2, \quad c = \sqrt{2}$.
15. $f(x) = x^3 + 2x^2 - 5x + 1, \quad c = -3$.
16. $f(x) = x^2 - 3x + 1, \quad c = i$.
17. Determine k so that $f(x) = x^3 - kx^2 + 3x + 7k$ is divisible by $x + 2$.

18. Determine all values of k such that $f(x) = k^2x^4 - 3kx^2 + 1$ is divisible by $x - 1$.
19. Use the Factor Theorem to show that $x - 3$ is a factor of $f(x) = x^4 - 2x^3 + x^2 - 8x - 12$.
20. Show that $x + 2$ is a factor of $f(x) = x^{10} - 1024$.
21. Prove that $f(x) = x^4 + 2x^2 + 1$ has no factors of the form $x - c$, where $c \in \mathbf{R}$.
22. Find the remainder if the polynomial $5x^{100} - 6x^{75} + 4x^{50} + 3x^{25} + 2$ is divided by $x + 1$.
23. Use the Factor Theorem to prove that $x - y$ is a factor of $x^n - y^n$, for all positive integers n. If n is even, show that $x + y$ is also a factor of $x^n - y^n$.
24. If n is an odd positive integer, prove that $x + y$ is a factor of $x^n + y^n$.

4 SYNTHETIC DIVISION

When applying the Remainder Theorem, it is necessary to divide by polynomials of the form $x - c$. The process referred to as *synthetic division* simplifies the work when divisors are of the above form. We shall illustrate this process by means of examples.

If the polynomial $3x^4 - 8x^3 + 9x + 5$ is divided by $x - 2$ in the usual way, we obtain

$$
\begin{array}{r|l}
 & 3x^3 - 2x^2 - 4x + 1 \\
\hline
x - 2 & 3x^4 - 8x^3 + 0x^2 + 9x + 5 \\
 & 3x^4 - 6x^3 \\
\hline
 & \quad - 2x^3 + 0x^2 \\
 & \quad - 2x^3 + 4x^2 \\
\hline
 & \qquad - 4x^2 + 9x \\
 & \qquad - 4x^2 + 8x \\
\hline
 & \qquad\qquad x + 5 \\
 & \qquad\qquad x - 2 \\
\hline
 & \qquad\qquad\quad 7
\end{array}
$$

where the term $0x^2$ has been inserted in the dividend so that all powers of x are accounted for.

Because the preceding technique of long division seems to involve a great deal of labor for so simple a problem, we look for a means of abbreviating the notation. After arranging the terms which involve like powers of x in vertical columns as above, it is seen that the repeated expressions $3x^4$, $-2x^3$, $-4x^2$, and x may be deleted without too much chance of confusion. Also, it appears unnecessary to "bring down" the terms $0x^2$, $9x$, and 5 from the dividend as indicated. With the elimination of these repetitions, our work takes on the form

$$
\begin{array}{r|rrrrr}
 & 3x^3 & -\ 2x^2 & -\ 4x & +\ 1 & \\
\cline{2-6}
x-2 & 3x^4 & -\ 8x^3 & +\ 0x^2 & +\ 9x & +\ 5 \\
 & & -\ 6x^3 & & & \\
\cline{2-3}
 & & -\ 2x^3 & & & \\
 & & & 4x^2 & & \\
\cline{3-4}
 & & & -\ 4x^2 & & \\
 & & & & 8x & \\
\cline{4-6}
 & & & & x & \\
 & & & & & -\ 2 \\
\cline{5-6}
 & & & & & 7
\end{array}
$$

If we take care to keep like powers of x under one another and account for missing terms by means of zero coefficients as above, some labor can be saved by omitting the symbol x. Doing this in the preceding expression, we obtain

$$
\begin{array}{r|rrrrr}
 & 3 & -2 & -4 & 1 & \\
\cline{2-6}
1-2 & 3 & -8 & 0 & 9 & 5 \\
 & & -6 & & & \\
\cline{2-3}
 & & -2 & & & \\
 & & & 4 & & \\
\cline{3-4}
 & & & -4 & & \\
 & & & & 8 & \\
\cline{4-5}
 & & & & 1 & \\
 & & & & & -2 \\
\cline{5-6}
 & & & & & 7
\end{array}
$$

Since the divisor is a polynomial of the form $x - c$, the two coefficients in the far left position are always $1 - c$, and with this in mind we shall discard the coefficient 1. Moreover, to make our notation more compact, let us move the numbers up in the following way:

$$
\begin{array}{r|rrrrr}
 & 3 & -2 & -4 & 1 & \\
\cline{2-6}
-2 & 3 & -8 & 0 & 9 & 5 \\
 & & -6 & 4 & 8 & -2 \\
\cline{2-6}
 & & -2 & -4 & 1 & 7
\end{array}
$$

If we now insert the leading coefficient 3 in the first position of the last row, the first four numbers of that row are the coefficients 3, -2, -4, and 1 of the quotient and the final number 7 is the remainder. Since there is no need to write the coefficients of the quotient two times, we discard the first row in our scheme, obtaining

$$\begin{array}{r|rrrrr} -2 & 3 & -8 & 0 & 9 & 5 \\ & & -6 & 4 & 8 & -2 \\ \hline & 3 & -2 & -4 & 1 & 7 \end{array} \tag{2.19}$$

where the top line has also been deleted since there is no longer any need for it.

There is a simple way of interpreting (2.19). Note that any number in the second row can be obtained by multiplying the number in the third row of the *preceding* column by -2. Moreover, a number in the third row can be found by subtracting the number above it in the second row from the corresponding number in the first row. This suggests a procedure for carrying out (2.19) without actually thinking of the division process. After arranging the terms of the polynomial in decreasing powers of x, write the coefficients in a row, supplying 0 for any missing term. Then write the "$-c$" (in our case -2) to the left of this row, as indicated in (2.19). Next, bring down the leading coefficient 3 to the third row. Multiply this number by -2 to obtain the first number, -6, in the second row. Subtract -6 from -8 to obtain the second number, -2, in the third row. Then multiply by $-c$ (in our case -2) to obtain the third number, 4, in the second row. Subtract to get the third number, -4, in the third row. Continue this process until the final number in the third row (the remainder) is obtained.

It is possible to avoid the subtractions performed above if the number c is used in place of $-c$ in the far left position of the first row. Then when the above process is used, the signs of the elements in the second row are changed and consequently, to find elements of the third row, we *add* the number above it in the second row to the corresponding number in the first row. With this change (2.19) becomes

$$\begin{array}{r|rrrrr} 2 & 3 & -8 & 0 & 9 & 5 \\ & & 6 & -4 & -8 & 2 \\ \hline & 3 & -2 & -4 & 1 & 7 \end{array}$$

This, then, is the process known as *synthetic division.*

EXAMPLE 1. Use synthetic division to find the quotient and remainder when $2x^4 + 5x^3 - 2x - 8$ is divided by $x + 3$.

Solution: Since we are to divide by $x + 3$, the c in the expression $x - c$ is -3. Hence the synthetic division takes the form

$$\begin{array}{r|rrrrr} -3 & 2 & 5 & 0 & -2 & -8 \\ & & -6 & 3 & -9 & 33 \\ \hline & 2 & -1 & 3 & -11 & 25 \end{array}$$

Therefore the quotient is $2x^3 - x^2 + 3x - 11$ and the remainder is 25.

Some authors prefer to place the number c in the upper right-hand corner instead of the upper left-hand corner of the array of numbers used for synthetic division. This is illustrated in the next two examples.

EXAMPLE 2. If $f(x) = 3x^5 - 38x^3 + 5x^2 - 1$, use synthetic division to find $f(4)$.

Solution: By the Remainder Theorem $f(4)$ is the remainder when $f(x)$ is divided by $x - 4$. Dividing synthetically, we have

$$\begin{array}{rrrrrr|l} 3 & 0 & -38 & 5 & 0 & -1 & 4 \\ & 12 & 48 & 40 & 180 & 720 & \\ \hline 3 & 12 & 10 & 45 & 180 & 719 & \end{array}$$

Hence $f(4) = 719$.

Sometimes synthetic division is employed to help find zeros of polynomials. By the method illustrated in the previous example, $f(c) = 0$ if and only if the remainder in the synthetic division by $x - c$ is 0.

EXAMPLE 3. Show that -11 is a zero of the polynomial

$$f(x) = x^3 + 8x^2 - 29x + 44.$$

Solution: Dividing synthetically by $x + 11$, we have

$$\begin{array}{rrrr|l} 1 & 8 & -29 & 44 & -11 \\ & -11 & 33 & -44 & \\ \hline 1 & -3 & 4 & 0 & \end{array}$$

Thus $f(-11) = 0$.

The preceding example shows that -11 is in the solution set of the *equation* $x^3 + 8x^2 - 29x + 44 = 0$. In Section 6 we shall use synthetic division in this way to find solutions of equations.

EXERCISES

In Exercises 1–10 use synthetic division to find the quotient and remainder if the first polynomial is divided by the second.

1. $2x^3 + x^2 - 3x + 5;\quad x - 2.$
2. $4x^3 - 3x^2 + x + 7;\quad x + 3.$
3. $x^3 - 8x + 7;\quad x + 5.$
4. $2x^3 - 4x^2 + 2;\quad x - 4.$
5. $x^5 + x^2 + 1;\quad x + 2.$
6. $x^4 - 6x - 7;\quad x - 2.$
7. $2x^4 - x^2 + 1;\quad x - 1/2.$
8. $(1/3)x^3 - (2/9)x^2 + (1/27)x + 1;\quad x + 1/3.$
9. $x^4 + 2ix^3 - ix + 5;\quad x - i.$
10. $x^3 - x^2;\quad x - 1 + i.$

Use synthetic division to work Exercises 11–14.

11. If $f(x) = x^4 - 5x^3 + 2x^2 + x - 5$, find $f(2)$ and $f(-2)$.
12. If $f(x) = 0.2x^3 - 0.03x + 0.015$, find $f(-0.2)$ and $f(0.2)$.
13. If $f(x) = x^3 - x^2 + 5$, find $f(1 + 2i)$ and $f(1 - 2i)$.
14. If $f(x) = x^2 + 4x - 2$, find $f(3 - i)$ and $f(3 + i)$.

In the following exercises use synthetic division to show that c is a zero of $f(x)$.

15. $f(x) = 2x^4 - 6x^3 + 4x^2 - 17x + 15, \quad c = 3.$
16. $f(x) = 3x^3 + 9x^2 - 11x + 4, \quad c = -4.$
17. $f(x) = 2x^4 - 5x^3 - x^2 + 3x + 1, \quad c = -1/2.$
18. $f(x) = 3x^5 - 2x^4 + 6x^3 - 4x^2 - 9x + 6, \quad c = 2/3.$
19. $f(x) = 2x^4 + 3x^3 - 3x^2 + 3x - 5 = 0, \quad c = -i.$
20. $f(x) = 3x^3 - 5x^2 + 4x + 2, \quad c = 1 + i.$

5 FACTORIZATION THEORY

The Factor Theorem (2.18) shows that there is a close relationship between the zeros of a polynomial $f(x)$ and the factors of $f(x)$. Indeed, if a number c can be found such that $f(c) = 0$, then a factor $x - c$ is immediately obtained. Unfortunately, except in special cases, zeros of polynomials are very difficult to find. For example, given the polynomial $f(x) = x^5 - 3x^4 + 4x^3 + 4x - 10$, there are no obvious zeros. Moreover, there exists no device such as the quadratic formula which can be applied to produce the zeros. In spite of the practical difficulty of determining zeros of polynomials, it is possible to make some headway concerning the *theory* of such zeros. The next result is basic for the development of this theory.

(2.20) Fundamental Theorem of Algebra

If $f(x)$ is a polynomial of degree greater than 0, with complex coefficients, then $f(x)$ has at least one complex zero.

Because the proof of this theorem requires advanced mathematical methods, it is impossible to give it here. We shall, however, accept the validity of (2.20) and use it in our work. Note that as a special case of (2.20), if all the coefficients are real, then $f(x)$ has at least one complex zero. We might remark that if $a + bi$ is a complex zero of a polynomial, it may happen that $b = 0$, in which case we refer to the number as a *real* zero.

If (2.20) is combined with the Factor Theorem, the following useful result is obtained.

(2.21) Corollary

Every polynomial of degree greater than 0 with complex coefficients has a factor of the form $x - c$, where c is a complex number.

Corollary (2.21) enables us, at least in theory, to express a polynomial as a product of polynomials of degree 1. In order to see this, let us suppose $f(x)$ is a polynomial of degree $n > 0$ in $F[x]$. Then by (2.21) we can write

$$f(x) = (x - c_1)f_1(x), \tag{2.22}$$

where c_1 is a complex number and $f_1(x) \in F[x]$. By (2.9) the sum of the degrees of $x - c_1$ and $f_1(x)$ is n and consequently, since $x - c_1$ has degree 1, $f_1(x)$ must have degree $n - 1$. If $n - 1 = 0$, then $f_1(x)$ is a nonzero element a of F and we have $f(x) = a(x - c_1)$. On the other hand, if $n - 1 > 0$, then by (2.21) again, there exists a complex number c_2 such that

$$f_1(x) = (x - c_2)f_2(x),$$

where $f_2(x)$ has degree $n - 2$. Substituting in (2.22), we have

$$f(x) = (x - c_1)(x - c_2)f_2(x).$$

If the degree $n - 2$ of $f_2(x)$ is 0, we write

$$f(x) = a(x - c_1)(x - c_2),$$

where $a \in F$ and the factorization is completed. Otherwise, using (2.21), we continue the factorization, obtaining a factor $x - c_3$ of $f_2(x)$. Since the degrees of the polynomials $f_i(x)$ decrease by one at each step and since the sum of the degrees of the polynomials in the factorization must equal n, then after n steps we reach a polynomial $f_n(x)$ of degree 0. Thus $f_n(x) = a$, where $a \in F$, and we have

$$f(x) = a(x - c_1)(x - c_2)\cdots(x - c_n), \tag{2.23}$$

where $c_1, c_2, \cdots, c_n$ are complex numbers and each c_i is a zero of $f(x)$. Evidently, the leading coefficient of the polynomial on the right in (2.23) is a. Since two polynomials are equal if and only if corresponding coefficients are equal, it follows that a is the leading coefficient of $f(x)$. This proves the following theorem.

(2.24) Theorem

If $f(x)$ is a polynomial of degree $n > 0$ with complex coefficients, then there exist n complex numbers $c_1, c_2, \cdots, c_n$ such that

$$f(x) = a(x - c_1)(x - c_2)\cdots(x - c_n),$$

where a is the leading coefficient of $f(x)$.

(2.25) Corollary

If $f(x)$ is a polynomial of degree $n > 0$ with complex coefficients, then $f(x)$ has at most n different complex zeros.

Proof: Suppose $f(x)$ has *more* than n different complex zeros. Choose $n + 1$ of these zeros and label them $c_1, c_2, \cdots, c_n$, and c. Now the c_i can be used as in the proof of (2.24) to obtain the factorization (2.23). Substituting c for x and using the fact that $f(c) = 0$, we obtain

$$0 = a(c - c_1)(c - c_2)\cdots(c - c_n).$$

However, each factor on the right side is different from zero, because $c \neq c_i$ for all i. Since the product of nonzero complex numbers cannot equal zero, we have a contradiction.

EXAMPLE 1. Find a polynomial $f(x)$ of degree 3 which has zeros 2, -1, and i.

Solution: By the Factor Theorem $f(x)$ has factors $x - 2$, $x + 1$, and $x - i$. No other factors of degree 1 exist, since by the Factor Theorem another linear factor $x - c$ would produce a fourth zero of $f(x)$ in violation of (2.25). Hence $f(x)$ has the form

$$f(x) = a(x - 2)(x + 1)(x - i),$$

where $a \in F$. Any nonzero value can be assigned to a. If we let $a = 1$, we obtain

$$f(x) = x^3 - (1 + i)x^2 + (-2 + i)x + 2i.$$

The complex numbers $c_1, c_2, \cdots, c_n$ in (2.24) are not necessarily all different. For example, the polynomial $f(x) = x^3 + x^2 - 5x + 3$ has the factorization

$$f(x) = (x + 3)(x - 1)(x - 1). \tag{2.26}$$

If a factor $x - c$ occurs k times in the factorization (2.24), then c is called a *zero of multiplicity* k. Thus, in (2.26), 1 is a zero of multiplicity 2 and -3 is a zero of multiplicity 1.

If a zero of multiplicity k is counted as k zeros, then (2.24) tells us that a polynomial $f(x)$ of degree $n > 0$ has *at least* n zeros (not necessarily all different). Combining this statement with (2.25), we have the following basic result.

(2.27) Theorem

If $f(x)$ is a polynomial of degree $n > 0$ with complex coefficients and if a zero of multiplicity k is counted k times, then $f(x)$ has precisely n complex zeros.

EXAMPLE 2. Express $f(x) = x^5 - 4x^4 + 13x^3$ as a product of linear factors and list the five zeros of $f(x)$.

Solution: We begin by removing the common factor x^3. Thus

$$f(x) = x^3(x^2 - 4x + 13).$$

By the quadratic formula, the zeros of the polynomial $x^2 - 4x + 13$ are given by

$$\frac{4 \pm \sqrt{16 - 52}}{2} = \frac{4 \pm \sqrt{-36}}{2} = 2 \pm 3i.$$

Hence by the Factor Theorem $x^2 - 4x + 13$ has factors $x - (2 + 3i)$ and $x - (2 - 3i)$. Consequently we may write

$$f(x) = x \cdot x \cdot x \cdot (x - 2 - 3i)(x - 2 + 3i),$$

which is the desired factorization. Since $x - 0$ occurs as a factor three times, the number 0 is a root of multiplicity three and the five zeros of $f(x)$ are 0, 0, 0, $2 + 3i$, and $2 - 3i$.

EXERCISES

In Exercises 1–4 find polynomials of degree 3 with the indicated zeros.

1. $-4, i, 1 - i.$
2. $i, 2i, 3i.$
3. $2 + i, 2 - i, 3.$
4. $\sqrt{3}, \sqrt{\pi}, -\sqrt{\pi}.$
5. Find a polynomial of degree 4 such that both 3 and -2 are zeros of multiplicity two.
6. Find a polynomial of degree 8 such that -1 is a zero of multiplicity three and 0 is a zero of multiplicity five.

In Exercises 7–14 find the zeros of the polynomials and state the multiplicity of each zero.

7. $f(x) = 3(x + 2)^2(x^2 + 2).$
8. $f(x) = (x - 7)^2(x + 1).$
9. $f(x) = x^4 - 4x^3 - 5x^2.$
10. $f(x) = (x^2 - 9)^2.$
11. $f(x) = x^3(x^2 - 4)^2.$
12. $f(x) = (x^2 - x - 12)^2.$
13. $f(x) = (x^2 + 2x - 3)(x^2 + 5x + 6).$
14. $f(x) = (2x + 5)^4.$
15. Show that 3 is a zero of multiplicity two of the polynomial $f(x) = x^4 - 6x^3 + 13x^2 - 24x + 36$ and express $f(x)$ as a product of linear factors.
16. Show that 2 is a zero of multiplicity two of the polynomial $f(x) = x^4 - 6x^3 + 9x^2 + 4x - 12$ and express $f(x)$ as a product of linear factors.
17. Show that 1 is a zero of multiplicity three of $f(x) = x^4 + x^3 - 9x^2 + 11x - 4$ and find the other zero.

18. Show that -1 is a zero of multiplicity four of $f(x) = x^6 + 4x^5 + x^4 - 16x^3 - 29x^2 - 20x - 5$ and find the other zeros.
19. Let $f(x)$ and $g(x)$ be polynomials of degree not greater than n, where n is a positive integer. Show that if $f(x)$ and $g(x)$ are equal in value for more than n distinct values of x, then $f(x)$ and $g(x)$ are identical, that is, coefficients of like powers are the same. [*Hint:* Write $f(x)$ and $g(x)$ as in (2.2) and consider $h(x) = f(x) - g(x) = (a_n - b_n)x^n + \cdots + (a_0 - b_0)$. Show that $h(x)$ has more than n distinct zeros and conclude from (2.25) that $a_i = b_i$, for all i.]
20. Determine real numbers A and B such that $A(2x + 3) + B(x - 7) = 3x - 2$ is an identity. [*Hint:* First write the equation in the form $(2A + B)x + (3A - 7B) = 3x - 2$. Then, by Exercise 19, $2A + B = 3$ and $3A - 7B = -2$. Now solve for A and B.]

Determine real numbers A, B, and C such that the following equations are identities (see Exercise 20).

21. $A(3x - 5) + B(2x - 1) + Cx^2 = 6 - 5x$.
22. $A(x - 3) + (Bx + C)(4x - 1) = 8x + 5$.
23. $A(2x + 1) + (Bx + C)(2x - 1) = (x + 1)(4x + 3)$.
24. $A(x - 2)(x^2 + 1) + B(x^2 + 1) + (Cx + D)(x - 2)^2 = 3x - 5$.

6 ZEROS OF POLYNOMIALS WITH REAL COEFFICIENTS

In this section we shall concentrate on polynomials with real coefficients. It will be necessary, however, to make use of certain properties of complex numbers. Recall that the complex number $a - bi$ is called the *conjugate* of the complex number $a + bi$, where $a, b \in \mathbf{R}$. To simplify the notation, if z is a complex number, then its conjugate will be denoted by $\bar{z}$. The next theorem contains some basic properties concerning conjugates.

(2.28) Theorem

If z and w are complex numbers, then

(i) $\overline{z + w} = \bar{z} + \bar{w}$,

(ii) $\overline{z \cdot w} = \bar{z} \cdot \bar{w}$,

(iii) $\overline{z^n} = \bar{z}^n$, for all positive integers n.

(iv) $\bar{z} = z$ if and only if z is real.

Proof: Let $z = a + bi$ and $w = c + di$, where $a, b, c, d \in \mathbf{R}$. Then $z + w = (a + c) + (b + d)i$ and, by definition of conjugate together with the properties of addition of complex numbers, we have

$$\begin{aligned}\overline{z + w} &= (a + c) - (b + d)i\\ &= (a - bi) + (c - di)\\ &= \bar{z} + \bar{w}.\end{aligned}$$

This proves (i).

By (1.10), $z \cdot w = (ac - bd) + (ad + bc)i$ and hence the conjugate is

$$\overline{z \cdot w} = (ac - bd) - (ad + bc)i,$$

which equals $(a - bi)(c - di)$, that is, $\bar{z} \cdot \bar{w}$. This proves (ii).

If we set $w = z$ in (ii), then $\overline{z \cdot z} = \bar{z} \cdot \bar{z}$, that is $\overline{z^2} = \bar{z}^2$. Also, $\overline{z^3} = \overline{z^2 \cdot z} = \overline{z^2} \cdot \bar{z} = \bar{z}^2 \cdot \bar{z} = \bar{z}^3$ (Supply reasons!). Continuing in this manner, it appears that $\overline{z^n} = \bar{z}^n$, for all positive integers n. A complete proof of (iii) requires the method of mathematical induction (see Appendix I).

Finally, to prove (iv), if $z = a + bi$ is real, then $b = 0$ and clearly $\bar{z} = z$. Conversely, if $\bar{z} = z$, then $a - bi = a + bi$ and, by the definition of equality of complex numbers, $-b = b$, which implies that $b = 0$; that is, z is real.

It is not difficult to extend (i) and (ii) of (2.28). Thus if z, w, and u are complex numbers, then, applying (i) twice, we have

$$\overline{(z + w) + u} = \overline{z + w} + \bar{u} = \bar{z} + \bar{w} + \bar{u},$$

and likewise for more than three complex numbers. In words, this is stated "the conjugate of a sum of complex numbers equals the sum of the conjugates." A similar result is true for products.

An interesting fact about polynomials with real coefficients is illustrated by Example 2 of the preceding section. Observe that the two complex zeros of $x^5 - 4x^4 + 13x^3$ are conjugates of one another. This is no accident. In fact, the following general result is true.

(2.29) Theorem

If $f(x)$ is a polynomial of degree $n > 0$ with real coefficients and if z is a complex zero of $f(x)$, then the conjugate $\bar{z}$ is also a zero of $f(x)$.

Proof: Write

$$f(x) = a_n x^n + a_{n-1} x^{n-1} + \cdots + a_1 x + a_0,$$

where $a_i \in \mathbf{R}$ and $a_n \neq 0$. If $f(z) = 0$, then we have

$$a_n z^n + a_{n-1} z^{n-1} + \cdots + a_1 z + a_0 = 0.$$

If two complex numbers are equal, then so are their conjugates. Consequently the conjugates of each side of this equation are also equal — that is,

$$\overline{a_n z^n + a_{n-1} z^{n-1} + \cdots + a_1 z + a_0} = \bar{0} = 0, \tag{2.30}$$

where the fact that $\bar{0} = 0$ follows from (iv) of (2.28). As pointed out above, the conjugate of a sum of complex numbers equals the sum of the conjugates. Hence (2.30) may be rewritten

$$\overline{a_n z^n} + \overline{a_{n-1} z^{n-1}} + \cdots + \overline{a_1 z} + \overline{a_0} = 0. \tag{2.31}$$

Using (2.28), we can write, for each i,

$$\overline{a_i z^i} = \overline{a_i} \cdot \overline{z^i} = \overline{a_i} \cdot \bar{z}^i = a_i \bar{z}^i.$$

Therefore (2.31) may be written

$$a_n \bar{z}^n + a_{n-1} \bar{z}^{n-1} + \cdots + a_1 \bar{z} + a_0 = 0,$$

which means that $f(\bar{z}) = 0$. This proves the theorem.

EXAMPLE 1. Find a polynomial $f(x)$ of degree 4 with real coefficients which has zeros $2 + i$ and $-3i$.

Solution: By (2.29), $f(x)$ must also have zeros $2 - i$ and $3i$ and consequently, by the Factor Theorem, it has factors $x - (2 + i)$, $x - (2 - i)$, $x - (-3i)$, and $x - (3i)$. Multiplying these factors, we obtain a polynomial of the required type. Thus

$$\begin{aligned} f(x) &= [x - (2 + i)][x - (2 - i)](x - 3i)(x + 3i) \\ &= (x^2 - 4x + 5)(x^2 + 9) \\ &= x^4 - 4x^3 + 14x^2 - 36x + 45. \end{aligned}$$

If a polynomial with real coefficients is factored as in (2.24) some of the factors $x - c_i$ may have a complex coefficient c_i. We can, however, obtain a factorization into polynomials with real coefficients, as stated in the next theorem.

(2.32) Theorem

Every polynomial of positive degree with real coefficients can be written as a product of linear and quadratic polynomials with real coefficients, where the quadratic factors have no real zeros.

Proof: If $f(x)$ has degree $n > 0$, then, by (2.27), $f(x)$ has precisely n complex zeros $c_1, c_2, \cdots, c_n$ and, as in (2.24), we obtain the factorization

$$f(x) = a(x - c_1)(x - c_2)\cdots(x - c_n).$$

Of course, some of the c_i may be real. In these cases we obtain the linear factors referred to in the statement of the theorem. If a zero c_i is not real, then by (2.29) the conjugate $\overline{c_i}$ is also a zero of $f(x)$ and hence must be one of the numbers in the set $c_1, c_2, \cdots, c_n$. This implies that both $x - c_i$ and $x - \overline{c_i}$ appear in the factorization of $f(x)$. If these factors are multiplied, we obtain $x^2 - (c_i + \overline{c_i})x + c_i\overline{c_i}$, which has *real* coefficients (Why?). Thus the complex zeros of $f(x)$ and their conjugates give rise to quadratic polynomials as stated in the theorem. Using the terminology on p. 65, the factors given by (2.32) are irreducible over **R**.

EXAMPLE 2. Express $x^4 - 2x^2 - 3$ as a product of linear and quadratic polynomials with real coefficients.

Solution: The zeros of the given polynomial can be found by solving the equation $x^4 - 2x^2 - 3 = 0$. This equation may be regarded as quadratic in x^2. Solving for x^2 by means of the quadratic formula, we obtain

$$x^2 = \frac{2 \pm \sqrt{4 + 12}}{2} = \frac{2 \pm 4}{2},$$

or $x^2 = 3$ and $x^2 = -1$. Thus the zeros are $\sqrt{3}$, $-\sqrt{3}$, i, and $-i$ and we obtain the factorization

$$x^4 - 2x^2 - 3 = (x - \sqrt{3})(x + \sqrt{3})(x - i)(x + i).$$

Multiplying the last two factors, we arrive at the desired factorization

$$x^4 - 2x^2 - 3 = (x - \sqrt{3})(x + \sqrt{3})(x^2 + 1).$$

The solution for this example could also have been obtained by factoring the original expression immediately, without first finding the zeros. Thus

$$x^4 - 2x^2 - 3 = (x^2 - 3)(x^2 + 1) = (x + \sqrt{3})(x - \sqrt{3})(x^2 + 1).$$

We have already pointed out the fact that in general it is very difficult to find zeros of polynomials of high degree. Most of our results up to this point have been primarily of theoretical value in the sense that they tell us that zeros and factorizations exist, but they do not indicate how to go about finding them. However, if the coefficients are all integers or rational numbers, there is a method for finding the rational zeros, if they exist. The method is a consequence of the following theorem.

(2.33) Theorem

Let $f(x) = a_n x^n + a_{n-1}x^{n-1} + \cdots + a_1 x + a_0$ be a polynomial with integer coefficients. If c/d is a rational zero of $f(x)$, where c and d have no common prime factors and $c > 0$, then c is a factor of a_0 and d is a factor of a_n.

Proof: We begin by showing that c is a factor of a_0. If $c = 1$, the theorem follows at once, since 1 is a factor of *any* number. Thus suppose $c \neq 1$. Then also $c/d \neq 1$, for if $c/d = 1$, we obtain $c = d$ and, since c and d have no prime factor in common, this implies that $c = d = 1$, a contradiction. Therefore in the following we have $c \neq 1$ and $c \neq d$.

Since $f(c/d) = 0$,

$$a_n(c^n/d^n) + a_{n-1}(c^{n-1}/d^{n-1}) + \cdots + a_1(c/d) + a_0 = 0.$$

Multiplying by d^n and then adding $-a_0d^n$ to both sides, we obtain

$$a_nc^n + a_{n-1}c^{n-1}d + \cdots + a_1cd^{n-1} = -a_0d^n,$$

which may be written

$$c(a_nc^{n-1} + a_{n-1}c^{n-2}d + \cdots + a_1d^{n-1}) = -a_0d^n.$$

This shows that c is a factor of the integer a_0d^n. Hence if c is factored into primes, say $c = p_1 p_2 \cdots p_k$, then each p_i is also a factor of a_0d^n. However, by hypothesis, none of the p_i are factors of d. This implies that each p_i is a factor of a_0, that is, c is a factor of a_0.

A similar argument may be used to prove that d is a factor of a_n.

The technique of using (2.33) for finding rational solutions of equations with integer coefficients is illustrated in the following example.

EXAMPLE 3. Find all rational solutions of the equation

$$3x^4 + 14x^3 + 14x^2 - 8x - 8 = 0.$$

Solution: The problem is equivalent to finding the rational zeros of the indicated polynomial. By (2.33), if c/d is a rational zero and $c > 0$, then c is a divisor of -8 and d is a divisor of 3. Hence the possible choices for c are 1, 2, 4, and 8, whereas the choices for d are ± 1 and ± 3. Therefore any rational roots must lie among the numbers ± 1, ± 2, ± 4, ± 8, $\pm 1/3$, $\pm 2/3$, $\pm 4/3$, $\pm 8/3$. Of these sixteen possibilities, not more than four can be zeros, by (2.25). It is necessary to check to see which of them, if any, are zeros. The method of synthetic division is recommended for doing this. We have

$$\begin{array}{rrrrr|r} 3 & 14 & 14 & -8 & -8 & -2 \\ & -6 & -16 & 4 & 8 & \\ \hline 3 & 8 & -2 & -4 & 0 & \end{array}$$

Thus -2 is a zero. Moreover, the synthetic division provides the coefficients of the quotient in the division of the polynomial by $x + 2$. Therefore we have the following factorization of the given polynomial:

$$(x + 2)(3x^3 + 8x^2 - 2x - 4).$$

The remaining solutions of the equation must be zeros of the second factor and consequently we may use the latter polynomial to check for solutions. Dividing by $x + 2/3$ synthetically gives us

$$\begin{array}{rrrr|r} 3 & 8 & -2 & -4 & -2/3 \\ & -2 & -4 & 4 & \\ \hline 3 & 6 & -6 & 0 & \end{array}$$

and we see that $-2/3$ is a zero.

The remaining zeros are in the solution set of the equation $3x^2 + 6x - 6 = 0$, or equivalently, $x^2 + 2x - 2 = 0$. By the quadratic formula this equation has solutions $-1 + \sqrt{3}$ and $-1 - \sqrt{3}$. Hence there are two rational roots -2 and $-2/3$ and two irrational roots.

Theorem (2.33) may also be applied to equations with rational coefficients. We merely multiply both sides of such an equation by the least common denominator of all the coefficients to obtain an equation with integer coefficients and then proceed as above.

EXAMPLE 4. Find all rational solutions of the equation

$$(2/3)x^4 + (1/2)x^3 - (5/4)x^2 - x - (1/6) = 0.$$

Solution: Multiplying both sides of the equation by 12 produces the equivalent equation

$$8x^4 + 6x^3 - 15x^2 - 12x - 2 = 0.$$

If, as in (2.33), c/d is a rational solution, then the choices for c are 1 and 2 and the choices for d are ± 1, ± 2, ± 4, and ± 8. Hence the only possible rational roots are ± 1, ± 2, $\pm 1/2$, $\pm 1/4$, and $\pm 1/8$. By trial we have

$$\begin{array}{rrrrr|r} 8 & 6 & -15 & -12 & -2 & -1/2 \\ & -4 & -1 & 8 & 2 & \\ \hline 8 & 2 & -16 & -4 & 0 & \end{array}$$

and hence $-1/2$ is a solution. Using sythetic division on the coefficients of the quotient, we have

$$\begin{array}{rrrr|r} 8 & 2 & -16 & -4 & -1/4 \\ & -2 & 0 & 4 & \\ \hline 8 & 0 & -16 & 0 & \end{array}$$

and consequently $-1/4$ is a solution. The last synthetic division gives us the quotient $8x^2 - 16$. Setting this equal to zero and solving, we obtain $x = \pm\sqrt{2}$. Thus the given equation has rational solutions $-1/2$, $-1/4$ and irrational solutions $\sqrt{2}$ and $-\sqrt{2}$.

The discussion in this section gives no information about finding the irrational or complex zeros of polynomials. The examples we have worked are not typical of problems encountered in applications. Indeed, it is not unusual for a polynomial with rational coefficients to have *no* rational zeros. Except in the simplest cases, the best that can be accomplished is to find decimal approximations to the irrational zeros. There exist methods which may be used to find some of these irrational zeros to any degree of accuracy. One rather primitive method is illustrated in Example 1 on p. 93. Another standard technique is to use what is called "Newton's Method." This is usually studied in the calculus and consequently we shall not discuss it in this book. In practice, computing machines have, to a large extent, taken over the task of approximating irrational solutions of equations.

If only rough approximations to the real solutions of an equation $f(x) = 0$ are required, the *graphical method* is available. To use this method, sketch the graph of $y = f(x)$ and estimate where $y = 0$, that is, approximate the x-intercepts. Needless to say, the accuracy of the approximation depends on the care with which the graph is sketched.

EXERCISES

1. Find a polynomial of degree 2 with real coefficients that has $3 - 2i$ as a zero.
2. Find a polynomial of degree 3 with real coefficients that has roots $5i$ and -2.
3. Find a polynomial of degree 4 with real coefficients that has roots $1 + i$ and $2 + 2i$.
4. Find a polynomial of degree 4 with real coefficients that has roots i and $2 - \sqrt{3}i$.

In Exercises 5–16 find all rational solutions of the given equations. If possible, find all the roots of the equations.

5. $2x^3 - 3x^2 - 17x + 30 = 0.$
6. $2x^3 - 3x^2 - 7x - 6 = 0.$
7. $6x^3 + 11x^2 - 4x - 4 = 0.$
8. $6x^3 + 11x^2 - 57x - 20 = 0.$
9. $3x^3 + 8x^2 - x - 20 = 0.$
10. $12x^3 - x^2 + 7x + 2 = 0.$
11. $x^4 + x^3 - 5x^2 - 15x - 18 = 0.$
12. $8x^4 + 2x^3 - 7x^2 + 2x - 15 = 0.$
13. $(9/4)x^4 - (15/4)x^3 - 20x^2 + (11/2)x + 1/2 = 0.$
14. $(1/3)x^4 - x^3 - x^2 + (13/3)x - 2 = 0.$
15. $x^4 + 3x^3 - 30x^2 - 6x + 56 = 0.$
16. $x^4 - 3x^3 - 43x^2 + 9x + 20 = 0.$

Prove that the equations in Exercises 17–20 have no rational roots.

17. $2x^4 + 2x^3 + 9x^2 - x - 5 = 0.$
18. $3x^4 - 9x^3 - 2x^2 - 15x - 5 = 0.$
19. $x^5 - 3x^3 + 4x^2 + x - 2 = 0.$
20. $2x^5 + 3x^3 + 7 = 0.$
21. If n is an odd positive integer, prove that a polynomial of degree n with real coefficients has at least one real zero.
22. Show that (2.29) is not necessarily true if $f(x)$ has complex coefficients.
23. Complete the proof of (2.33) by showing that d is a factor of a_n.
24. If a polynomial of the form $x^n + a_{n-1}x^{n-1} + \cdots + a_1x + a_0$, where $a_i \in \mathbf{Z}$, has a rational root r, show that $r \in \mathbf{Z}$ and that r is a factor of a_0.

7 REVIEW EXERCISES

Oral

Define or discuss each of the following.

1. Polynomial.
2. Polynomial function.
3. The degree of a polynomial.
4. Linear polynomial.
5. Irreducible polynomial.
6. The division algorithm for polynomials.
7. The Remainder Theorem.
8. The Factor Theorem.
9. The Fundamental Theorem of Algebra.
10. Conjugate of a complex number.
11. The multiplicity of a zero of a polynomial.
12. The relation between rational zeros and coefficients of a polynomial with integer coefficients.

Written

1. Find $f(x) + g(x)$, $f(x) - g(x)$, and $f(x) \cdot g(x)$ if:
   (a) $f(x) = 2x^3 - x + 5, \quad g(x) = x^2 + x + 2.$
   (b) $f(x) = 7x^4 + x^2 - 1, \quad g(x) = 7x^4 - x^3 + 4x.$
   (c) $f(x) = 3, \quad g(x) = x.$
   (d) $f(x) = 0, \quad g(x) = 2x^2 - 7x + 4.$

2. Sketch the graph of the polynomial function f defined as follows:
   (a) $f(x) = x^3 - 4x.$ (b) $f(x) = -x^2 + 8x - 16.$

3. Express each of the following polynomials as a product of polynomials which are irreducible over **R**:
   (a) $x^4 - 16.$ (b) $8x^3 - 27.$ (c) $2x^4 - x^3 - 2x^2 + x.$

4. In each of the following find the quotient and remainder if $f(x)$ is divided by $g(x)$:
   (a) $f(x) = -3x^4 - x^3 + x^2 - 4x + 3, \quad g(x) = 2x^3 + 4x - 1.$
   (b) $f(x) = 9x^2 + 4x - 1, \quad g(x) = x^3.$
   (c) $f(x) = 6x - 5, \quad g(x) = 2x + 3.$

5. If $f(x) = -2x^4 - 5x^3 + 4x^2 - 9$:
   (a) Use the Remainder Theorem to find $f(2)$.
   (b) Prove that $x + 3$ is a factor of $f(x)$.

6. In each of the following use synthetic division to find the quotient and remainder if $f(x)$ is divided by $g(x)$:
   (a) $f(x) = 3x^5 - x^2 + 3, \quad g(x) = x + 2.$
   (b) $f(x) = x^4 + 2ix^3 - ix + 5, \quad g(x) = x - i.$
7. Find polynomials of degree 3 with the following zeros:
   (a) 1, 2, 3. (b) $3 + 2i$, $3 - 2i$, 3. (c) $2i$, $-i$, 1.
8. Find a polynomial of degree 7 such that:
   (a) 1 is a zero of multiplicity 3 and 0 is a zero of multiplicity 4.
   (b) 0 is a zero of multiplicity 4 and 1 is a zero of multiplicity 3.
9. Show that -2 is a zero of multiplicity 3 of the polynomial $x^5 + 8x^4 + 21x^3 + 14x^2 - 20x - 24$ and express this polynomial as a product of linear factors.
10. Find all rational solutions of the following equations:
    (a) $x^4 - 3x^3 - 15x^2 + 19x + 30 = 0.$
    (b) $30x^3 + 107x^2 - 5x - 42 = 0.$
    (c) $x^4 + 2x^3 + 6x^2 + x + 10 = 0.$

Supplementary Questions

1. Prove that the only polynomials in $F[x]$ which have multiplicative inverses are the polynomials of degree 0. Which polynomials in $\mathbf{Z}[x]$ (see Exercise 13 p. 60) have multiplicative inverses?
2. Let S denote the subset of $F[x]$ consisting of the polynomials (2.1) with $a_0 = 0$. Prove that S is closed relative to addition, subtraction, and multiplication. Which of the field properties (1.3)–(1.7) are true in S?
3. Let S denote the subset of $F[x]$ consisting of all polynomials such that the odd powers of x have zero coefficients. Prove that S is closed relative to addition, subtraction, and multiplication. Which of the field properties (1.3)–(1.7) are true in S? Are the same properties true if the word "odd" is replaced by the word "even"?
4. Given the equation $ax^2 + bx + c = 0$, $a \neq 0$, derive the quadratic formula by writing $x^2 + (b/a)x = (-c/a)$ and then adding $b^2/4a^2$ to both sides and taking square roots. State conditions on a, b, and c which imply that the equation has unequal rational roots.
5. The polynomials in $F[x]$ are sometimes *defined* as the set S of all infinite sequences $(a_0, a_1, a_2, \cdots, a_n, \cdots)$ where $a_i \in F$ and all but a finite number of the a_i are zero. Define equality, addition and multiplication in S in a manner analogous to the corresponding definitions in $F[x]$. What are the identity elements relative to addition and multiplication? What is the additive inverse of $(a_0, a_1, \cdots, a_n, \cdots)$? Prove that addition is commutative and associative. What element of S corresponds to the polynomial x? How should one define the *degree* of an element of S? State several advantages and disadvantages of this approach to polynomials.

chapter three Additional Topics on Functions

In this chapter we shall discuss several advanced topics about functions. Our point of view will, for the most part, be intuitive. A more complete and rigorous development of this material is reserved for books on the calculus.

1 INCREASING OR DECREASING FUNCTIONS

If $a, b \in \mathbf{R}$ and $a < b$, then the set of all real numbers between a and b is called the *open interval* from a to b. It is customary to use the ordered pair (a, b) to denote this interval. Although we have also used ordered pairs to denote points in a coordinate plane, there is little chance for confusion, since it should always be clear from the discussion whether the symbol (a, b) represents an interval or a point. To reiterate, the open interval (a, b) is defined by

$$(a, b) = \{x \mid a < x < b\}.$$

Since there is a one-to-one correspondence between real numbers and points on a coordinate line l, we may define the graph of an open interval in a natural way, namely as the set of points on l which correspond to the numbers in the interval. Thus if A and B are the points on l corresponding to a and b, respectively, then the graph of the open interval (a, b) is the set of all points on l between A and B, but not including A or B. The graph may be illustrated as in Fig. 3.1 by darkening an appropriate part of l. The parentheses in the figure indicate that the points A and B are not to be included.

Figure 3.1

We shall also employ *closed intervals*, denoted by $[a, b]$, and *half-open intervals*, denoted by $[a, b)$ or $(a, b]$. These are defined as follows:

$$[a, b] = \{x \mid a \le x \le b\},$$
$$[a, b) = \{x \mid a \le x < b\},$$
$$(a, b] = \{x \mid a < x \le b\}.$$

Occasionally, we shall refer to the following *unbounded* sets of real numbers:

$$(a, \infty) = \{x \mid x > a\},$$
$$[a, \infty) = \{x \mid x \ge a\},$$
$$(-\infty, a) = \{x \mid x < a\},$$
$$(-\infty, a] = \{x \mid x \le a\},$$
$$(-\infty, \infty) = \mathbf{R}.$$

These are sometimes called *infinite intervals.* A portion of the graph of (a, ∞) is pictured in Fig. 3.2, where the arrow indicates that the graph

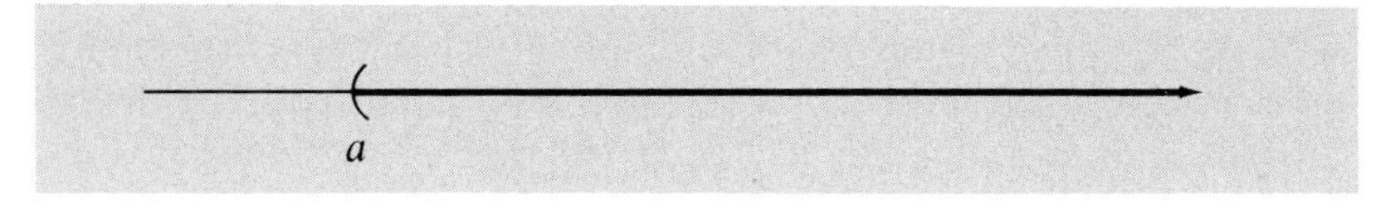

Figure 3.2

continues indefinitely to the right. It is, of course, impossible to sketch the entire graph. In spite of this, we often refer to such a drawing as the graph of the given interval. The symbol ∞ (translated "infinity") is merely a notational device and is *never* to be interpreted as representing a real number.

As an aid to describing the behavior of functions, we introduce a new terminology. In the following definition I denotes an arbitrary interval. Thus I might be open, closed, half-open, infinite, and so on.

(3.1) Definition of Increasing or Decreasing Function

A function f is *increasing* in an interval I if $f(x_1) < f(x_2)$ whenever $x_1 < x_2$ in I. A function f is *decreasing* in an interval I if $f(x_1) > f(x_2)$ whenever $x_1 < x_2$ in I.

If we consider the graph of an increasing function, then as abscissas of points on the graph increase so do the ordinates. This means that the graph "rises" as x increases. On the other hand, if a function is decreasing, then ordinates of points decrease as abscissas increase — that is, the graph "falls" as x increases.

As an illustration of (3.1) let us consider the function f defined by $f(x) = x^2 - 3$ (see Example 2 on p. 35). The graph of f is sketched in

Fig. 1.18. It appears from the graph that f is decreasing in the interval $(-\infty, 0]$ and increasing in $[0, \infty)$. This can be proved algebraically as follows: If a and b are positive real numbers and $a < b$, then also $a^2 < b^2$. Hence $a^2 - 3 < b^2 - 3$, that is, $f(a) < f(b)$. Thus f increases in the set of positive real numbers. Next let us consider negative real numbers. If $a < b < 0$, then $-a > -b > 0$, whence $(-a)^2 > (-b)^2$. Therefore $a^2 > b^2$, or $a^2 - 3 > b^2 - 3$, and we have proved that $f(a) > f(b)$; that is, f decreases in the set of negative real numbers. It follows that $f(x)$ takes on its least value when $x = 0$. This smallest value, -3, is called the minimum value of f, according to (3.2) below. The corresponding point is the lowest point on the graph. Clearly $f(x)$ does not take on a maximum (largest) value.

(3.2) Definition of Maximum or Minimum Value

Let a function f be defined on an interval I and let u and v be numbers in I.

(i) If $f(x) \leq f(v)$ for all $x \in I$, then $f(v)$ is called the *maximum value* of f on I.

(ii) If $f(x) \geq f(u)$ for all $x \in I$, then $f(u)$ is called the *minimum value* of f on I.

If $f(v)$ is the maximum value of f on I, we say that f *takes on* its maximum value at v. Graphically, if we restrict our attention to points with abscissas in I, then the point $P(v, f(v))$ is the highest point on the graph, as illustrated in Fig. 3.3. Similarly, if $f(u)$ is the minimum value of f on I we say that f *takes on* its minimum value at u. The point $Q(u, f(u))$ is the lowest point on the part of the graph corresponding to I (see Fig. 3.3).

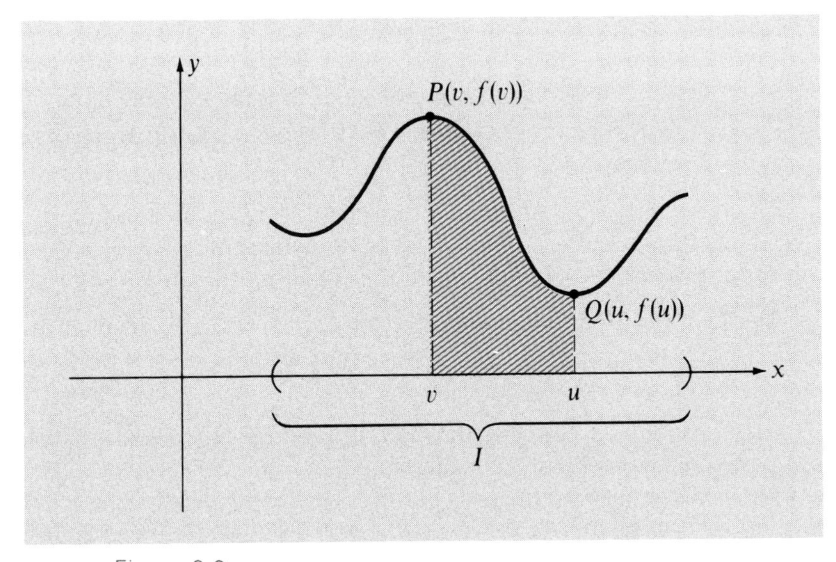

Figure 3.3

Maximum or minimum values are sometimes called *extreme values* or *extrema* of f.

Certain functions may have a maximum value but no minimum value, and vice versa. Other functions may have neither a maximum nor minimum value. It is worth noting that if a function f is increasing in an interval $(a, c]$ and decreasing in $[c, b)$, then f takes on a maximum value at c in the interval (a, b). Similarly, if a function f is decreasing in $(a, c]$ and increasing in $[c, b)$, then f takes on a minimum value at c.

As another illustration, consider the function f defined by $f(x) = \sqrt{x - 1}$ (see Example 4 on p. 36). From the graph of f as sketched in Fig. 1.20 we see that f is increasing throughout the interval $[1, \infty)$. Evidently, in this interval f takes on a minimum value at 1, but f has no maximum value. If we restrict our attention to a finite closed interval such as $[2, 3]$, then f takes on a minimum value at 2 and a maximum value at 3 (Why?).

The function f defined by $f(x) = 1/x$, whose graph is sketched in Fig. 1.21, decreases in the interval $(-\infty, 0)$ and also in $(0, \infty)$. There is neither a maximum nor minimum value for f in either of these intervals.

The determination of maximum and minimum values of functions is extremely important in mathematics and its applications. Our present methods for finding these values rely heavily on our ability to obtain an accurate description of the graph of f. If f is defined in a complicated manner, it may be very difficult to sketch the graph merely by plotting points. In such cases more advanced mathematical techniques are needed in order to find maximum and minimum values of functions.

EXERCISES

1. Express each of the following intervals in terms of inequalities:
   (a) $(-3, 5)$. (b) $[-2, 3]$. (c) $(0, 5)$.
   (d) $(-3, \infty)$. (e) $(-\infty, -1)$.

2. Prove that if a function f is increasing in an interval $(a, c]$ and decreasing in $[c, b)$, then f takes on a maximum value at c in the interval (a, b).

3. If f is the linear function defined by $f(x) = mx + b$, where $m, b \in \mathbf{R}$, prove that (a) if $m > 0$ then f is increasing in $(-\infty, \infty)$; (b) if $m < 0$ then f is decreasing in $(-\infty, \infty)$.

4. Prove that a constant function f takes on an infinite number of maximum and minimum values.

5–30. Determine the intervals in which the functions defined in Exercises 5–30 on pp. 39–40 are increasing and the intervals in which they are decreasing.

2 CONTINUOUS FUNCTIONS

In the process of sketching the graphs of certain functions by plotting many points and then connecting them with a smooth curve we tacitly make use of a property called *continuity*. This property guarantees that there are no sharp "jumps" or "breaks" in the graph and hence, roughly speaking, when considering a small portion of the graph, the sketch can be made without lifting the pencil from the paper. Due to the technical nature of the definition we shall not attempt to give a formal treatment of continuity here. A rigorous approach to this topic will be found in texts on the calculus. In the discussion to follow we shall proceed intuitively, relying on graphs for illustrations.

Let c be a number in the domain of a function f and consider the point $P(c, f(c))$ on the graph of f. If, as illustrated in Fig. 3.4, $Q(x, f(x))$ is

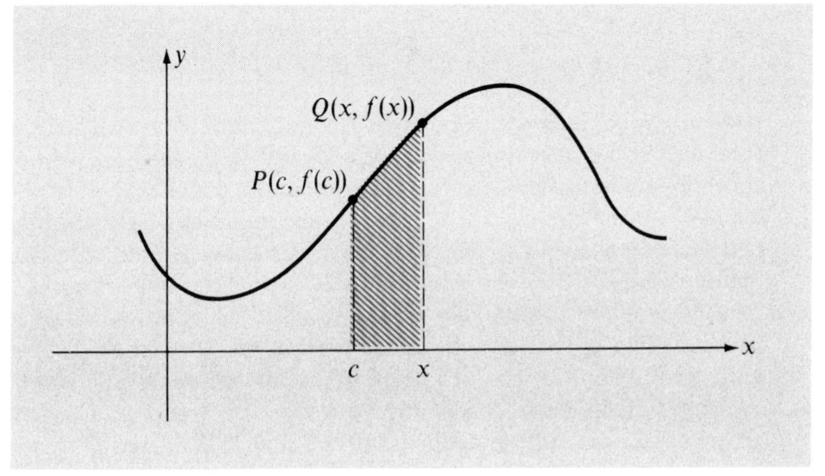

Figure 3.4

another, arbitrary, point on the graph, then the function f is said to be *continuous* at c if, whenever x is "close to" c, then Q is "close to" P. Since the values of f may be interpreted as ordinates of points on the graph, the preceding statement is equivalent to saying that $f(x)$ is close to $f(c)$ whenever x is close to c. This implies that for points whose abscissas lie in a small interval containing c, the sketch of the graph has a "connected" or "unbroken" appearance. The lack of preciseness in this discussion is due to the phrase "close to" which we have used. It would be somewhat more accurate to say that a function f is continuous at c if the number $|f(x) - f(c)|$ can be made as small as desired for any choice of x sufficiently close to c — that is, by making $|x - c|$ sufficiently small. Fig. 3.5 illustrates the graph of a function which is *not* continuous at certain numbers. A break in the graph appears at the point corresponding to a, and the function is not continuous there (we say that f is *discontinuous*, or has a *discontinuity*, at a). Note that if x is close to a and $x > a$, then $f(x)$ is *not*

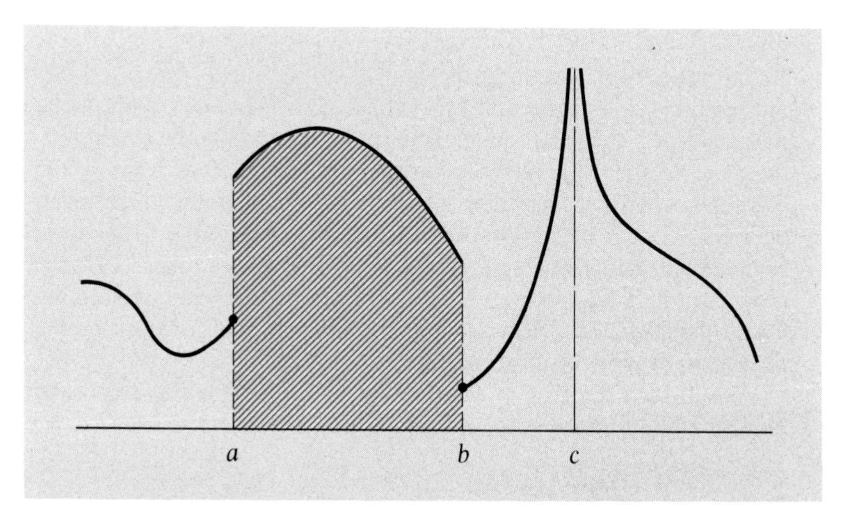

Figure 3.5

close to $f(a)$ as required. Similarly, f is not continuous at b. However, if we restrict our attention to the *open* interval (a, b), then f is continuous at every number in this interval. Discontinuities of the type illustrated at a and b are sometimes called "jump" discontinuities. The graph of the step function sketched in Fig. 1.23 illustrates many jump discontinuities. Indeed, there is such a discontinuity at every integer value of x.

The graph of the function sketched in Fig. 3.5 indicates that as x gets closer and closer to c, $f(x)$ increases without bound, and consequently f cannot be continuous at c. This type of discontinuity is also illustrated by the graph of the function defined by $f(x) = 1/x$ (see Fig. 1.21). As a matter of fact, the latter function is not continuous at 0 since 0 is not in the domain of f.

A function is said to be *continuous on* (*or in*) *an interval I* if it is continuous at every number in the interval. Thus the function f sketched in Fig. 3.5 is continuous in the *open* interval (a, b) and in the half-open interval $[b, c)$. The graph indicates that f is also continuous in the infinite intervals $(-\infty, a]$ and (c, ∞). It can be shown that a polynomial function is continuous at *every* real number c, and hence in the infinite interval $(-\infty, \infty)$.

Functions which are continuous on *closed* intervals are important in advanced courses in mathematics. We shall conclude this section by stating (without proof) several important properties of such functions.

(3.3) The Intermediate Value Theorem

If a function f is continuous on a closed interval $[a, b]$ and if $f(a) \neq f(b)$, then f takes on every value between $f(a)$ and $f(b)$.

The theorem states that if w is any number between $f(a)$ and $f(b)$, then there is a number c between a and b such that $f(c) = w$. If we regard the graph of the continuous function f as extending in an unbroken manner from the point $P(a, f(a))$ to the poi[illegible] $f(b))$, as illustrated in Fig. 3.6, then for any number w between [illegible] ...pears that a

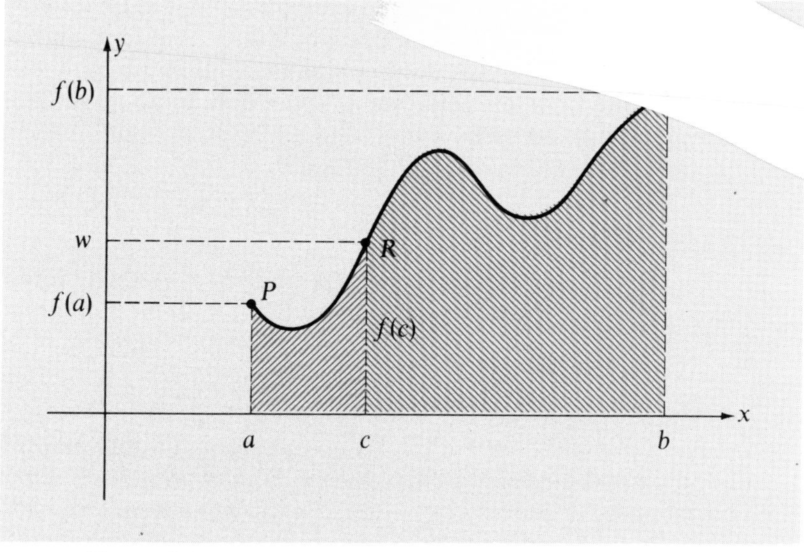

Figure 3.6

horizontal line with y-intercept w should intersect the graph in at least one point R. The abscissa c of R is a number such that $f(c) = w$.

As a consequence of (3.3) we see that if $f(a)$ and $f(b)$ have opposite signs, then there is a number c between a and b such that $f(c) = 0$; that is, f has a zero at c. Geometrically, this implies that if the point $P(a, f(a))$ on the graph of a continuous function lies below the x-axis, and the point $Q(b, f(b))$ lies above the x-axis, or vice versa, then the graph crosses the x-axis at some point $R(c, 0)$, where $a < c < b$. This fact may be used to approximate zeros of functions, as illustrated by the next example.

EXAMPLE 1. Show that $f(x) = x^3 - 3x - 1$ has a zero c between 1 and 2. Isolate c between (a) successive tenths; (b) successive hundredths.

Solution: Since $f(1) = -3$ and $f(2) = 1$, it follows from our previous remarks that there is a number c between 1 and 2 such that $f(c) = 0$. The table below contains several values of $f(x)$ when x is between 1 and 2.

x	1.6	1.7	1.8	1.9
$f(x)$	-1.704	-1.187	-0.568	0.159

Since $f(1.8)$ and $f(1.9)$ have opposite signs, the zero c is between 1.8 and 1.9. Next, calculating $f(x)$ when x is between 1.8 and 1.9 we find that $f(1.87) = -0.070797$ and $f(1.88) = 0.004672$. Consequently c is between 1.87 and 1.88.

The method of successive approximation illustrated in the preceding example is very tedious unless a calculating machine is available. However, it is an elementary process which can be used to approximate zeros of continuous functions to any number of decimal places.

Maximum and minimum values were defined on page 89. For continuous functions we have the following important result.

(3.4) Theorem

If a function f is continuous on a closed interval $[a, b]$, then f takes on a maximum value $f(v)$ and a minimum value $f(u)$ for some numbers u, v in $[a, b]$.

The hypothesis that the interval be closed is essential. For example, the function defined by $f(x) = 1/x$ is continuous on the *open* interval $(0, 1)$ but it does not take on a maximum value there (Why not?). Similarly, f is continuous on the interval $(-1, 0)$ but it does not take on a minimum value. On the other hand, if $[a, b]$ is a *closed* interval not containing 0, maximum and minimum values for f exist (Why?).

EXERCISES

1. Show that the function considered in Example 1 has a zero c between -1 and 0. Isolate c between (a) successive tenths; (b) successive hundredths.
2. Same as Exercise 1 for the numbers -2 and -1.
3. Prove that the equation $x^3 + 5x^2 - 8x - 13 = 0$ has three irrational roots and isolate them between successive integers.
4. Prove that the equation $x^{20} + 4x^{14} + 2x^6 + 1 = 0$ has no real roots.
5. Prove that the linear function f defined by $f(x) = mx + b$, $m \neq 0$, takes on every real value.
6. Sketch the graph of a function f which is defined on the closed interval $[-1, 1]$ by the three conditions: $f(-1) = 2, f(1) = -2$, and f has no zero between -1 and 1. Why doesn't this contradict (3.3)?
7. Define a function f which is continuous on the closed interval $[0, 1]$ and has neither a maximum nor minimum on the open interval $(0, 1)$. Why doesn't this contradict (3.4)?

8. Sketch the graph of a function f defined on a closed interval $[a, b]$ which has neither a maximum nor a minimum on $[a, b]$. Can f be continuous? Explain.
9. Use the graph of f to discuss the discontinuities of the functions defined in Exercises 15, 21, and 27 on p. 40.
10. Same as Exercise 9 for Exercises 18, 28, and 30 on p. 40.

3 RATIONAL FUNCTIONS

A function f is said to be *rational* if it is the quotient of two polynomial functions — that is, if one can write

$$f(x) = \frac{g(x)}{h(x)}, \tag{3.5}$$

where $g(x)$ and $h(x)$ are polynomials. Since division by zero is not permissible, the domain of f is $\{a \in \mathbf{R} \mid h(a) \neq 0\}$. It can be shown that a rational function is continuous at every number in its domain. The behavior of f near a zero of h needs special consideration. We shall once more proceed in an intuitive manner. If $h(c) = 0$ but $g(c) \neq 0$, then it follows from the continuity of h and g that when x is very close to c, the denominator in (3.5) is very close to 0 and the numerator is very close to $g(c)$. In this event the quotient $g(x)/h(x)$ is numerically very large, the sign depending on the signs of $g(x)$ and $h(x)$. We say that f increases or decreases without bound as x approaches c, if, by choosing x sufficiently close to c, $f(x)$ can be made arbitrarily large and positive or arbitrarily large and negative, respectively. Another terminology which is used is to say that f becomes positively infinite (or negatively infinite) as x approaches c. The function whose graph is sketched in Fig. 3.5 becomes positively infinite as x approaches c. The behavior of certain rational functions may differ when x is restricted to values greater than c or values less than c. As a simple illustration let us consider the function f defined by $f(x) = 1/x$ (see Fig. 1.21). In this case we say that f increases without bound as x approaches 0 through positive values, whereas f decreases without bound as x approaches 0 through negative values.

If a function f becomes positively or negatively infinite as x approaches c (through values greater than c or less than c), then the vertical line l with x-intercept c is called a *vertical asymptote* for the graph of f. Roughly speaking, the point $P(x, f(x))$ on the graph gets closer and closer to l as x takes on values closer and closer to c. For the graph sketched in Fig. 3.5 the vertical line through $R(c, 0)$ is a vertical asymptote. Likewise, in Fig. 1.21, the y-axis is a vertical asymptote for the graph. In general, if $h(c) = 0$ and $g(c) \neq 0$, then the graph of the rational function f defined by (3.5) has a vertical asymptote with x-intercept c.

A horizontal line l with equation $y = k$ is called a *horizontal asymptote* for the graph of a function f if $f(x)$ approaches k as x increases (or decreases) without bound. In this case the point $P(x, f(x))$ on the graph is very close to l when x is large and positive or numerically large and negative. For the graph sketched in Fig. 1.21 the x-axis is a horizontal asymptote.

Graphs of rational functions often have vertical or horizontal asymptotes. This is illustrated in the following example.

EXAMPLE 1. Sketch the graph of the rational function f defined by

$$f(x) = \frac{x^2}{x^2 - 4}.$$

Solution: The domain of f consists of all real numbers except 2 and -2. At the latter numbers the denominator vanishes, and hence by our previous remarks there are vertical asymptotes for the graph with x-intercepts 2 and -2. More specifically, if $x > 2$ then $x^2 > 4$ and hence $x^2 - 4 > 0$. Thus when $x > 2$ the expression $x^2/(x^2 - 4)$ is positive, and as x approaches 2 through values *greater* than 2, $f(x)$ *increases* without bound. On the other hand, if $0 < x < 2$ then $x^2/(x^2 - 4)$ is negative, and hence as x approaches 2 through values *less* than 2, $f(x)$ becomes *negatively* infinite. Similarly, one can argue that as x approaches -2 through values less than -2, $f(x)$ becomes positively infinite, whereas if x approaches -2 through values greater than -2, $f(x)$ becomes negatively infinite. If the reader is not convinced of this he should calculate functional values such as $f(-2.1)$, $f(-2.01)$, $f(-1.9)$, $f(-1.99)$, etc.

The x-intercepts are found by solving the equation $f(x) = 0$. This gives us $x^2/(x^2 - 4) = 0$, or $x = 0$. The y-intercept $f(0)$ is also 0.

There are several excluded regions for the graph. If $x > 2$ then $x^2 > 4$ and hence $x^2 - 4 > 0$. It follows that $f(x)$ is positive if $x > 2$, and the corresponding part of the graph lies above the x-axis. As a matter of fact, $f(x)$ is greater than 1 if $x > 2$, since the numerator x^2 is greater than the denominator $x^2 - 4$. The same thing is true if $x < -2$. On the other hand, if $-2 < x < 2$ then $x^2 - 4 < 0$, and hence

$$\frac{x^2}{x^2 - 4} \leq 0.$$

Therefore the part of the xy-plane above the y-axis from $x = -2$ to $x = 2$ is an excluded region.

Dividing the numerator and denominator of the given quotient by x^2 gives us

$$f(x) = \frac{1}{1 - (4/x^2)}$$

As x increases or decreases without bound, the denominator $1 - (4/x^2)$ approaches 1, and therefore so does $f(x)$. It follows that the horizontal line with y-intercept 1 is a horizontal asymptote for the graph.

If we augment this discussion by plotting a sufficient number of points, we obtain the sketch shown in Fig. 3.7. The verification of this is

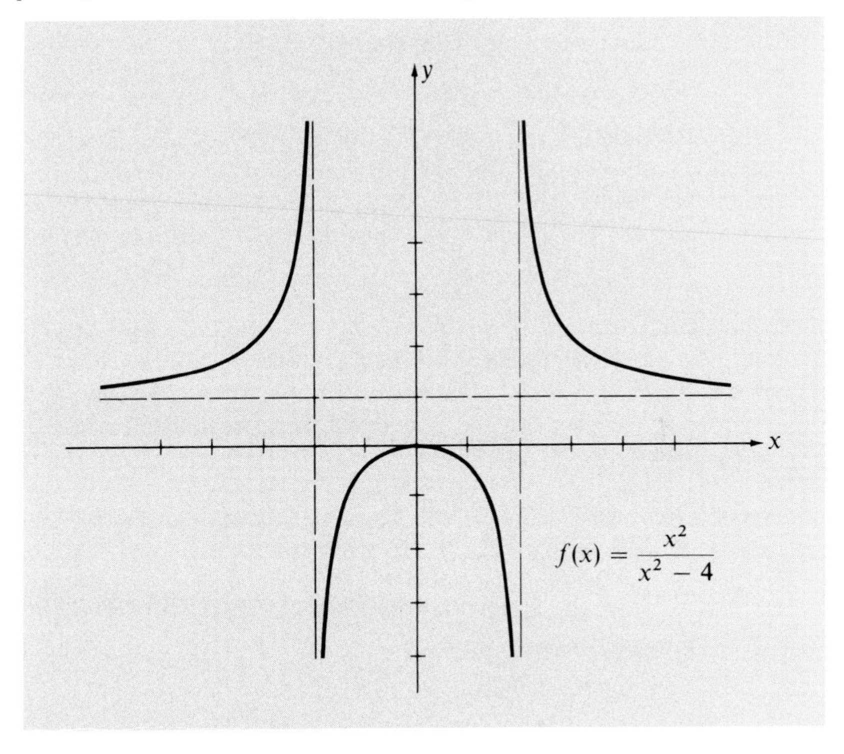

Figure 3.7

left to the reader. The asymptotes are indicated by means of dashed lines. They are not part of the graph of f but are only included as a guide to sketching the graph.

EXERCISES

Sketch the graph of the rational function f defined as indicated.

1. $f(x) = \dfrac{1}{(x - 1)^2}$.
2. $f(x) = \dfrac{x^2}{(x - 1)^2}$.
3. $f(x) = \dfrac{x}{4 - x^2}$.
4. $f(x) = \dfrac{2x}{x^2 + x - 2}$.
5. $f(x) = \dfrac{x^2 - 1}{x}$.
6. $f(x) = \dfrac{x^2}{x^2 - 1}$.

4 IMPLICIT FUNCTIONS

An equation in two variables often determines one or more functions. Thus if p and q are algebraic expressions in x and y, then the solution set S of the equation $p = q$ may be expressed as

$$S = \{(x, y) \mid p = q\},$$

where the notation refers to the set of all ordered pairs (x, y) that are solutions of $p = q$. If, for each x in some subset D of $\mathbf{R}$, there is exactly one y such that $(x, y) \in S$, then a function f is determined, where $f(x) = y$ for all x in D. The solution set S may then be expressed in the form

$$S = \{(x, y) \mid y = f(x)\},$$

where a suitable domain D is chosen for f. A function f determined in this way is said to be defined *implicitly* by the equation $p = q$ and is referred to as an *implicit function.* Sometimes an implicit function may be found by solving the equation for y in terms of x, as illustrated in the next example.

EXAMPLE 1. Find an implicit function which is determined by the equation $x(x + 1) - 2y = 3(x - y)$.

Solution: Each of the following equations is equivalent to the given one:

$$\begin{aligned} x^2 + x - 2y &= 3x - 3y \\ -2y + 3y &= 3x - x - x^2 \\ y &= 2x - x^2. \end{aligned}$$

Thus the solution set S of the given equation is

$$S = \{(x, y) \mid y = 2x - x^2\}$$

and the equation determines implicitly the function f, where $f(x) = 2x - x^2$ for all $x \in \mathbf{R}$. Incidentally, the graph of the implicit function f is shown in Fig. 1.12.

If, in Example 1, we substitute $f(x)$ for y in the given equation we obtain

$$x(x + 1) - 2f(x) = 3(x - f(x)),$$

or

$$x(x + 1) - 2(2x - x^2) = 3(x - (2x - x^2)).$$

The latter equation reduces to the identity

$$3x^2 - 3x = 3x^2 - 3x.$$

This is characteristic of every function f defined implicitly by an equation in x and y, and may be used as a check to tell whether a function which has been found is an implicit function.

Equations in x and y may determine several implicit functions, as illustrated in the next example.

Example 2. Find two implicit functions determined by the equation $y^2 = x$.

Solution: The equation is the same as that considered in Example 5 on page 24. It was pointed out there that the solution set is the *union* of the solution sets of the two equations $y = \sqrt{x}$ and $y = -\sqrt{x}$. If we define functions f and g by

$$f(x) = \sqrt{x}, \qquad g(x) = -\sqrt{x},$$

where $x \geq 0$, then when $f(x)$ (or $g(x)$) is substituted for y in the given equation we obtain the identity $x = x$. Thus f and g are implicit functions.

The graph of the equation $y^2 = x$ is sketched in Fig. 1.13. Notice that the graph of f corresponds to the part above the x-axis, whereas the graph of g is the part below the x-axis. Actually, the equation $y^2 = x$ determines many other implicit functions. For example, if h is defined as follows:

$$h(x) = \begin{cases} -\sqrt{x} & \text{if } 0 \leq x < 3 \\ \sqrt{x} & \text{if } x \geq 3, \end{cases}$$

then h is an implicit function, since substitution of $h(x)$ for y in the equation leads to the identity $x = x$. The graph of h is sketched in Fig. 3.8. Note that h is not continuous at $x = 3$.

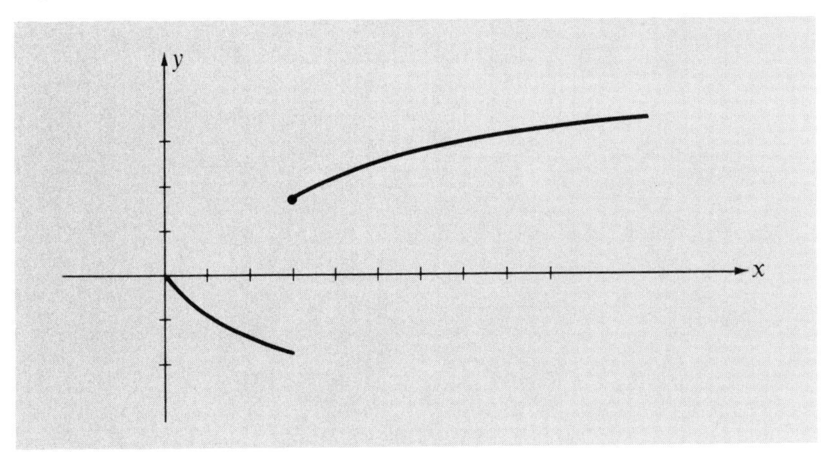

Figure 3.8

The method for finding implicit functions illustrated in Examples 1 and 2 is not always easy to apply. For example, it is very difficult to solve the equation $y^4 + 3y - 4x^3 = 5x + 1$ for y in terms of x. Nevertheless, this equation may still determine an implicit function f in the sense that, when $f(x)$ is substituted for y, then

$$[f(x)]^4 + 3[f(x)] - 4x^3 = 5x + 1$$

is true for all x in the domain of f. It is possible to state conditions under which such an implicit function exists. Results of this nature and investigations of properties of implicit functions are found in more advanced courses in mathematics.

EXERCISES

Find at least one implicit function f determined by each of the following equations and state the domain of f.

1. $3x - 2y + 4 = 2x^2 + 3y - 7x$.
2. $x^2 + 4xy = 3x - 2$.
3. $x^2 + y^2 - 1 = 0$.
4. $3x^2 - 4y^2 = 12$.
5. $x^3 - xy + 4y = 1$.
6. $4xy^2 - 2x + 3 = x^2$.
7. $x^2 - 2xy + y^2 = x$.
8. $3x^2 - 4xy + y^2 = 0$.
9. $\sqrt{x} + \sqrt{y} = 1$.
10. $|x - y| = 2$.

5 REVIEW EXERCISES

Oral

Define or discuss each of the following.

1. Open interval.
2. Closed interval.
3. Half-open interval.
4. Increasing function.
5. Decreasing function.
6. Maximum value of a function.
7. Minimum value of a function.
8. Continuous function.
9. Discontinuity of a function.
10. The Intermediate Value Theorem.
11. Rational function
12. Vertical asymptote.
13. Horizontal asymptote.
14. Implicit function.

Written

1. Prove that if a function f is decreasing in an interval $[a, c]$ and increasing in $[c, b]$, then $f(c)$ is the minimum value of f in the interval $[a, b]$.

2. If f is a constant function and $c \in \mathbf{R}$, prove that $f(c)$ is both a maximum and minimum value of f on every interval containing c.

3. If $f(x) = x^3$ and $[a, b]$ is any closed interval, show that f takes on a minimum value at a and a maximum value at b.

4. Find the intervals in which f is increasing and the intervals in which f is decreasing if:
   (a) $f(x) = 1/x^2$.
   (b) $f(x) = x^2 - 4x$.
   (c) $f(x) = |x - 5|$.
   (d) $f(x) = x^3 + 4$.
   (e) $f(x) = (x - 5)(x - 1)$.

5. Show that the equation $x^3 + x^2 - 5x + 2 = 0$ has an irrational root c in the open interval $(1, 2)$ and isolate c between successive tenths.

6. Sketch the graph of the function f defined by:

   (a) $f(x) = \dfrac{x}{x - 9}$. (b) $f(x) = \dfrac{x}{x^2 - 9}$.

   (c) $f(x) = \dfrac{x^2}{x - 9}$. (d) $f(x) = \dfrac{1}{x^3 - 9x}$.

7. Describe a function f which is defined on $[0, 1]$ and has a maximum value but no minimum value. Can f be continuous? Explain.
8. Find at least one implicit function determined by each of the following equations and state the domain of f:

   (a) $2xy + y - x^2 + 5 = 0$. (b) $6x^2 - 4y^3x + 9 = 0$.
   (c) $\sqrt[3]{x} + \sqrt[3]{y} = 1$. (d) $x^4y^2 + 3y = 5x$.

Supplementary Questions

1. If $S = (a, b)$ and $T = (c, d)$ are open intervals such that $S \cap T \neq \varnothing$, prove that $S \cap T$ and $S \cup T$ are open intervals. Is the analogous statement true for closed intervals? For half-open intervals?
2. If f is the quadratic function defined by $f(x) = ax^2 + bx + c$ where $a > 0$, prove that f takes on a minimum value at $-b/2a$. What is true if $a < 0$?
3. Let f be the function defined by the conditions: $f(x) = 0$ if x is rational and $f(x) = 1$ if x is irrational. Is f increasing on any interval? Is f decreasing? Does f take on a maximum or minimum value? Is f continuous at any number?
4. If f and g are increasing in an interval $[a, b]$ is $f + g$ increasing? Is fg increasing? Is $f - g$ increasing?
5. Is the composite function of two increasing functions increasing? What is true if both functions are decreasing? What if one function is increasing and the other is decreasing?
6. If a function f is continuous and increasing in an interval $[a, b]$ prove that f is $1-1$. Is the inverse function f^{-1} increasing in $[f(a), f(b)]$? What can be said if f is continuous and decreasing in $[a, b]$?

chapter four

Exponential and Logarithmic Functions

The functions considered in the preceding chapters were mainly algebraic functions — that is, functions which can be defined by equations of the form $f(x) = p$, where p is an algebraic expression in x. Functions which are not algebraic are termed *transcendental*. In this chapter we shall define two important families of transcendental functions and investigate some of their properties.

1 EXPONENTIAL FUNCTIONS

Exponents are used to denote products of a real number a with itself. Thus we write $a^2 = a \cdot a$, $a^3 = a \cdot a \cdot a$, and in general, if n is a positive integer $a^n = a \cdot a \cdot \cdots \cdot a$, where there are n factors, all equal to a, on the right-hand side. For example, $(\frac{1}{2})^5 = \frac{1}{2} \cdot \frac{1}{2} \cdot \frac{1}{2} \cdot \frac{1}{2} \cdot \frac{1}{2} = \frac{1}{32}$, $(-3)^3 = (-3) \cdot (-3) \cdot (-3) = -27$, and so on. If $a \neq 0$, the extension to non-positive exponents is made by defining $a^0 = 1$ and, if $n \in \mathbf{N}$, $a^{-n} = 1/a^n$. The *laws of exponents* are: $a^n \cdot a^m = a^{n+m}$, $(a^n)^m = a^{nm}$, $(ab)^n = a^n \cdot b^n$, $(a/b)^n = a^n/b^n$, and $a^m/a^n = a^{m-n}$, where $a, b \in \mathbf{R}$ and $m, n \in \mathbf{Z}$ (provided no zero denominators occur). These may be proved by mathematical induction (see Appendix I). As illustrations we have

$$(2a^2b^3c)^4 = 2^4(a^2)^4(b^3)^4c^4 = 16a^8b^{12}c^4,$$

$$\left(\frac{2x}{y^4}\right)^2 (x^2y)^3 = \left(\frac{2^2x^2}{y^8}\right)(x^6y^3) = 4(x^2x^6)\left(\frac{y^3}{y^8}\right) = \frac{4x^8}{y^5},$$

$$\frac{3u^3v^{-5}}{(2uv^{-2})^{-2}} = \frac{3u^3v^{-5}}{2^{-2}u^{-2}v^4} = 3 \cdot 2^2 \cdot u^{3-(-2)}v^{-5-4} = 12u^5v^{-9}.$$

If a and b are nonnegative real numbers and n is a positive integer, or if a and b are both negative and n is an odd positive integer, then

$$\sqrt[n]{a} = b \quad \text{means that} \quad b^n = a.$$

The number $\sqrt[n]{a}$ is called the *principal nth root of a*. If $n = 2$, it is customary to write $\sqrt{a}$ instead of $\sqrt[2]{a}$ and to call $\sqrt{a}$ the (principal) *square root* of a. The number $\sqrt[3]{a}$ is referred to as the (principal) *cube root* of a. For example,

$$\begin{aligned} \sqrt[5]{\tfrac{1}{32}} &= \tfrac{1}{2} &\quad \text{since} \quad (\tfrac{1}{2})^5 &= \tfrac{1}{32}, \\ \sqrt[4]{81} &= 3 &\quad \text{since} \quad 3^4 &= 81, \\ \sqrt[3]{-8} &= -2 &\quad \text{since} \quad (-2)^3 &= -8, \\ \sqrt{16} &= 4 &\quad \text{since} \quad 4^2 &= 16. \end{aligned}$$

Complex numbers are needed to define $\sqrt[n]{a}$ if $a < 0$ and n is an *even* positive integer, since for all real numbers b, $b^n \geq 0$ whenever n is even. We shall not define such roots at this time.

It is important to observe that when $\sqrt[n]{a}$ is real, it is a *unique* real number. More generally, if $b^n = a$ for a positive integer n, then b is called *an* nth root of a. For example, both 4 and -4 are square roots of 16, since $4^2 = 16$ and also $(-4)^2 = 16$. However, the *principal* square root of 16 is 4. In elementary arithmetic one sometimes sees the expression $\sqrt{16} = \pm 4$, meaning that either $\sqrt{16} = 4$ or $\sqrt{16} = -4$, that is, $\sqrt{a}$ is used to denote *all* square roots of a. We wish to emphasize that this is *not* done in advanced mathematics.

To complete our terminology, any symbol $\sqrt[n]{a}$ is called a *radical*, a is called the *radicand*, and n is called the *index* of the radical. The symbol $\sqrt{\ }$ is called a *radical sign*. It is possible to state laws for radicals which are analogous to the laws of exponents, however we shall not do so here.

Radicals may be used to define rational exponents. First, if $a \in \mathbf{R}$ and $n \in \mathbf{N}$ we define $a^{1/n} = \sqrt[n]{a}$, provided $\sqrt[n]{a}$ is a real number. Next, if m/n is a rational number, where n is positive and the integers m and n have no common factor except 1 or -1, we define

$$a^{m/n} = (\sqrt[n]{a})^m = \sqrt[n]{a^m}.$$

The laws of exponents are also true for rational exponents. As illustrations we have

$$(-27)^{2/3}(4)^{-5/2} = (\sqrt[3]{-27})^2(\sqrt{4})^{-5} = (-3)^2(2)^{-5} = 9/32,$$

$$\left(\frac{2x^{2/3}}{y^{1/2}}\right)^2 \left(\frac{3x^{-5/6}}{y^{1/3}}\right) = \left(\frac{4x^{4/3}}{y}\right)\left(\frac{3x^{-5/6}}{y^{1/3}}\right) = \frac{12x^{1/2}}{y^{4/3}} = \frac{12x^{1/2}y^{2/3}}{y^2}.$$

In order to avoid complex numbers when considering numbers such as $a^{1/2}$, $a^{3/4}$, $a^{7/6}$, etc., we shall assume throughout the remainder of this section that *a is a positive real number*. It is then possible to define a unique real number a^x for *every* real number x (rational or irrational) in such a way that the laws of exponents remain valid. Given a number such as a^π, if we use the decimal representation $3.14159\cdots$ for π and consider the numbers a^3, $a^{3.1}$, $a^{3.14}$, $a^{3.141}$, $a^{3.1415}, \cdots$, we might expect that each successive power gets closer to what we wish to consider as a^π. This is precisely what occurs when a^x is properly defined. However, the definition requires deeper concepts than are available to us, and consequently it is better to leave it for a more advanced course in mathematics.

Although we omit definitions and proofs, we shall assume, henceforth, that formulas such as $a^x a^y = a^{x+y}$, etc., are valid whether x or y are rational or irrational.

Before continuing with the discussion of real exponents, let us establish several results about rational exponents. We show first that if $a > 1$, then $a^r > 1$ for every positive rational number r. If $a > 1$, then multiplying both sides by a, we obtain $a^2 > a$ and hence $a^2 > a > 1$. Multiplying by a again, we obtain $a^3 > a^2 > a > 1$. Continuing, we see that $a^n > 1$ for all positive integers n. (A complete proof requires mathematical induction.) Similarly, if $0 < a < 1$ it follows that $a^n < 1$ for all positive integers n. Now consider $a > 1$ and $r = p/q$, where p and q are positive integers. If $a^{p/q} \leq 1$, then from above, $(a^{p/q})^q \leq 1$, or $a^p \leq 1$, which contradicts the fact that $a^n > 1$ for all positive integers n. Hence $a^r > 1$ for every positive rational number r.

(4.1) Theorem

Suppose $a > 1$. If r and s are rational numbers such that $r < s$, then $a^r < a^s$.

Proof: If $r < s$, then $s - r$ is a positive rational number, and therefore $a^{s-r} > 1$. Multiplying both sides by a^r, we obtain $a^s > a^r$.

Since for each real number x there corresponds a unique real number a^x, we can define a function.

(4.2) Definition of Exponential Function

If $a > 0$, then the *exponential function f with base a* is defined by

$$f(x) = a^x,$$

for all $x \in \mathbf{R}$.

It is possible to extend Theorem (4.1) to the case of *real* exponents r and s. Thus if $a > 1$ and x_1, x_2 are real numbers such that $x_1 < x_2$, then it can be shown that $a^{x_1} < a^{x_2}$, that is, $f(x_1) < f(x_2)$. This says that if $a > 1$, then the exponential function f with base a is an *increasing* function for all real numbers. On the other hand, if $0 < a < 1$, then f *decreases* for all real numbers.

EXAMPLE 1. Sketch, on the same coordinate plane, the graph of f when

$$\text{(a) } f(x) = 2^x, \text{ (b) } f(x) = 3^x, \text{ (c) } f(x) = 4^x.$$

Solution: As shown below, let us make one table for all of the functions.

x	-2	-1	0	1	2	3
2^x	$\frac{1}{4}$	$\frac{1}{2}$	1	2	4	8
3^x	$\frac{1}{9}$	$\frac{1}{3}$	1	3	9	27
4^x	$\frac{1}{16}$	$\frac{1}{4}$	1	4	16	64

The sketches are shown in Fig. 4.1, where for clarity we have indicated the graph of $y = 3^x$ by using dashes.

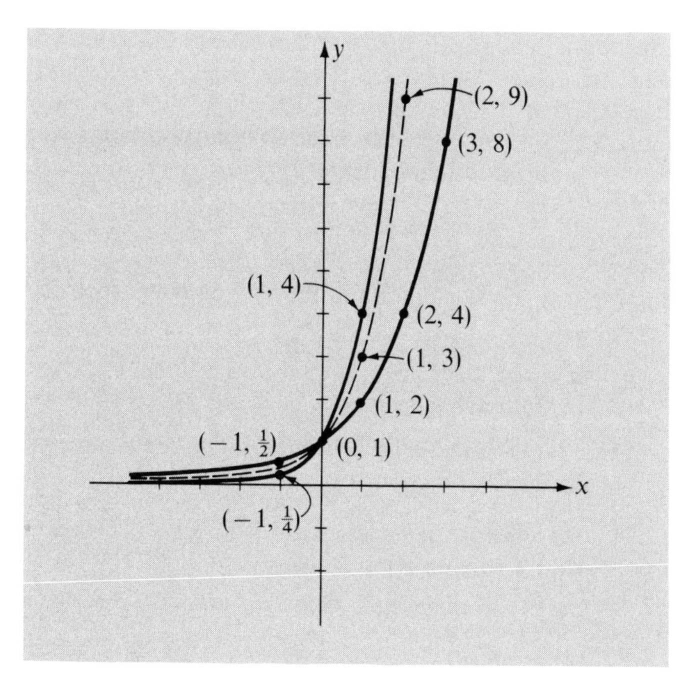

Figure 4.1

In the preceding example notice that as the base a is increased the graph rises more sharply for positive values of x, but for negative values the graph comes closer to the x-axis. It can be shown that if $1 < a < b$, then $a^x < b^x$ for positive values of x and $b^x < a^x$ for negative values of x. In particular, this would tell us that the graph of the function given by (b) lies "between" the graphs of the functions given by (a) and (c), as shown in Fig. 4.1. Similarly, the graph of the function defined by $f(x) = (2.5)^x$ lies "between" the graphs given by (a) and (b).

EXAMPLE 2. Sketch the graph of f if $f(x) = (\frac{1}{2})^x$.

Solution: We have the the table shown below. A portion of the graph is

x	-3	-2	-1	0	1	2	3
$f(x)$	8	4	2	1	$\frac{1}{2}$	$\frac{1}{4}$	$\frac{1}{8}$

sketched in Fig. 4.2. Since $(\frac{1}{2})^x = 2^{-x}$, the graph is the same as the graph of the function defined by $f(x) = 2^{-x}$.

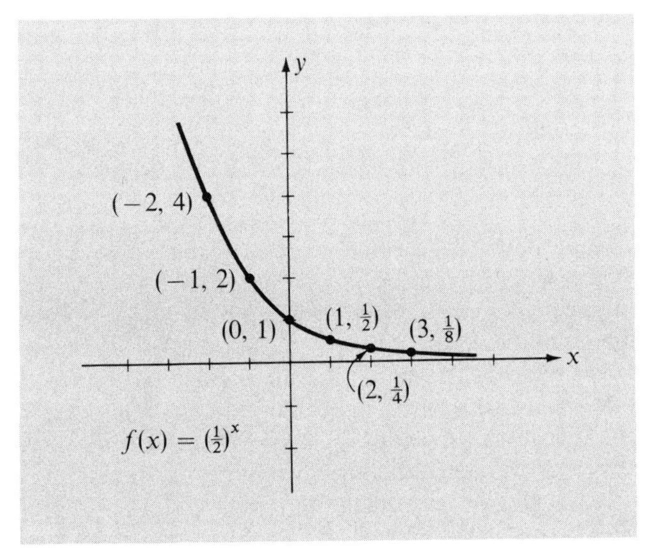

Figure 4.2

In advanced mathematics and applications, it is not unusual to work with functions f defined by $f(x) = a^p$, where p is an algebraic expression in x. We do not intend to study such functions in detail; as an illustration, however, let us consider one example.

EXAMPLE 3. Sketch the graph of f if $f(x) = 2^{-x^2}$.

Solution: Since $f(x) = 1/2^{x^2}$, it is clear that when x is very large numerically, $f(x)$ is close to zero and hence the corresponding point on the curve is close to the x-axis. The maximum value of $f(x)$ occurs when the denominator 2^{x^2} has its smallest value, which is at $x = 0$. Tabulating some coordinates of points on the graph, we have the table shown below.

x	-2	-1	0	1	2
$f(x)$	$\frac{1}{16}$	$\frac{1}{2}$	1	$\frac{1}{2}$	$\frac{1}{16}$

Part of the graph is sketched in Fig. 4.3. Functions of this type arise in the study of the branch of mathematics called *probability*.

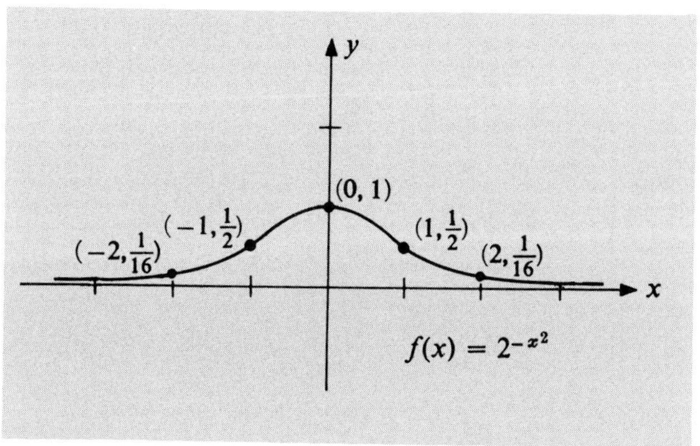

Figure 4.3

Many physical quantities vary *exponentially* in the sense that they may be described by means of an exponential function. Among the many examples of this are population growth, radioactive decay, charge and discharge of an electrical condenser, and rates at which certain chemicals dissolve in liquids. As an illustration, it might be observed experimentally that the number of bacteria in a certain culture doubles every hour. If there are 1000 bacteria present at the start of the experiment, then the experimenter would have the readings shown below, where t is the time in hours and $f(t)$ is the bacteria count at time t. Thus it appears that $f(t) = (1000)2^t$. This formula makes it possible to predict the number of bacteria present at *any* time t.

t	0	1	2	3	4
$f(t)$	1000	2000	4000	8000	16000

EXERCISES

Eliminate negative exponents and simplify.

1. $\dfrac{2^{-2} - 3^{-3}}{(-2)^2 + (-3)^3}$.

2. $4^0 + 0^4$.

3. $\dfrac{(6x^3)^2}{(2x^2)^3}$.

4. $(3y^3)^4(4y^2)^{-3}$.

5. $(5x^2y^{-3})(4x^{-5}y^4)$.

6. $(-2r^2s)^5(3r^{-1}s^3)^2$.

7. $\left(\dfrac{3x^5y^4}{x^0y^{-3}}\right)^2$.

8. $\dfrac{(3x^{-3}y)^{-2}}{(x^2y^{-2})^3}$.

9. $\left(\dfrac{5a^{-1}b^2}{2c^{-3}}\right)^{-1}\left(\dfrac{ab^{-1}}{c^2}\right)^4$.

10. $\left(\dfrac{3x^3y^{-2}}{7x^{-5}y^8}\right)^0$.

11. $((ab^2)^{-2})^{-2}$.

12. $a^0 + (b^0 + c)^0$.

13. $\dfrac{x^{-1}}{y^{-1}} - \left(\dfrac{x}{y}\right)^{-1}$.

14. $(r + s)^{-1}(r^{-1} + s^{-1})$.

15. $(-0.008)^{2/3}$.

16. $(0.008)^{-2/3}$.

17. $(3x^{5/6})(8x^{2/3})$.

18. $(27a^6)^{-2/3}$.

19. $\dfrac{(x^6y^3)^{-1/3}}{(x^4y^2)^{-1/2}}$.

20. $\left(\dfrac{a^{2/3}}{b^{-2}}\right)^{-1}\left(\dfrac{a^{1/2}}{b^{1/3}}\right)^3$.

In Exercises 21–30 sketch the graph of the function f defined by the given expression.

21. $f(x) = (\frac{3}{2})^x$.
22. $f(x) = (\frac{2}{3})^x$.
23. $f(x) = 4^{x/2}$.
24. $f(x) = 4^{-(x/2)}$.
25. $f(x) = 2^{x+1}$.
26. $f(x) = 2^{x-1}$.
27. $f(x) = 2^{1-x}$.
28. $f(x) = 2^{-x-1}$.
29. $f(x) = 2^{-(x+3)^2}$
30. $f(x) = 2^{3-x^2}$.
31. If $a > 1$, how does the graph of $y = a^x$ compare with the graph of $y = a^{-x}$?
32. How does the graph of $y = a^x$ compare with the graph of $y = -a^x$?
33. Use (4.1) to prove that if $0 < a < 1$ and r and s are rational numbers such that $r < s$, then $a^r > a^s$.
34. Why has $a = 0$ been ruled out in the discussion of a^x?
35. The half-life of radium is 1600 years — that is, given any quantity, one-half of it will disintegrate in 1600 years. If one begins with an amount q_0 milligrams, then the quantity q remaining after t years is given by $q = q_0 2^{kt}$. Find k.
36. The number of bacteria in a certain culture is given by $Q = 2(3^t)$, where t is measured in hours and Q in thousands. What is the initial number of bacteria? What is the number after 10 minutes? After 30 minutes? After 1 hour?

2 LOGARITHMS

Suppose a is any positive real number different from 1. If we examine the graph of $y = a^x$ (see Fig. 4.4 for the case $a > 1$), it seems evident that

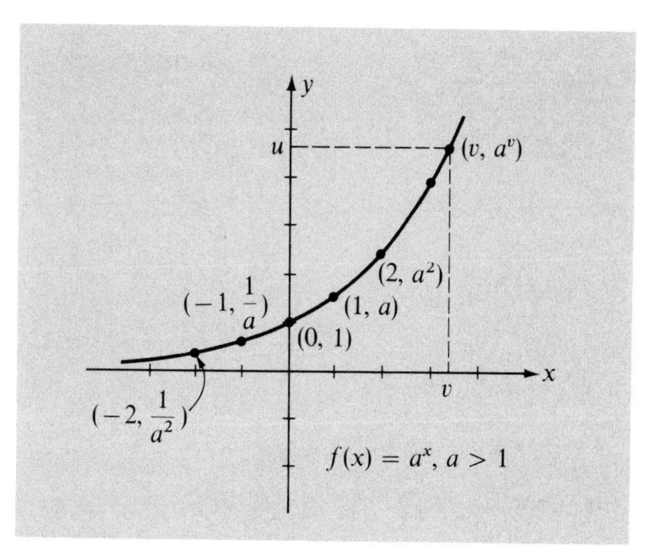

Figure 4.4

if u is any positive real number, then u is the ordinate of some point on the graph — that is, there is a number v such that $u = a^v$. Moreover, the number u can occur as an ordinate only once, since, as we have observed, if $x_1 < x_2$, then $a^{x_1} < a^{x_2}$. This means that the exponential function is a one-to-one function from the set **R** onto the set of positive real numbers. Geometrically, a horizontal line $y = u$, $u > 0$, intersects the graph in precisely one point. A similar situation exists when $0 < a < 1$. The preceding discussion makes the following theorem seem plausible.

(4.3) Theorem

If $a > 0$ and $a \neq 1$, then for each positive real number u there is a unique real number v such that $a^v = u$.

We wish to emphasize that (4.3) has not been proved, as our discussion has been strictly on an intuitive level. A proof could be supplied by showing that the exponential function is continuous (see p. 91) and then using the Intermediate Value Theorem (3.3); however, this is beyond the scope of our work.

Theorem (4.3) enables us to state the following definition.

(4.4) Definition of Logarithm

Let $a > 0$ and $a \neq 1$. If u is any positive real number, then the (unique) exponent v such that $a^v = u$ is called the *logarithm of u to the base a* and is denoted by $\log_a u$.

A convenient way of stating (4.4) is to write equivalent equations as follows:

(4.5) $$v = \log_a u \quad \text{if and only if} \quad a^v = u,$$

where it is assumed that a and u are restricted as in (4.4). In the following discussion *we shall always assume that a is a positive real number different from* 1 *and that* $u > 0$.

As illustrations of (4.5), we may write

$$3 = \log_2 8, \qquad \text{since} \quad 2^3 = 8,$$
$$-2 = \log_5 \tfrac{1}{25}, \qquad \text{since} \quad 5^{-2} = \tfrac{1}{25},$$
$$4 = \log_{10} 10{,}000, \qquad \text{since} \quad 10^4 = 10{,}000.$$

Since $\log_a u$ is the exponent v such that $a^v = u$, we have the following important identity:

(4.6) $$a^{\log_a u} = u.$$

Also, since $a^x = a^x$ for every real number x, we may use (4.5) with $v = x$ and $u = a^x$ to obtain

(4.7) $$x = \log_a a^x.$$

Letting $x = 1$ in (4.7), we have $1 = \log_a a$. Another important formula is

(4.8) $$\log_a 1 = 0.$$

This is true since $a^0 = 1$.

EXAMPLE 1. (a) Find $\log_4 2$. (b) Find u if $\log_3 u = 4$.

Solution: (a) We let $v = \log_4 2$. By (4.5) this is equivalent to $4^v = 2$ and we see that $v = \frac{1}{2}$. Hence $\log_4 2 = \frac{1}{2}$.

(b) If $\log_3 u = 4$, then $3^4 = u$. Hence $u = 81$.

EXAMPLE 2. (a) Solve the equation $\log_4 (5 + x) = 3$. (b) Solve the inequality $\log_{10} x > 2$.

Solution: (a) If $\log_4 (5 + x) = 3$, then by (4.5) we have $5 + x = 4^3$, which is equivalent to $x = 59$. Hence the solution set is $\{59\}$.

(b) If $2 < \log_{10} x$, then since the exponential function with base 10 is increasing, we have $10^2 < 10^{\log_{10} x}$. By (4.6) this implies $100 < x$. Hence the solution set is the infinite interval $(100, \infty)$.

The three statements (i), (ii), and (iii) given in the following theorem are called the *laws of logarithms*.

(4.9) Theorem

Let $a > 0$ and $a \neq 1$. If u and w are positive real numbers, then

$$\begin{aligned} &\text{(i)} \quad \log_a (uw) = \log_a u + \log_a w, \\ &\text{(ii)} \quad \log_a (u/w) = \log_a u - \log_a w, \\ &\text{(iii)} \quad \log_a (u^c) = c \log_a u, \quad \text{for every real number } c. \end{aligned}$$

Proof: Let

$$r = \log_a u \quad \text{and} \quad s = \log_a w. \tag{4.10}$$

By (4.5) this implies $a^r = u$ and $a^s = w$. Consequently

$$a^r a^s = uw,$$

and, by a law of exponents,

$$a^{r+s} = uw.$$

From (4.5) this is equivalent to

$$r + s = \log_a (uw).$$

Substituting for r and s from (4.10), we obtain (i).

To prove (ii), we begin with (4.10) as above and *divide* a^r by a^s, obtaining

$$\frac{a^r}{a^s} = \frac{u}{w},$$

or

$$a^{r-s} = \frac{u}{w}.$$

From (4.5) the latter equation is equivalent to

$$r - s = \log_a (u/w),$$

which, by (4.10), gives us (ii).

Finally, if c is a real number, then, using the same notation as above, we have $(a^r)^c = u^c$, which, by a law of exponents, gives us $a^{rc} = u^c$, or $a^{cr} = u^c$. By the definition of logarithm, this implies $cr = \log_a u^c$, or, substituting for r, $c \log_a u = \log_a u^c$.

EXAMPLE 3. If $\log_a 3 = 0.4771$ and $\log_a 2 = 0.3010$, find

$$\text{(a) } \log_a 6, \quad \text{(b) } \log_a (3/2), \quad \text{(c) } \log_a \sqrt{2}, \quad \text{(d) } (\log_a 3)/(\log_a 2).$$

Solution: (a) Writing $6 = 2 \cdot 3$ and using (i) of (4.9) we have $\log_a 6 = \log_a (2 \cdot 3) = \log_a 2 + \log_a 3 = 0.4771 + 0.3010 = 0.7781$.

(b) By (ii) of (4.9), $\log_a (3/2) = \log_a 3 - \log_a 2 = 0.4771 - 0.3010 = 0.1761$.

(c) Using (iii) of (4.9), we may write $\log_a \sqrt{2} = \log_a 2^{1/2} = (1/2) \log_a 2 = (1/2)(0.3010) = 0.1505$.

(d) There is no law of logarithms which will enable us to change the form of $(\log_a 3)/(\log_a 2)$. It is necessary to *divide* 0.4771 by 0.3010, obtaining approximately 1.585. It is important to notice the difference between this and part (b).

EXAMPLE 4. Solve the following equations:

(a) $\log_5 (2x + 3) = \log_5 11 + \log_5 3$.
(b) $2 \log_7 x = \log_7 36$.
(c) $\log_4 (x + 6) - \log_4 10 = \log_4 (x - 1) - \log_4 2$.

Solution: (a) Using (i) of (4.9) we may write the given equation in the form

$$\log_5 (2x + 3) = \log_5 (11 \cdot 3) = \log_5 33.$$

Consequently $2x + 3 = 33$, or $2x = 30$. Therefore, the solution set is $\{15\}$.

(b) In order for $\log_7 x$ to exist, x must be positive. If $x > 0$, then by (iii) of (4.9) the given equation is equivalent to $\log_7 x^2 = \log_7 36$. It follows that $x^2 = 36$ and hence the solution set is $\{6\}$.

(c) The equation may be written as

$$\log_4 (x + 6) - \log_4 (x - 1) = \log_4 10 - \log_4 2.$$

Applying (ii) of (4.9) gives us

$$\log_4 \left(\frac{x + 6}{x - 1}\right) = \log_4 \frac{10}{2} = \log_4 5.$$

Therefore $(x + 6)/(x - 1) = 5$. From the latter equation we have $x + 6 = 5x - 5$, or $4x = 11$. This gives us the solution set $\{11/4\}$.

Sometimes it is necessary to *change the base* of a logarithm. That is, given $\log_a u$, we may wish to find $\log_b u$ for some $b > 0$, $b \neq 1$. This can be accomplished as follows. Let $v = \log_b u$. Then $b^v = u$. Hence $\log_a b^v = \log_a u$ or, by (iii) of (4.9), $v \log_a b = \log_a u$. Solving for v (that is, $\log_b u$), we have

$$\log_b u = \frac{1}{\log_a b} \cdot \log_a u. \tag{4.11}$$

An interesting case occurs when $u = a$ in (4.11). Since $\log_a a = 1$, we obtain

$$\log_b a = \frac{1}{\log_a b}. \tag{4.12}$$

As a final remark, note that there is no law for expressing $\log_a (u + w)$ in terms of simpler logarithms. It is evident that this does not always equal $\log_a u + \log_a w$, since the latter expression equals $\log_a (uw)$.

The laws of logarithms are sometimes used as in the following example.

EXAMPLE 5. Express $\log_a x^3\sqrt{y}/z^2$ in terms of the logarithms of x, y, and z.

Solution: Using (4.9) we may write

$$\begin{aligned}\log_a \frac{x^3\sqrt{y}}{z^2} &= \log_a (x^3\sqrt{y}) - \log_a z^2\\ &= \log_a x^3 + \log_a \sqrt{y} - \log_a z^2\\ &= 3\log_a x + (1/2)\log_a y - 2\log_a z.\end{aligned}$$

EXERCISES

Use (4.5) to change the equations in Exercises 1–4 to logarithmic form and the equations in Exercises 5–8 to exponential form.

1. $3^2 = 9$.
2. $2^4 = 16$.
3. $10^{-3} = 0.001$.
4. $u^v = w$.
5. $\log_4 64 = 3$.
6. $\log_3 (1/27) = -3$.
7. $\log_a 1 = 0$.
8. $\log_s p = t$.

In Exercises 9–16 find each number.

9. $\log_2 (1/8)$.
10. $\log_3 (81)$.
11. $\log_{10} (10{,}000)$.
12. $\log_{1/2} 4$.
13. $3^{\log_3 5}$.
14. $\log_{10} (\frac{1}{10})$.
15. $\log_4 \sqrt[3]{4}$.
16. $10^{5 \log_{10} 2}$.

Find the solution sets in Exercises 17–36.

17. $\log_3 (x + 5) = -1$.
18. $\log_7 x = 2/3$.
19. $\log_{10} (x^2) = -5$.
20. $\log_2 (x^2 - x - 2) = 2$.
21. $\log_x 5 = 2$.
22. $10^{\log_{10} x} = 2/13$.
23. $\log_2 (x + 1) < 4$.
24. $\log_3 (2 - 5x) > 1$.
25. $2 \le \log_{10} x \le 3$.
26. $0 < \log_{10} x < 1$.
27. $\log_3 2x = \log_3 3 + \log_3 5$.
28. $\log_5 2x = \log_5 8 - \log_5 3$.
29. $\log_6 (3x + 1) = \log_6 10 - \log_6 2$.
30. $3 \log_2 x = 2 \log_2 8$.
31. $\log_4 x - \log_4 (x - 1) = 2 \log_4 3$.
32. $2 \log_5 \sqrt{x} = 3$.
33. $\log_{10} x^2 - \log_{10} 5 = \log_{10} 7 + \log_{10} 2x$.
34. $\log_{10} x^2 = \log_{10} x$.
35. $\log_5 (x - 1) + \log_5 (x - 2) = 2 \log_5 \sqrt{6}$.
36. $(1/3) \log_8 (x + 1) = 2 \log_8 3 - (2/3) \log_8 (x + 1)$.

In Exercises 37–40 express the logarithm in terms of logarithms of x, y, and z.

37. $\log_a \dfrac{xy^3}{z^2}$.
38. $\log_a \dfrac{\sqrt[3]{xz^2}}{y}$.
39. $\log_a \sqrt{x\sqrt{y}}$.
40. $\log_a \sqrt[4]{\dfrac{x}{y^2z^5}}$.

In Exercises 41 and 42 write the expression as one logarithm.

41. $2 \log_a x + (1/5) \log_a (x + 1) - 3 \log_a (x - 1)$.
42. $\log_a (y/x) + 2 \log_a x - \log_a (xy^2)$.

43. If $\log_a x = 3$, find $\log_{1/a} x$ and $\log_a (1/x)$.
44. Prove that $\log_{1/a} x = \log_a (1/x)$ for all $x > 0$.
45. Prove that $\log_a \left(\dfrac{x - \sqrt{x^2 - 1}}{x + \sqrt{x^2 - 1}}\right) = 2 \log_a (x - \sqrt{x^2 - 1})$.
46. Prove that $\log_a (\sqrt{1 + x^2} + x) = -\log_a (\sqrt{1 + x^2} - x)$.

3 LOGARITHMIC FUNCTIONS

The concept of logarithm can be used to introduce a new function having the set **P** of positive real numbers as domain.

(4.13) Definition of Logarithmic Function

Let $a \in \mathbf{P}$, $a \neq 1$. The function f from **P** to **R** defined by

$$f(x) = \log_a x,$$

for all $x \in \mathbf{P}$, is called the *logarithmic function with base a*.

The graph of f is the same as the graph of the equation $y = \log_a x$, which, by (4.5), is equivalent to

$$x = a^y. \tag{4.14}$$

Equation (4.14) can be graphed, as in Section 1, by merely interchanging the roles of x and y. In order to find some pairs in the solution set of (4.14), we substitute for y and find the corresponding values of x. If $a > 1$, then we obtain the curve sketched in Fig. 4.5. Note that f is an

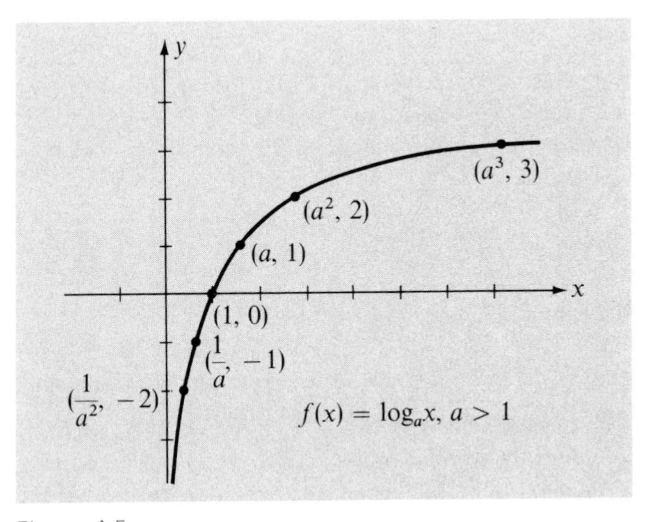

Figure 4.5

increasing function throughout its domain. If $0 < a < 1$, then the graph has the general shape shown in Fig. 4.6, whence f is a decreasing function. Note that for every a under consideration, the region to the left of the y-axis is excluded. There is no y-intercept and the x-intercept is 1.

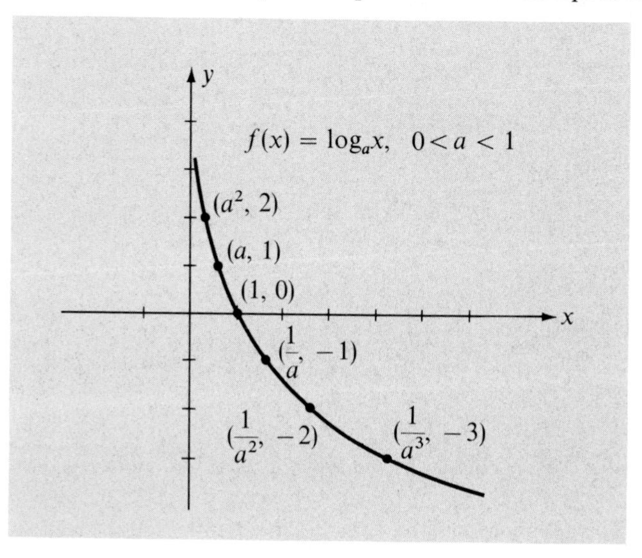

Figure 4.6

In mathematics one often encounters functions defined by expressions of the form $\log_a p$, where p is an algebraic expression in x. Such functions are classified as members of the logarithmic family. However, the graph may differ from those given in Figs. 4.5 and 4.6, as is illustrated in the following examples.

EXAMPLE 1. Sketch the graph of f if $f(x) = \log_3(-x)$, $x < 0$.

Solution: If $x < 0$, then $-x > 0$, and hence $\log_3(-x)$ is defined. We wish to graph the equation $y = \log_3(-x)$, which is equivalent to $3^y = -x$, or $x = -(3^y)$. We make the table shown below.

y	-2	-1	0	1	2
x	$-\frac{1}{9}$	$-\frac{1}{3}$	-1	-3	-9

Part of the graph is shown in Fig. 4.7.

EXAMPLE 2. Sketch the graph of the equation $y = \log_3|x|$, $x \neq 0$.

Solution: Since $|x| > 0$ for all $x \neq 0$, there will be points on the graph corresponding to negative values of x as well as to positive values. If $x > 0$, then $|x| = x$ and therefore the graph to the right of the y-axis coincides with the graph of $y = \log_3 x$. When $x < 0$, then $|x| = -x$

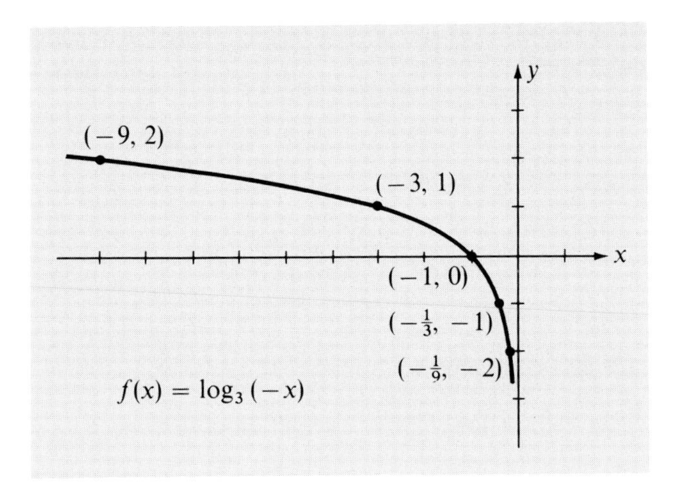

Figure 4.7

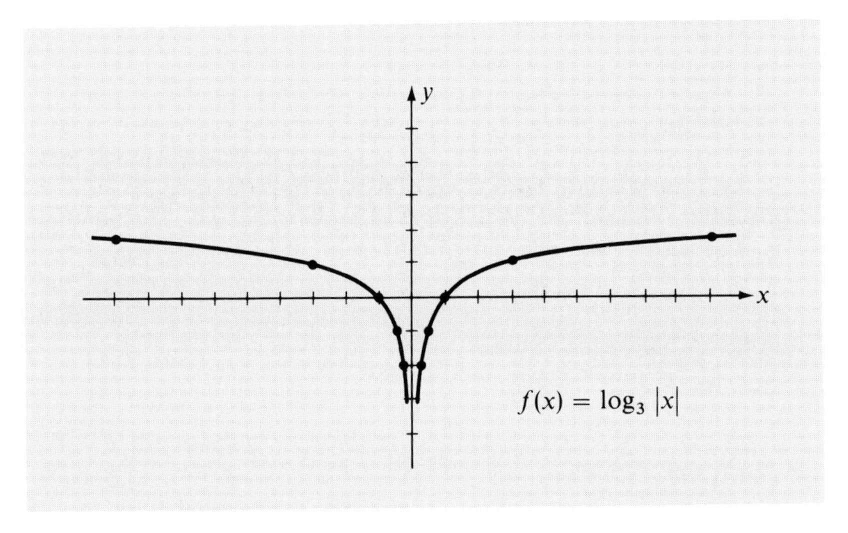

Figure 4.8

and the graph is the same as that of $y = \log_3(-x)$ (see Example 1). The graph is shown in Fig. 4.8.

There is a close relationship between exponential and logarithmic functions. This is to be expected since the logarithmic function has been defined in terms of the exponential function. A precise description of the relationship is stated in the following theorem.

(4.15) Theorem

If $a > 0$ and $a \neq 1$, then the exponential and logarithmic functions with base a are inverse functions of one another.

Proof: Let $f(x) = a^x$ and $g(x) = \log_a x$, where the domains and ranges of f and g are as given previously. By (1.26) we must show that

$$f(g(x)) = x \quad \text{and} \quad g(f(x)) = x.$$

For all positive real numbers x we have

$$\begin{aligned} f(g(x)) &= a^{g(x)} && \text{(definition of } f) \\ &= a^{\log_a x} && \text{(definition of } g) \\ &= x && (4.6). \end{aligned}$$

Also, for all real numbers x we have

$$\begin{aligned} g(f(x)) &= \log_a f(x) && \text{(definition of } g) \\ &= \log_a a^x && \text{(definition of } f) \\ &= x && (4.7). \end{aligned}$$

This proves the theorem.

In calculus and its applications, an irrational number, denoted by e, is used for the logarithmic base. To five decimal places, $e \doteq 2.71828$. This base arises naturally in the theory of certain mathematical concepts and in the description of many types of physical phenomena. For this reason logarithms to the base e are called *natural logarithms*. The notation $\ln x$ is often used as an abbreviation for $\log_e x$. We shall not have occasion to work with natural logarithms. For computational purposes, the base 10 is used. This base gives us the system of common logarithms discussed in the sections which follow.

EXERCISES

In Exercises 1–8 sketch the graph of the function f defined by the given expression.

1. $f(x) = \log_2 (1 + x)$.
2. $f(x) = \log_2 (1 - x)$.
3. $f(x) = \log_2 |1 + x|$.
4. $f(x) = \log_2 |1 - x|$.
5. $f(x) = 2 \log_2 (-x)$.
6. $f(x) = -\log_2 (x^2)$.
7. $f(x) = \log_3 \sqrt{x}$.
8. $f(x) = |\log_3 x|$.
9. Sketch, on the same rectangular coordinate system, the graphs of $y = \log_2 x$, $y = \log_3 x$, and $y = \log_4 x$. Formulate a conjecture based on your graphs. If $e \doteq 2.7$, how does the graph of $y = \log_e x$ compare with these graphs?
10. Sketch, on the same rectangular coordinate system, the graphs of $y = \log_a x$ and $y = a^x$, for $a = 2$. Is there any relationship between the graphs? (cf. p. 49).

4 COMMON LOGARITHMS

Logarithms can be employed to simplify certain types of numerical computations. When used for this purpose, the base 10 is usually chosen and the logarithms are referred to as *common logarithms*. For convenience, we shall use the symbol "log x" as an abbreviation for $\log_{10} x$. The reason for choosing base 10 is because every positive real number x can be written in the form $x = a \cdot 10^k$, where $1 \leq a < 10$ and k is an integer. This is called the *scientific form* for the number x. As illustrations we have $513 = (5.13)10^2$, $2375 = (2.375)10^3$, $720000 = (7.2)10^5$, $0.641 = (6.41)10^{-1}$, $0.00000438 = (4.38)10^{-6}$ and $4.601 = (4.601)10^0$. If x is any positive real number, then writing $x = a \cdot 10^k$, where $1 \leq a < 10$ and k is an integer, we have, by (i) of (4.9),

(4.16) $$\log x = \log a + \log 10^k.$$

Applying (4.7), we obtain

$$\log x = \log a + k, \qquad 1 \leq a < 10, \qquad k \text{ an integer.}$$

Thus, in order to find $\log x$ for any positive real number x, it is sufficient to know the logarithms of numbers between 1 and 10. In (4.16) the number $\log a$, $1 \leq a < 10$, is called the *mantissa* and the integer k is called the *characteristic* of $\log x$.

Since $\log x$ increases as x increases, if $1 \leq a < 10$, then

$$\log 1 \leq \log a < \log 10,$$

or, by (4.8) and (4.7),

$$0 \leq \log a < 1.$$

Hence the mantissa of a logarithm is some number between 0 and 1. It is usually necessary to approximate logarithms when working numerical problems. For example, it can be shown that

$$\log 2 = 0.3010299957\cdots,$$

where the decimal is nonrepeating and nonterminating. In our work we shall round off logarithms of numbers to four decimal places and write

$$\log 2 \doteq 0.3010.$$

The symbol "$\doteq$," which may be read "is approximately equal to," is used to indicate that the expression on the right is an *approximation* of log 2 to the nearest ten-thousandth. When a number between 0 and 1 is written as a finite decimal, it is referred to as a *decimal fraction*. Thus, to summarize, (4.16) tells us that if x is a positive real number, then $\log x$ *can be approximated as the sum of a positive decimal fraction* (*the mantissa*) *and an integer k* (*the characteristic*). We shall refer to this representation as the *standard form* for $\log x$.

Logarithms of many of the numbers between 1 and 10 have been calculated. Table 1 contains four-decimal-place approximations for logarithms of numbers between 1.00 and 9.99 at intervals of 0.01. This table can be used to find the logarithm of any three-digit number to four-decimal-place accuracy. There exist far more extensive tables which provide logarithms of many additional numbers to much greater accuracy than four decimal places.

The use of Table 1 is illustrated in the following examples.

EXAMPLE 1. Find

$$\text{(a)}\ \log 43.6, \qquad \text{(b)}\ \log 43{,}600, \qquad \text{(c)}\ \log 0.0436.$$

Solution: (a) Since $43.6 = (4.36)10^1$, the characteristic of log 43.6 is 1. From Table 1 we see that the mantissa, log 4.36, approximated to four decimal places is 0.6395. Hence

$$\log 43.6 \doteq 0.6395 + 1 = 1.6395.$$

(b) Writing $43{,}600 = (4.36)10^4$, we see that the mantissa is the same as in part (a) and the characteristic is 4. Hence

$$\log 43{,}600 \doteq 0.6395 + 4 = 4.6395.$$

(c) If we write $0.0436 = (4.36)10^{-2}$, then

$$\log 0.0436 = \log 4.36 + (-2).$$

Hence

$$\log 0.0436 \doteq 0.6395 + (-2). \tag{4.17}$$

We could combine 0.6395 and -2 and write

$$\log 0.0436 \doteq -1.3605.$$

However, this would not be the standard form, since $-1.3605 = -0.3605 + (-1)$, a number in which the decimal fraction is *negative*. A common error is to write (4.17) as -2.6395. This is incorrect since $-2.6395 = -0.6395 + (-2)$, which is not the same as $0.6395 + (-2)$.

When a logarithm has negative characteristic, as above, it is customary either to leave it in standard form or to rewrite the logarithm, keeping the decimal part positive. As an illustration of the latter technique, let us add and subtract 8 on the right side of (4.17). This gives us

$$\log 0.0436 \doteq 0.6395 + (8 - 8) + (-2),$$

or

$$\log 0.0436 \doteq 8.6395 - 10.$$

Clearly one could also write

$$\log 0.0436 \doteq 18.6395 - 20 = 43.6395 - 45,$$

and so on, as long as the *integral part* of the logarithm is -2.

EXAMPLE 2. Find

(a) $\log (0.00652)^2$, (b) $\log (0.00652)^{-2}$,
(c) $\log (0.00652)^{1/2}$

Solution: (a) By (iii) of (4.9) we can write

$$\log (0.00652)^2 = 2 \log (0.00652).$$

Since $0.00652 = (6.52)10^{-3}$, we have

$$\begin{aligned} \log 0.00652 &= \log 6.52 + (-3) \\ &\doteq 0.8142 + (-3). \end{aligned}$$

Hence

$$2 \log 0.00652 \doteq 1.6284 + (-6).$$

The standard form is $0.6284 + (-5)$.

(b) Proceeding in the same manner,

$$\begin{aligned} \log (0.00652)^{-2} &= -2 \log 0.00652 \\ &\doteq -2[0.8142 + (-3)] \\ &= -1.6284 + 6. \end{aligned}$$

It is important to note that -1.6284 means $-0.6284 + (-1)$ and that hence the decimal part is negative. To obtain the standard form, we could write

$$\begin{aligned} -1.6284 + 6 &= 6.0000 - 1.6284 \\ &= 4.3716, \end{aligned}$$

which shows that the mantissa is 0.3716 and the characteristic is 4.

(c) Again using (iii) of (4.9),

$$\begin{aligned} \log (0.00652)^{1/2} &= (1/2) \log 0.00652 \\ &\doteq (1/2)[0.8142 + (-3)]. \end{aligned}$$

If we multiply through by 1/2, the standard form will not be obtained, since neither number in the resulting sum will be the characteristic. In order to avoid this, we adjust the expression within brackets by adding and subtracting a suitable number. If we use 1 in this capacity, we obtain

$$(1/2)[1.8142 + (-4)] = 0.9021 + (-2),$$

which is in standard form. We could also have used numbers other than 1, for example,

$$(1/2)[17.8142 + (-20)] = 8.9021 + (-10).$$

Table 1 can be used to find an approximation to x if $\log x$ is given. This is illustrated in the following example.

EXAMPLE 3. Find a decimal approximation to x if (a) $\log x = 1.7959$, (b) $\log x = -3.5918$.

Solution: (a) The mantissa 0.7959 determines the sequence of digits in x and the characteristic determines the position of the decimal point. Referring to the *body* of Table 1, we see that the mantissa 0.7959 is associated with the number 6.25. Since the characteristic is 1, x lies between 10 and 100 and consequently we have $x \doteq 62.5$.

(b) In order to find x from Table 1, $\log x$ must be written in standard form. To change $\log x = -3.5918$ to standard form, we add and subtract 4, obtaining

$$\begin{aligned}\log x &= (4 - 3.5918) - 4 \\ &= 0.4082 - 4.\end{aligned}$$

From Table 1 the mantissa 0.4082 is associated with 2.56. Since the characteristic of $\log x$ is -4, we have $x \doteq 0.000256$.

In each of the previous examples the mantissas appeared in Table 1. When this is not the case, x can be approximated by the method of interpolation discussed in the next section.

EXERCISES

In Exercises 1–14 use Table 1 and the laws of logarithms to approximate the common logarithms of each of the given numbers.

1. 6.71; 671; 0.00671.
2. 39.4; 3940; 0.394.
3. 97; 970,000; 0.97.
4. 1.02; 102; 10,200.
5. 0.0000707; 70.7; 707.
6. 0.0008; 0.8; 8.
7. $(88.5)^2$; $(88.5)^{1/2}$; $(88.5)^{-2}$.
8. $(5470)^4$; $(5470)^{40}$; $(5470)^{1/4}$.
9. $(0.243)^3$; $(0.243)^{-3}$; $(0.243)^{1/3}$.
10. $(0.014)^{10}$; $10^{0.014}$; $10^{1.632}$.
11. $\dfrac{437}{26.1}$.
12. $\dfrac{(6.93)(30.1)}{537}$.
13. $\sqrt{10.2}\,(412)^3$.
14. $\dfrac{(0.0061)^{10}}{\sqrt{0.29}}$.

In Exercises 15–24 use Table 1 to find a decimal approximation for x. If $\log x$ does not appear in the table, find the entry in the table which most nearly approximates x.

15. $\log x = 3.8102$.
16. $\log x = 1.4183$.
17. $\log x = 0.0212$.
18. $\log x = 7.7774 - 10$.
19. $\log x = 8.8789 - 10$.
20. $\log x = 5.5386$.
21. $\log x = -1.1399$.
22. $\log x = -3.6421$.
23. $\log x = -0.0191$.
24. $\log x = -5.2235$.

5 LINEAR INTERPOLATION

The only logarithms that can be found *directly* from Table 1 are logarithms of numbers that contain at most three nonzero digits. If *four* nonzero digits are involved, then it is possible to obtain an approximation by using the method of linear interpolation described in this section. In order to illustrate this process and at the same time give some justification for it, let us consider as a specific example log 12.64. Since the logarithmic function to the base 10 is increasing, this number lies between log 12.60 $\doteq$ 1.1004 and log 12.70 $\doteq$ 1.1038. Examining the graph of $y = \log x$, we have the situation shown in Fig. 4.9, where we have distorted the units on

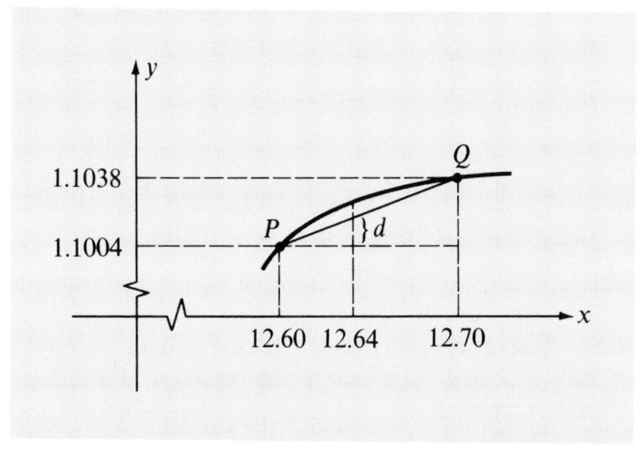

Figure 4.9

the x- and y-axes and also the portion of the graph shown. A more accurate drawing would show that the graph of $y = \log x$ is much closer to the line segment joining $P(12.60, 1.1004)$ to $Q(12.70, 1.1038)$ than we have pictured. Thus it appears that log 12.64, which is the ordinate of the point on the graph having abscissa 12.64, can be approximated by finding the ordinate of the point on the *line segment* from P to Q with abscissa 12.64. If we let d be as in Fig. 4.9, then this ordinate is $1.1004 + d$. The number d can be approximated by using similar triangles. Thus, using Fig. 4.10, where the graph of $y = \log x$ has been deleted, we may form the following proportion

$$\frac{d}{0.0034} = \frac{0.04}{0.1}.$$

Hence

$$d = \frac{(0.04)(0.0034)}{0.1} = 0.00136.$$

When operating in this manner, we always round off decimals to the same number of places as appear in the body of the table; therefore $d \doteq 0.0014$.

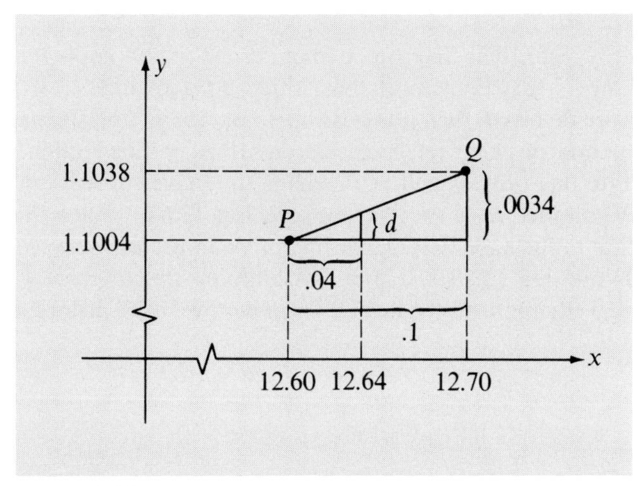

Figure 4.10

Then

$$\log 12.64 \doteq 1.1004 + 0.0014 = 1.1018.$$

This process is referred to as *linear interpolation* since a straight line is used to approximate the graph of $y = \log x$. The process is not restricted to tables of logarithms — it can be used for any table in which entries vary in a manner similar to the behavior of a linear function.

Hereafter we shall not sketch a graph when interpolating. Rather we shall use a scheme such as is illustrated in the following example.

EXAMPLE 1. Approximate log 572.6.

Solution: The proportion involved can easily be seen if we arrange our work as follows:

$$10\left\{\begin{array}{l} 6\left\{\begin{array}{l}\log 572.0 \doteq 2.7574 \\ \log 572.6 = \ ?\end{array}\right\} d \\ \quad \log 573.0 \doteq 2.7582 \end{array}\right\} 0.0008,$$

where we have indicated differences by appropriate symbols alongside of the braces. We have the proportion

$$\frac{d}{0.0008} = \frac{6}{10},$$

or

$$d = (6/10)(0.0008) = 0.00048 \doteq 0.0005.$$

Hence $\log 572.6 \doteq 2.7574 + 0.0005 = 2.7579$. Another way of working this type of problem is to reason that, since 572.6 is 6/10 of the way from 572.0 to 573.0, then log 572.6 is (approximately) 6/10 of the way from 2.7574 to 2.7582. Hence $\log 572.6 \doteq 2.7574 + (6/10)(0.0008) \doteq 2.7574 + 0.0005 = 2.7579$.

EXAMPLE 2. Approximate log 0.003678.

Solution: We write

$$10\left\{\begin{array}{l} 8\left\{\begin{array}{l}\log 0.003670 \doteq 0.5647 + (-3) \\ \log 0.003678 = ?\end{array}\right\}d \\ \log 0.003680 \doteq 0.5658 + (-3)\end{array}\right\}0.0011.$$

Since we are interested only in ratios, we use the numbers 8 and 10 on the left side because their ratio is the same as the ratio of 0.000008 to 0.000010. Thus we have

$$\frac{d}{0.0011} = \frac{8}{10} = 0.8,$$

and $d = (0.0011)(0.8) = 0.00088 \doteq 0.0009$. Hence

$$\begin{aligned}\log 0.003678 &\doteq [0.5647 + (-3)] + 0.0009 \\ &= 0.5656 + (-3).\end{aligned}$$

If a number x is written in the form $x = a \cdot 10^k$, where $1 \leq a < 10$, then, before using Table 1 to find log x by interpolation, a should be rounded off to three decimal places. Another way of saying this is that x should be rounded off to four *significant figures.* Some examples will help to clarify the procedure. Thus if $x = 36.4635$, round off to 36.46 before approximating log x; a number such as 684,279 should be written as 684,300; for decimals such as 0.096202 write 0.09620; and so on. The reason for this is that Table 1 does not guarantee more than four-digit accuracy, since the mantissas which appear in it are approximations. Of course, this means that if *more* than four-digit accuracy is required in a problem, then Table 1 cannot be used. If in more extensive tables the logarithm of a number containing n digits can be found *directly*, then interpolation is allowed for numbers involving $n + 1$ digits, and numbers should be rounded off accordingly.

The method of interpolation can also be used to find x when given log x. If we use Table 1, then x may be found to four significant figures. In effect, here we are given the *ordinate* of a point and are asked to find the *abscissa.* A geometric argument similar to the one given earlier can be supplied to justify the procedure, which is illustrated in the next example.

EXAMPLE 3. Find x to four significant figures if $\log x = 1.7949$.

Solution: The mantissa 0.7949 does not appear in Table 1, but it can be isolated between adjacent entries, namely the mantissas corresponding to 6.230 and 6.240. We arrange the work as follows:

$$0.1\left\{\begin{array}{l} r\left\{\begin{array}{l}\log 62.30 \doteq 1.7945 \\ \log x = 1.7949\end{array}\right\}0.0004 \\ \log 62.40 \doteq 1.7952\end{array}\right\}0.0007.$$

Then

$$\frac{r}{0.1} = \frac{0.0004}{0.0007} = \frac{4}{7}$$

and $r = (0.1)(4/7) \doteq 0.06$. Hence $x \doteq 62.30 + 0.06 = 62.36$.

If we are given $\log x$, then the number x is called the *antilogarithm* of $\log x$. In Example 3 the antilogarithm of $\log x = 1.7949$ is $x \doteq 62.36$. Sometimes the notation antilog $(1.7949) \doteq 62.36$ is used.

EXERCISES

Use the method of linear interpolation to approximate the common logarithms of the numbers in Exercises 1–8.

1. 35.74.
2. 293.8.
3. 7463.
4. 0.8464.
5. 0.001693.
6. 55,550.
7. 200,700.
8. 0.01234.

In Exercises 9–16 use linear interpolation to approximate x.

9. $\log x = 2.4397$.
10. $\log x = 4.8165$.
11. $\log x = 0.4461$.
12. $\log x = 0.0186$.
13. $\log x = 8.6312 - 10$.
14. $\log x = 2.9392 - 5$.
15. $\log x = -8.6312$.
16. $\log x = -1.9382$.

6 COMPUTATIONS WITH LOGARITHMS

Owing to the development of computing machines and portable calculators, the importance of logarithms for numerical computations has diminished in recent years. However, since mechanical devices are not always available, it is worthwhile to have some familiarity with the use of logarithms for solving arithmetic problems. At the same time the practice gained in working numerical problems will lead to a deeper understanding of the theoretical aspects of logarithms. We illustrate some of the computational techniques by means of the following examples.

EXAMPLE 1. Use logarithms to approximate

$$N = \frac{(59700)(0.0163)}{41.7}.$$

Solution: Using (4.9) and Table 1, we have

$$\begin{aligned} \log N &= \log 59700 + \log (0.0163) - \log 41.7 \\ &\doteq 4.7760 + (0.2122 - 2) - (1.6201) \\ &= 4.9882 - 3.6201 \\ &= 1.3681. \end{aligned}$$

Finding the antilogarithm, we have, to three significant figures,

$$N \doteq 23.3.$$

EXAMPLE 2. Approximate $N = \sqrt[3]{56.11}$ to four significant figures.

Solution: Writing $N = (56.11)^{1/3}$, we have, by (iii) of (4.9),

$$\log N = (1/3) \log (56.11).$$

Interpolating from Table 1, we have

$$10\left\{1\left\{\begin{matrix}\log 56.10 \doteq 1.7490 \\ \log 56.11 = \ ? \end{matrix}\right\}d \atop \log 56.20 \doteq 1.7497\right\}0.0007,$$

$$\frac{d}{0.0007} = \frac{1}{10},$$

or $d = 0.00007 \doteq 0.0001$. Hence $\log 56.11 \doteq 1.7491$ and we have

$$\log N \doteq (1/3)(1.7491) \doteq 0.5830.$$

We now find the antilogarithm by interpolation from Table 1.

$$0.01\left\{r\left\{\begin{matrix}\log 3.820 \doteq 0.5821 \\ \log N \doteq 0.5830 \end{matrix}\right\}9 \atop \log 3.830 \doteq 0.5832\right\}11$$

$$\frac{r}{0.01} = \frac{9}{11}.$$

Therefore $r \doteq 0.008$ and $N \doteq 3.828$.

If we were interested in only *three* significant figures, then the above interpolations could have been avoided. In the remaining examples we shall, for simplicity, work with three-digit numbers.

EXAMPLE 3. Approximate

$$N = \frac{(1.32)^{10}}{\sqrt[5]{0.0268}}.$$

Solution:

$$\begin{aligned}\log N &= 10 \log (1.32) - (1/5) \log (0.0268) \\ &\doteq 10(0.1206) - (1/5)(3.4281 - 5) \\ &= 1.206 - 0.6856 + 1 \\ &= 1.5204.\end{aligned}$$

Finding the antilogarithm (to three significant figures), we have

$$N \doteq 33.1.$$

EXAMPLE 4. Find x if $x^{2.1} = 6.5$.

Solution: Taking the common logarithm of both sides and using (iii) of (4.9), we have

$$2.1 \log x = \log 6.5.$$

Hence

$$\log x = \frac{\log 6.5}{2.1},$$

or by (4.5),

$$x = 10^{\log 6.5/2.1}.$$

If an approximation to x is desired, then from the second equation we have

$$\log x \doteq \frac{0.8129}{2.1} \doteq 0.3871.$$

From Table 1, the antilogarithm (to three significant figures) is

$$x \doteq 2.44.$$

EXAMPLE 5. Approximate

$$N = \frac{69.3 + \sqrt[3]{56.1}}{\log 807}.$$

Solution: Since we have no formula for the logarithm of a sum, the two terms in the numerator must be *added* before the logarithm can be found. From Example 2 we have $\sqrt[3]{56.1} \doteq 3.8$ and hence the numerator is (approximately) 73.1. Moreover, from Table 1, $\log 807 \doteq 2.9069$. Thus, writing the numerator and denominator of N to three significant figures, we have

$$N \doteq \frac{73.1}{2.91}.$$

This problem is too easy to bother with logarithms. By long division we find $N \doteq 25.1$.

EXERCISES

Use logarithms to approximate each of Exercises 1–20 to three significant figures.

1. (42.7)(0.364).
2. (0.00639)(0.0127).
3. $\dfrac{74{,}600}{9{,}230}$.
4. $\dfrac{2.07}{58.1}$.

5. $(5.17)^4$.
6. $(0.162)^5$.
7. $\sqrt[4]{0.267}$.
8. $\sqrt[6]{46.3}$.
9. $\dfrac{(6.23)(17.4)^2}{81.5}$.
10. $\dfrac{(7.06)^3}{(31.3)\sqrt{1.05}}$.
11. $\sqrt[3]{(4.61)(1.29)^2}$.
12. $\left[\dfrac{(13.1)^4}{\sqrt{2.41}}\right]^{-1/2}$.
13. $\dfrac{(-6.43)^2(80.6)^{-1/3}}{-74.3}$.
14. $\sqrt[5]{\dfrac{(127)^2(69.2)}{42{,}700}}$.
15. $10^{-4.623}$.
16. $(10^{0.764})(0.764)^{10}$.
17. $(2.07)^{0.36}$.
18. $\sqrt[3]{1.62 + \sqrt[3]{1.62}}$.
19. $\dfrac{\log 46.3}{\log 3.14}$.
20. $\dfrac{8.31 + \log 9.42}{\sqrt[10]{0.666}}$.
21. The area A of a triangle with sides a, b, and c may be calculated from the formula $A = \sqrt{s(s-a)(s-b)(s-c)}$, where s is one-half the perimeter. Approximate the area of a triangle with sides 12.6, 18.2, and 14.1.
22. The volume V of a right circular cone of altitude h and radius of base r is $V = \frac{1}{3}\pi r^2 h$. Approximate the volume of a cone of radius 2.43 and altitude 7.28.
23. The formula used in physics to approximate the period T (seconds) of a simple pendulum of length L (feet) is $T = 2\pi\sqrt{L/(32.2)}$. Approximate the period of a pendulum 33 inches long.
24. The pressure p (pounds per cubic foot) and volume v (cubic feet) of a certain gas are related by the formula $pv^{1.4} = 600$. Approximate the pressure if $v = 8.22$ cubic feet.

7 EXPONENTIAL AND LOGARITHMIC EQUATIONS AND INEQUALITIES

In certain equations and inequalities variables appear as exponents or in logarithmic expressions. Some of these may be solved as illustrated in the examples below.

EXAMPLE 1. Solve the equation $3^x = 21$.

Solution: Taking the common logarithm of both sides and using (iii) of (4.9), we obtain

$$x \log 3 = \log 21.$$

Hence the solution set is determined by

$$x = \frac{\log 21}{\log 3}.$$

A common error is to *subtract* log 3 from log 21. This is incorrect since x equals the *quotient* of log 21 by log 3. If an approximation for x is desired, we may use Table 1 to write

$$x \doteq \frac{1.3222}{0.4771} \doteq 2.77,$$

where the last number is obtained by *dividing* 1.3222 by 0.4771. A partial check on the solution is to note that since $3^2 = 9$ and $3^3 = 27$, then x should lie between 2 and 3, somewhat closer to 3 than to 2.

EXAMPLE 2. Solve $5^{2x+1} = 6^{x-2}$.

Solution: Taking the common logarithm of both sides produces

$$(2x + 1) \log 5 = (x - 2) \log 6,$$

which in turn leads to the following chain of equivalent equations:

$$\begin{aligned} 2x \log 5 + \log 5 &= x \log 6 - 2 \log 6 \\ x(2 \log 5 - \log 6) &= -\log 5 - \log 6^2 \\ x &= \frac{-(\log 5 + \log 36)}{\log 5^2 - \log 6} = \frac{-\log 180}{\log (25/6)}. \end{aligned}$$

Thus the solution set consists of one real number. If an approximation to the solution is desired, we may proceed as in Example 1.

EXAMPLE 3. Solve $(2/5)^x > 0.7$.

Solution: Writing $0.7 < (2/5)^x$ and using the fact that the logarithmic function to the base 10 is an increasing function, we have

$$\log (0.7) < x \log (2/5).$$

Since log (2/5) is negative, if we divide both sides by this number the sense of the inequality is reversed. Thus the solution set is

$$\left\{x \mid x < \frac{\log (0.7)}{\log (2/5)}\right\}.$$

An approximate solution may be obtained by using Table 1.

EXAMPLE 4. Solve $\log (5x - 1) - \log (x - 3) = 2$.

Solution: The given equation may be written as

$$\log \frac{5x - 1}{x - 3} = 2.$$

The definition of logarithm gives us

$$\frac{5x - 1}{x - 3} = 10^2,$$

which is equivalent to

$$5x - 1 = 100x - 300.$$

Solving for x we obtain the solution set $\{299/95\}$.

EXAMPLE 5. Solve $\dfrac{a^x - a^{-x}}{2} = 3$ for x, where $a > 0$.

Solution: Multiplying both sides of the given equation by $2a^x$, gives us

$$a^{2x} - 1 = 6a^x,$$

which may be written

$$(a^x)^2 - 6(a^x) - 1 = 0.$$

If we let $u = a^x$ we obtain a quadratic equation in the variable u. Applying the quadratic formula, we obtain

$$a^x = \frac{6 \pm \sqrt{36 + 4}}{2} = 3 \pm \sqrt{10}.$$

Since a^x is never negative, the number $3 - \sqrt{10}$ must be discarded. Taking the logarithm of both sides to the base a, we have

$$x \log_a a = \log_a (3 + \sqrt{10}),$$

or, since $\log_a a = 1$,

$$x = \log_a (3 + \sqrt{10}).$$

EXERCISES

In Exercises 1–18 find the solution sets.

1. $5^x = 10$.
2. $2^{3-x} = 4$.
3. $3^{1-2x} = 2^{x+5}$.
4. $5^{3x+2} = 4^{x-1}$.
5. $2^{x^2} = 6$.
6. $4^{-x^2} = 3$.
7. $\log x = 1 - \log (x - 9)$.
8. $\log (x + 3) = 1 + \log (3x - 10)$.
9. $\log (x^2 + 1) - \log (x + 1) = 1 + \log (x - 1)$.
10. $\log (x + 2) - \log (4x + 3) = \log (1/x)$.
11. $(1/2)^x > 2$.
12. $(1.8)^x \leq 3$.
13. $\log (x^3) = (\log x)^3$.
14. $\sqrt{\log_2 x} = \log_2 \sqrt{x}$.

15. $x^{\sqrt{\log x}} = 10^8.$

16. $\log \sqrt{x^2 - 1} = 2.$

17. $\log (\log x) = 2.$

18. $\log (\log (\log x)) = 2.$

Use logarithms to the base a to solve the equations in Exercises 19 and 20 for x in terms of y.

19. $\dfrac{a^x + a^{-x}}{2} = y.$

20. $\dfrac{a^x - a^{-x}}{a^x + a^{-x}} = y.$

21. The current i in a certain electrical circuit is given by

$$i = \frac{E}{R}(1 - e^{-Rt/L}).$$

Use base e to solve for t in terms of the remaining symbols.

22. The formula $Q = Q_0e^{-kt}$ is used in the study of decay of a radioactive substance. Use base e to solve for t in terms of the remaining symbols.

8 REVIEW EXERCISES

Oral

Define or discuss each of the following.

1. The Laws of Exponents.
2. The principal nth root of a.
3. The exponential function to the base a.
4. The logarithm of u to the base a.
5. The logarithmic function to the base a.
6. Linear interpolation.

Written

1. Find the following numbers:
   (a) $\log_2 64$. (b) $15^{\log_{15} 7}$. (c) $\log_8 \sqrt[5]{8}$.
   (d) $10^{3 \log 2}$.

2. Sketch the graph of the function f defined by:
   (a) $f(x) = 5^x$. (b) $f(x) = 5^{-x}$. (c) $f(x) = 5^{-x^2}$.
   (d) $f(x) = -5^{x+2}$. (e) $f(x) = 5^{2-x}$.

3. Find the solution set of:
   (a) $\log_4 (x - 3) = 2$. (b) $\log_{10} x^2 = -2$.
   (c) $\log_5 (x + 1) < 2$. (d) $|\log x| < 1$.
   (e) $2 \log_3 (x + 1) - \log_3 (x + 4) = 2 \log_3 2$.

4. Sketch the graph of the function f if:
   (a) $f(x) = \log_3 x$. (b) $f(x) = \log_3 x^2$.
   (c) $f(x) = 2 \log_3 x$. (d) $f(x) = \log_3 2x$.

5. (a) Express $\log_a (z^2\sqrt{x}/y^3)$ in terms of logarithms of x, y, and z.
   (b) Express $\log_a (yx^2) + 5 \log_a (y/x) - 3 \log_a y\sqrt[3]{x}$ as one logarithm.

6. Use linear interpolation to approximate:
   (a) log 6,483. (b) log (0.001769).
   (c) log (0.8888).

7. Use linear interpolation to approximate x if:
   (a) $\log x = 4.6312$. (b) $\log x = 9.0186 - 10$.
   (c) $\log x = -1.6312$.

8. Use logarithms to approximate:

   (a) $\dfrac{(27.4)^2}{\sqrt[3]{948}}$. (b) $\sqrt[5]{\dfrac{64.7}{86.1}}$. (c) $(1.89)^{3.4}$.

9. Find the solution sets of each of the following:

   (a) $8^{3-x} = 2$. (b) $6^x = 2$.
   (c) $2^{1-x} = 3^{x+5}$. (d) $\log(x - 15) = 2 - \log x$.
   (e) $(\frac{1}{3})^x < 3$. (f) $\log x^2 = (\log x)^2$.

10. Solve for x in terms of y if:

   (a) $y = \dfrac{10^x + 10^{-x}}{2}$. (b) $y = \dfrac{10^x - 10^{-x}}{2}$.

Supplementary Questions

1. Prove the laws of exponents by means of mathematical induction (see Appendix I).
2. If $0 < a < 1$ and if r, s are rational numbers such that $r < s$, prove that $a^r > a^s$.
3. The function f defined by $f(x) = (1 + 1/x)^x$, $x > 0$, has values close to e (see p. 118) when x is large. Sketch the graph of f, using logarithms to approximate ordinates of points which have large abscissas.
4. The *hyperbolic sine* and *hyperbolic cosine* functions, denoted by sinh and cosh respectively, are defined by

$$\sinh x = \frac{e^x - e^{-x}}{2} \quad \text{and} \quad \cosh x = \frac{e^x + e^{-x}}{2},$$

   where e is the base for natural logarithms (see p. 118). If $e \doteq 2.7$, approximate the graphs of these functions.
5. Prove that the following are true for all $x, y \in \mathbf{R}$:

   (a) $\cosh x + \sinh x = e^x$.
   (b) $\cosh x - \sinh x = e^{-x}$.
   (c) $(\cosh x)^2 - (\sinh x)^2 = 1$.
   (d) $\sinh(x + y) = \sinh x \cosh y + \cosh x \sinh y$.
   (e) $\cosh(x + y) = \cosh x \cosh y + \sinh x \sinh y$.
   (f) $\sinh 2x = 2 \sinh x \cosh x$.
   (g) $\cosh 2x = (\cosh x)^2 + (\sinh x)^2$.
6. Use logarithms to the base e to obtain the inverse of the hyperbolic sine function.

chapter five The Trigonometric Functions

In this chapter we lay the foundation for work in the area of mathematics called *trigonometry*. The approach we use disguises the fact that the origins of trigonometry had to do with measurement of angles and triangles. Indeed, angles and triangles are not considered until Section 5. Our reason for using this approach is motivated by modern applications in which the notion of angle either is secondary or does not enter into the picture. Thus our main objective in Sections 1 through 4 is to study a certain collection of transcendental functions that behave in a special way — the so-called *trigonometric* or *circular functions*. Although numerous applications of these functions have been made to important problems involving natural phenomena, our early work deals mainly with the theory and not with applications. In Sections 5 through 7 our discussion turns to the angular aspects of trigonometry. These ideas have many practical uses and also serve to give additional insight into the nature of the trigonometric functions.

1 THE WRAPPING FUNCTION

Consider a unit circle U — that is, a circle of radius 1 — with center at the origin of a rectangular coordinate system. We shall assume that it is possible to measure the length of arc connecting points on U. In particular, if A is the point with coordinates $(1, 0)$ and if P is any other point on U, then measuring in a *counterclockwise direction* from A, as illustrated in

Fig. 5.1, there is a unique positive real number t called the *length of the arc* $\widehat{AP}$. Since the circumference of U is 2π, we have $0 < t < 2\pi$. If we let $t = 0$ when A and P are the same point, then for each point P on U there is associated precisely one real number in the half-open interval $[0, 2\pi)$.

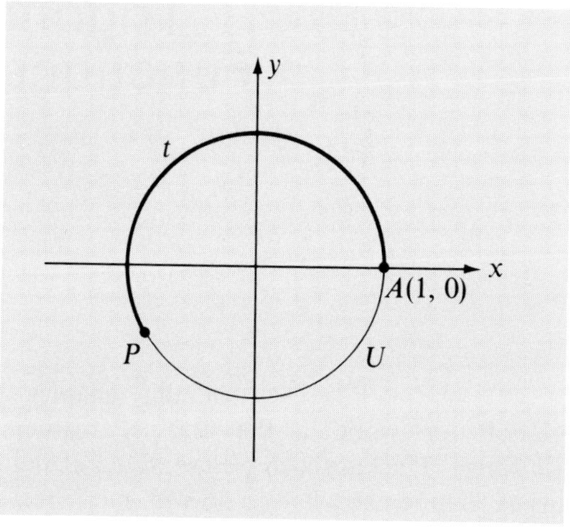

Figure 5.1

It can be shown, conversely, that for each real number t, $0 \leq t < 2\pi$, there is one and only one point P on U such that the length of $\widehat{AP}$ (measured in a counterclockwise direction) is t. Since the position of P depends on t, we sometimes use the functional notation and denote it by $P(t)$. This establishes a one-to-one correspondence between the real numbers in the interval $[0, 2\pi)$ and the points on U.

The discussion in the preceding paragraph can be extended so that with *every* real number there corresponds one and only one point on U. A convenient way to demonstrate this fact is to consider, as in Fig. 5.2, a real axis w with origin at the point $A(1, 0)$, tangent to U at A, with positive direction upward. Let us regard the w-axis as perfectly flexible and think of wrapping the positive part of the w-axis in a counterclockwise manner about U, as one would wrap thread about a spool. This is illustrated in Fig. 5.3, where the dashed line indicates the initial position of w. One complete revolution about U gives us our one-to-one correspondence between the real numbers in the interval $[0, 2\pi)$ and the points on U. If the wrapping process is continued, a second revolution of the w-axis about U produces a one-to-one correspondence between the numbers in the interval $[2\pi, 4\pi)$ and the points on U. Thus for each point on U there correspond two real numbers in $[0, 4\pi)$ and these two numbers differ numerically by 2π units. A third revolution associates with each number in $[4\pi, 6\pi)$ a unique point on U, and for each point P on U there correspond *three* numbers in the interval $[0, 6\pi)$. Continuing this wrapping procedure, we see that every positive real number t can be associated

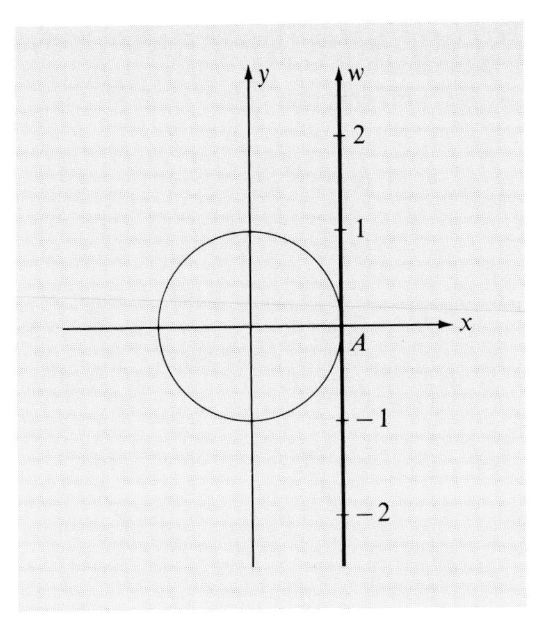

Figure 5.2

with a unique point $P(t)$ on U. Moreover, since the circumference of U has length 2π, two such real numbers t_1 and t_2 are associated with the same point if and only if $t_2 - t_1$ is a multiple of 2π — that is,

$$t_2 = t_1 + 2\pi n \tag{5.1}$$

for some integer n.

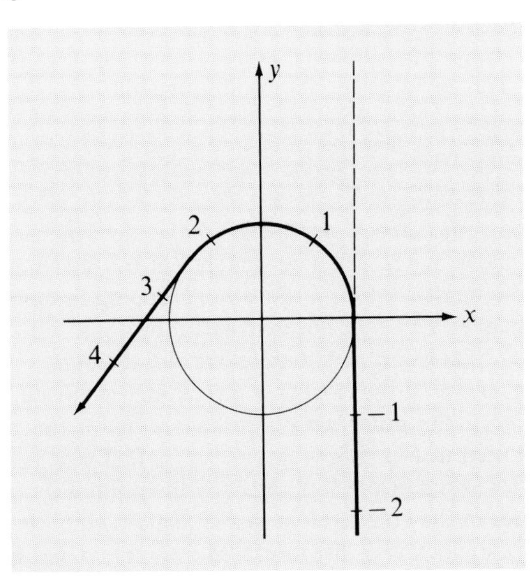

Figure 5.3

A similar correspondence may be obtained between the negative real numbers and points on U. In this case we wrap the negative part of the w-axis in a *clockwise* direction about U, as shown in Fig. 5.4. In this

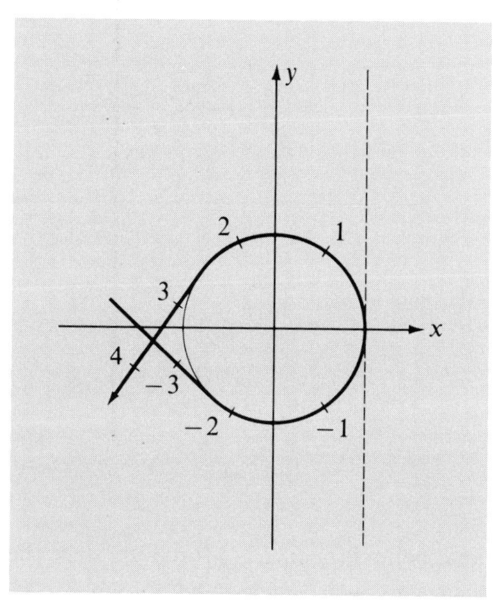

Figure 5.4

manner the real numbers in the interval $[-2\pi, 0)$ are associated with specific points on U; and so on.

The preceding discussion leads to the fact that to each real number t there corresponds a unique point $P(t)$ on the unit circle U. Moreover, t_1 and t_2 correspond to the same point if and only if (5.1) is true for some integer n. This gives us the important formula

$$P(t) = P(t + 2\pi n) \tag{5.2}$$

for every $t \in \mathbf{R}$ and $n \in \mathbf{Z}$.

The correspondence described above determines a function with domain $\mathbf{R}$ and range U which we shall refer to as the *wrapping function.* To reiterate, the point $P(t)$ on U that the wrapping function associates with the real number t can be located as follows. If $t > 0$, measure off a distance t around U in the counterclockwise direction. If $t < 0$, measure $|t|$ units around U in the clockwise direction. Of course, if $t = 0$, then we take $P(t) = A$, the point with coordinates $(1, 0)$. Several of the points $P(t)$ are plotted in Fig. 5.5.

If we let t range from 0 to 2π, then the point $P(t)$ traces out the unit circle U once in the counterclockwise direction, whereas $P(-t)$ traces out U once in the clockwise direction. Moreover, we see geometrically that $P(t)$ and $P(-t)$ are on the same vertical line for all $t \in \mathbf{R}$. Thus $P(-1)$ is

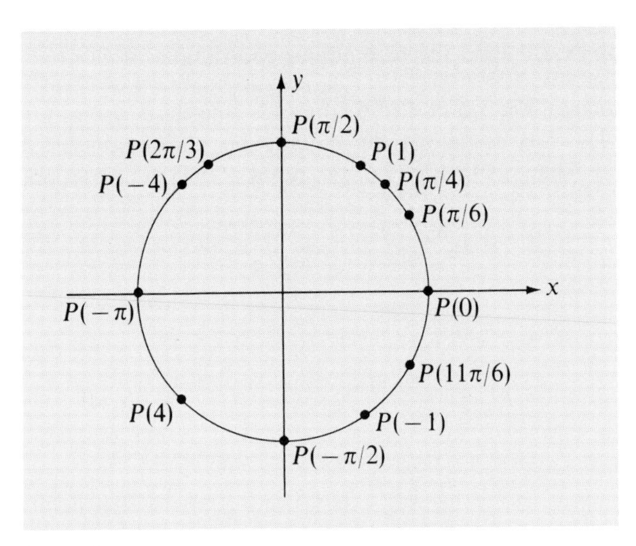

Figure 5.5

directly under $P(1)$, $P(-4)$ is directly above $P(4)$, and so on. This implies that the abscissas of $P(t)$ and $P(-t)$ are the same and the ordinates are negatives of one another. Therefore *if* $P(t)$ *has coordinates* (x, y), *then* $P(-t)$ *has coordinates* $(x, -y)$.

In the next section we shall make use of the rectangular coordinates (x, y) of $P(t)$ to define functions from **R** to **R**. If only elementary methods are available, then in order to find the rectangular coordinates of $P(t)$ it is necessary to specialize t. As an illustration, since $P(\pi)$ is the point π units from $A(1, 0)$ measured along U, it is one-half the way around the circle and hence has rectangular coordinates $(-1, 0)$. Similarly, the point $P(\pi/2)$ has coordinates $(0, 1)$. Since, by (5.2), $P(\pi/2) = P(\pi/2 + 2\pi n)$, we see that many other values of t lead to the point with coordinates $(0, 1)$; for example, we could use $t = 5\pi/2$, $9\pi/2$, or $-3\pi/2$.

Example 1. Find the rectangular coordinates of $P(\pi/4)$, $P(3\pi/4)$, $P(5\pi/4)$, and $P(7\pi/4)$.

Solution: Since the point $P(\pi/4)$ lies one-half the way from $P(0)$ to $P(\pi/2)$ (cf. Fig. 5.5), it follows from plane geometry that the rectangular coordinates of $P(\pi/4)$ are (c, c) for some positive real number c. Since P lies on U and an equation for U is $x^2 + y^2 = 1$, we must have $c^2 + c^2 = 1$, or $2c^2 = 1$. Solving for c, we have $c = \sqrt{1/2} = \sqrt{2}/2$. Consequently the rectangular coordinates of $P(\pi/4)$ are $(\sqrt{2}/2, \sqrt{2}/2)$.

Since $P(3\pi/4)$ lies one-half the way from $P(\pi/2)$ to $P(\pi)$, a similar argument gives us the rectangular coordinates $(-\sqrt{2}/2, \sqrt{2}/2)$. In like manner, the coordinates of $P(5\pi/4)$ are $(-\sqrt{2}/2, -\sqrt{2}/2)$, whereas the coordinates of $P(7\pi/4)$ are $(\sqrt{2}/2, -\sqrt{2}/2)$.

EXAMPLE 2. Find the rectangular coordinates of $P(\pi/6)$.

Solution: The point $P(\pi/6)$ on U is one-third the way from $A(1, 0)$ to $B(0, 1)$. Let us denote the rectangular coordinates of this point P by (c, d). Then, as in Fig. 5.6, the point P' which corresponds to $-\pi/6$ has

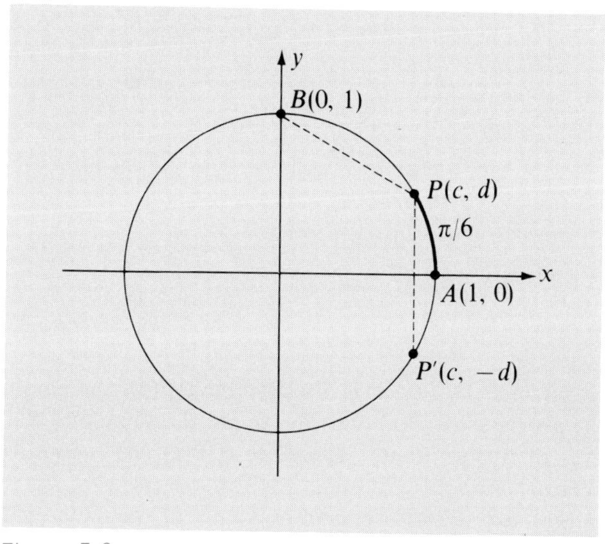

Figure 5.6

coordinates $(c, -d)$. The length of $\widehat{P'P}$ is the same as the length of $\widehat{PB}$, and hence $d(P', P) = d(P, B)$. Employing the distance formula gives us

$$2d = \sqrt{(c - 0)^2 + (d - 1)^2}.$$

Squaring both sides, we have

$$4d^2 = c^2 + d^2 - 2d + 1. \tag{5.3}$$

Since (c, d) are coordinates of a point on U, we have $c^2 + d^2 = 1$. Substituting 1 for $c^2 + d^2$ in (5.3) leads to

$$4d^2 + 2d - 2 = 0,$$

which may be written

$$2(2d - 1)(d + 1) = 0.$$

Since d is positive, this gives us $d = 1/2$. Using $c = \sqrt{1 - d^2}$, we obtain $c = \sqrt{3/4} = \sqrt{3}/2$, and consequently $P(\pi/6)$ has rectangular coordinates $(\sqrt{3}/2, 1/2)$.

EXAMPLE 3. Find the rectangular coordinates of $P(\pi/3)$.

Solution: An argument similar to that given in Example 2 may be used. Referring to Fig. 5.7, we see that if the rectangular coordinates of $P(\pi/3)$

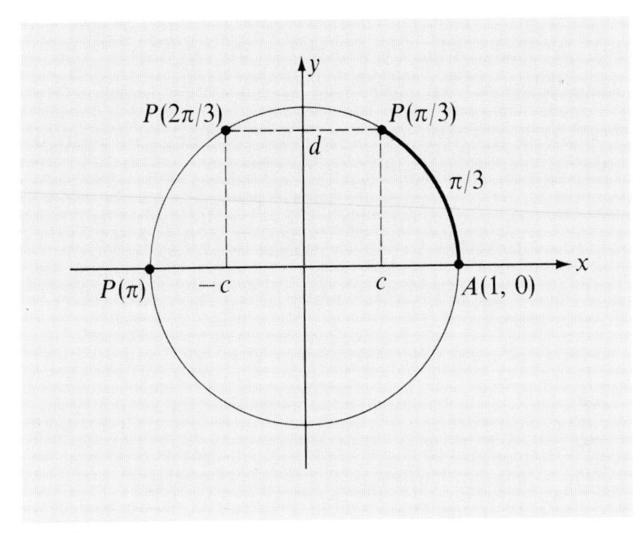

Figure 5.7

are (c, d), then the rectangular coordinates of $P(2\pi/3)$ are $(-c, d)$. Moreover, since

$$d(P(\pi/3), P(2\pi/3)) = d(A, P(\pi/3)),$$

we have, by the distance formula,

$$2c = \sqrt{(c - 1)^2 + d^2}.$$

We shall leave it as an exercise to show that this implies that $c = 1/2$ and $d = \sqrt{3}/2$. Hence the rectangular coordinates of $P(\pi/3)$ are $(1/2, \sqrt{3}/2)$.

EXERCISES

1. Complete the solution of Example 3.
2. Find the rectangular coordinates of $P(2\pi/3)$ by using the method given in Example 3.

In each of Exercises 3 and 4 determine the quadrant in which the indicated points lie and plot their approximate positions on the unit circle U.

3. (a) $P(3)$, (b) $P(-3)$, (c) $P(4.5)$, (d) $P(11\pi/6)$, (e) $P(-5\pi/4)$.
4. (a) $P(6)$, (b) $P(-2)$, (c) $P(5)$, (d) $P(5\pi/3)$, (e) $P(13\pi/4)$.

In each of Exercises 5–10 find the rectangular coordinates of the given point.

5. (a) $P(5\pi)$, (b) $P(7\pi/2)$, (c) $P(-8\pi)$, (d) $P(-5\pi/2)$.
6. (a) $P(27\pi)$, (b) $P(-11\pi/2)$, (c) $P(70\pi)$, (d) $P(23\pi/2)$.
7. $P(4\pi/3)$.
8. $P(7\pi/3)$.
9. $P(-5\pi/6)$.
10. $P(-3\pi/4)$.
11. Find two positive and two negative values of t such that $P(t)$ has rectangular coordinates $(\sqrt{3}/2, 1/2)$. [*Hint:* Use Example 2 and (5.2).]
12. Same as Exercise 11 for $(-\sqrt{2}/2, -\sqrt{2}/2)$.
13. If $P(t)$ has rectangular coordinates $(4/5, 3/5)$, find the rectangular coordinates of

    (a) $P(t + \pi)$, (b) $P(-t)$, (c) $P(t - \pi)$, (d) $P(-\pi - t)$.
14. Same as Exercise 13 for $(-3/5, 4/5)$.
15. Generalize Exercise 13 to the case where $P(t)$ is any point on U. [*Hint:* The coordinates of $P(t)$ can be written as (a, b), where either $b = \sqrt{1 - a^2}$ or $b = -\sqrt{1 - a^2}$.]

2 THE TRIGONOMETRIC FUNCTIONS

The wrapping function can be used to define a new class of functions from **R** to **R**. These functions are called either the *trigonometric functions* or the *circular functions*. There are six trigonometric functions and they are referred to as the *sine*, *cosine*, *tangent*, *cotangent*, *secant*, and *cosecant functions*. As an aid in specifying them, three letters instead of one are used, namely *sin*, *cos*, *tan*, *cot*, *sec*, and *csc*, respectively. If t is a real number, then the number which the sine function associates with t will be denoted by either $\sin(t)$ or $\sin t$, and similarly for the other five functions

(5.4) Definition of the Trigonometric Functions

If t is any real number, let $P(t)$ be the point on the unit circle U that the wrapping function associates with t. If the rectangular coordinates of $P(t)$ are (x, y), then

$$\sin t = y, \qquad \csc t = 1/y \quad (\text{if } y \neq 0),$$
$$\cos t = x, \qquad \sec t = 1/x \quad (\text{if } x \neq 0),$$
$$\tan t = y/x \quad (\text{if } x \neq 0), \qquad \cot t = x/y \quad (\text{if } y \neq 0).$$

It is important to remember the signs of the functional values for the trigonometric functions when $P(t)$ is in various quadrants. We see from

(5.4) that if $P(t)$ is in quadrant I, all functional values are positive. If $P(t)$ is in quadrant II, then y is positive and x is negative, and consequently the sine and cosecant functional values are positive and all others are negative. If $P(t)$ is in quadrant III, then both x and y are negative, and therefore the tangent and cotangent have positive values and all other functions have negative values. Finally, if $P(t)$ is in quadrant IV, the cosine and secant values are positive and all others are negative.

In order to use (5.4) to find the values of the trigonometric functions that correspond to a real number t, it is necessary to determine the rectangular coordinates (x, y) of the point $P(t)$ and then substitute in the appropriate formula. For special values of t these may be found as in the preceding section.

EXAMPLE 1. Find the values of the trigonometric functions for $t = 0$, $\pi/6$, $\pi/4$, $\pi/3$, and $\pi/2$.

Solution: The rectangular coordinates of $P(0)$ are $(1, 0)$. Since the ordinate y of $P(0)$ is 0, we see from (5.4) that the cosecant and cotangent functions are undefined. The remaining functional values may be found by substituting 1 for x and 0 for y in (5.4). Thus $\sin 0 = 0$, $\cos 0 = 1$, $\tan 0 = 0/1 = 0$, and $\sec 0 = 1/1 = 1$.

From Example 2 of the preceding section, the rectangular coordinates of $P(\pi/6)$ are $(\sqrt{3}/2, 1/2)$. The functional values for $t = \pi/6$ may therefore be obtained by substituting $\sqrt{3}/2$ for x and $1/2$ for y in (5.4). These are tabulated below.

To find the values of the trigonometric functions for $t = \pi/4$, the coordinates $(\sqrt{2}/2, \sqrt{2}/2)$ of $P(\pi/4)$, as calculated in Example 1 of the preceding section, may be used. In like manner, the values for $t = \pi/3$ may be determined from the coordinates $(1/2, \sqrt{3}/2)$. Finally, the values for $t = \pi/2$ are found by using $(0, 1)$. These values are arranged in tabular form below. The reader should check each entry in the table. A dash indicates that the function is undefined for the indicated number t. We shall leave verifications for other special values of t as exercises.

t	$\sin t$	$\cos t$	$\tan t$	$\csc t$	$\sec t$	$\cot t$
0	0	1	0	—	1	—
$\pi/6$	$1/2$	$\sqrt{3}/2$	$\sqrt{3}/3$	2	$2\sqrt{3}/3$	$\sqrt{3}$
$\pi/4$	$\sqrt{2}/2$	$\sqrt{2}/2$	1	$\sqrt{2}$	$\sqrt{2}$	1
$\pi/3$	$\sqrt{3}/2$	$1/2$	$\sqrt{3}$	$2\sqrt{3}/3$	2	$\sqrt{3}/3$
$\pi/2$	1	0	—	1	—	0

The domain of each trigonometric function can be determined from (5.4). The domain of the sine and cosine functions is all of **R**. For the tangent and secant functions we must exclude all real numbers t such that the abscissa x of $P(t)$ is 0. This set of numbers can be specified by

$$\{\pi/2 + n\pi \mid n \in \mathbf{Z}\}.$$

Thus the domain of the tangent and secant functions consists of all real numbers *except* $\pm\pi/2$, $\pm 3\pi/2$, $\pm 5\pi/2$, and so on. We often say that the tangent and secant are *undefined* for these numbers. Similarly, the domain of the cotangent and cosecant functions is the set of all real numbers except those numbers t for which the ordinate y of $P(t)$ is 0. The latter include the numbers 0, $\pm\pi$, $\pm 2\pi$, $\pm 3\pi$, and, in general, all numbers in the set $\{n\pi \mid n \in \mathbf{Z}\}$.

Let us now investigate the range of each trigonometric function. Since the pair (x, y) in (5.4) gives coordinates of a point on the unit circle U, we have $|x| \leq 1$ and $|y| \leq 1$. Consequently, since $\sin t = y$ and $\cos t = x$, the range of both the sine and cosine functions is the set of all numbers in the closed interval $[-1, 1]$. On the other hand, the range of the cosecant and secant functions consists of all real numbers of absolute value greater than or equal to 1 (Why?).

To determine the range of the tangent function, let us note first that if (x, y) are the coordinates of the point $P(t)$ on U, then $x^2 + y^2 = 1$ and hence either $y = \sqrt{1 - x^2}$ or $y = -\sqrt{1 - x^2}$. If we restrict $P(t)$ to the first or second quadrants, then $\tan t = \sqrt{1 - x^2}/x$. If a is *any* real number, there exists a real number x such that $\sqrt{1 - x^2}/x = a$. Specifically, let x equal $1/\sqrt{1 + a^2}$ or $-1/\sqrt{1 + a^2}$ according to whether a is positive or negative. This shows that $\tan t$ takes on all real values and hence that the range of the tangent function is all of $\mathbf{R}$. Similarly, the range of the cotangent function is $\mathbf{R}$.

We see from (5.4) that the following equations are true for all values of t for which denominators are not zero:

(5.5) $$\csc t = \frac{1}{\sin t},$$

(5.6) $$\sec t = \frac{1}{\cos t},$$

(5.7) $$\cot t = \frac{1}{\tan t},$$

(5.8) $$\tan t = \frac{\sin t}{\cos t},$$

(5.9) $$\cot t = \frac{\cos t}{\sin t}.$$

These formulas may be proved by substituting the equivalent forms in terms of x and y from (5.4). For example, $\tan t = y/x = \sin t/\cos t$.

Each of the equations (5.5)–(5.9) is called an *identity* since it is true for every t in the domains of the functions involved in the equation (cf. page 10).

Another useful relationship is a consequence of the fact that if (x, y) are coordinates of a point on U, then $y^2 + x^2 = 1$. Since $y = \sin t$ and $x = \cos t$, this gives us the identity

$$(\sin t)^2 + (\cos t)^2 = 1.$$

Except for $n = -1$, powers such as $(\cos t)^n$ are written in the form $\cos^n t$. The symbol $\cos^{-1} t$ is reserved for a special situation to be discussed in the next chapter; the same is true for the other trigonometric functions. With this agreement on notation, the previous identity may be written

$$\sin^2 t + \cos^2 t = 1. \tag{5.10}$$

Two other very important identities may be established easily. If $\cos t \neq 0$, then by dividing both sides of (5.10) by $\cos^2 t$ we obtain

$$\frac{\sin^2 t}{\cos^2 t} + 1 = \frac{1}{\cos^2 t}.$$

By (5.8) and (5.6), this is equivalent to

$$\tan^2 t + 1 = \sec^2 t. \tag{5.11}$$

Similarly, dividing both sides of (5.10) by $\sin^2 t$ leads to the identity

$$1 + \cot^2 t = \csc^2 t. \tag{5.12}$$

Identities (5.5)–(5.12) are often referred to as *fundamental identities* and are the basis for much of our future work in trigonometry. Other relationships involving the trigonometric functions are true; these will be discussed in later sections.

EXAMPLE 2. If $P(t)$ is in the second quadrant and $\sin t = 3/5$, find the values of the other trigonometric functions.

Solution: From (5.10) and the fact that $\cos t$ is negative when $P(t)$ is in quadrant II, we have $\cos t = -\sqrt{1 - \sin^2 t} = -\sqrt{1 - 9/25} = -\sqrt{16/25} = -4/5$. Hence $\tan t = \sin t/\cos t = (3/5)/(-4/5) = -3/4$. By (5.5)–(5.7), the remaining values are $\csc t = 5/3$, $\sec t = -5/4$, and $\cot t = -4/3$.

EXAMPLE 3. Use the fundamental identities to simplify

$$\sec t - \sin t \tan t,$$

where $t \neq \pi/2 + n\pi$, $n \in \mathbf{Z}$.

Solution:

$$\begin{aligned}
\sec t - \sin t \tan t &= \frac{1}{\cos t} - \sin t \left(\frac{\sin t}{\cos t}\right) && \text{(5.6) and (5.8)} \\
&= \frac{1 - \sin^2 t}{\cos t} && \text{(Why?)} \\
&= \frac{\cos^2 t}{\cos t} && \text{(5.10)} \\
&= \cos t && \text{(Why?).}
\end{aligned}$$

Example 3 illustrates manipulations which are often used to simplify problems involving the trigonometric functions. In the next chapter much more will be done along these lines.

We shall end this section with three useful identities. As mentioned on page 139, if the point $P(t)$ has coordinates (x, y), then $P(-t)$ has coordinates $(x, -y)$. Hence, by the definition of the trigonometric functions,

$$\sin(-t) = -y, \quad \cos(-t) = x, \quad \text{and} \quad \tan(-t) = -y/x.$$

The previous three equalities may be rewritten

$$\begin{aligned}
\sin(-t) &= -\sin t, \\
\cos(-t) &= \cos t, \\
\tan(-t) &= -\tan t,
\end{aligned} \tag{5.13}$$

where t is any real number in the domain of the indicated function. Similar formulas are true for the other trigonometric functions (see Exercise 3).

Incidentally, according to Exercise 17 of Section 5 in Chapter One, (5.13) implies that the sine and tangent are odd functions and the cosine is an even function.

EXAMPLE 4. Find $\sin(-\pi/6)$ and $\cos(-\pi/4)$.

Solution: Using (5.13) and the table on p. 143,

$$\sin(-\pi/6) = -\sin \pi/6 = -1/2$$

and

$$\cos(-\pi/4) = \cos \pi/4 = \sqrt{2}/2.$$

EXERCISES

1. Verify all entries in the table on p. 143.
2. Write out the proofs of (5.5)–(5.9).

3. Use (5.13) to establish the following identities:
   (a) $\csc(-t) = -\csc t$, (b) $\sec(-t) = \sec t$,
   (c) $\cot(-t) = -\cot t$.
4. Show that the range of the cotangent function is **R**.

In each of Exercises 5–8 find the quadrant in which $P(t)$ lies if the indicated conditions are true.

5. $\sin t < 0$ and $\tan t < 0$.
6. $\cos t > 0$ and $\tan t < 0$.
7. $\sec t < 0$ and $\cot t > 0$.
8. $\csc t > 0$ and $\sec t < 0$.

In Exercises 9–14 find the values of all six trigonometric functions if the given conditions are true.

9. $\sin t = -2/3$ and $\sec t > 0$.
10. $\cos t = -1/5$ and $\csc t < 0$.
11. $\tan t = 3/4$ and $\sin t < 0$.
12. $\csc t = 8$ and $\tan t < 0$.
13. $\sec t = -13/5$ and $\tan t > 0$.
14. $\cos t = -1/2$ and $\sin t > 0$.

In Exercises 15–18 use the rectangular coordinates of $P(t)$ to find the values of the six trigonometric functions for each of the indicated values of t.

15. (a) $t = -3\pi$, (b) $t = 7\pi/2$.
16. (a) $t = 21\pi$, (b) $t = -9\pi/2$.
17. (a) $t = 5\pi/6$, (b) $t = -11\pi/4$.
18. (a) $t = 5\pi/3$, (b) $t = 33\pi/4$.

In Exercises 19–24 use the fundamental identities to transform the first expression into the second.

19. $\cos t \csc t$, $\cot t$.
20. $\sec t/\tan t$, $\csc t$.
21. $\sin t(\csc t - \sin t)$, $\cos^2 t$.
22. $\tan t \sin t + \cos t$, $\sec t$.
23. $\sec x/\csc x$, $\tan x$.
24. $\tan^2 x/(\sec x + 1)$, $\sec x - 1$.

3 VALUES OF THE TRIGONOMETRIC FUNCTIONS

In the previous section several values of the trigonometric functions were calculated. We now wish to discuss the problem of determining *all* values. Let us consider, for the moment, only the sine function. We know from (5.4) that for every real number t there corresponds a unique real number, denoted by $\sin t$, which lies between -1 and 1. In general, the methods used to determine this number are beyond the scope of this book, since we do not have the processes necessary for measuring arc length. Thus we cannot, by elementary methods, find points $P(t)$ on the unit circle U that the wrapping function associates with values of t such as 0.74619, $-4/7$, $\sqrt{5}$, $\sqrt[3]{-11}$, or, for that matter, integers such as 1, 2, 3, 4, and so on. We may, however, make some general observations. If t varies from 0 to $\pi/2$, then the coordinates of $P(t)$ vary from (1, 0) to (0, 1). In particular, the ordinate — that is, the value of the sine function — increases from 0 to 1. Moreover, this function takes on *all* values between 0 and 1. If we let t range from $\pi/2$ to π, then the coordinates of $P(t)$ vary from $(0, 1)$ to $(-1, 0)$ and hence the sine function [the ordinate of $P(t)$] decreases from 1 to 0. In similar fashion, we see that as t varies from π to $3\pi/2$, $\sin t$ decreases from 0 to -1, and as t varies from $3\pi/2$ to 2π, $\sin t$ increases from -1 to 0. If we let t range through the interval $[2\pi, 4\pi)$, the identical pattern for $\sin t$ is repeated. The same is true for other intervals of length 2π. Indeed, using (5.2), we have the identity

(5.14) $$\sin(t + 2\pi n) = \sin t$$

for every $t \in \mathbf{R}$ and $n \in \mathbf{Z}$. According to the next definition, this implies that the sine function is periodic.

(5.15) Definition of Periodic Function

A function f with domain X is *periodic* if there exists a positive real number k such that

$$f(x + k) = f(x)$$

for every $x \in X$. If a least such positive real number k exists, it is called the *period* of f.

If a function f has period k, then the ordinate of the point with abscissa x is the same as the ordinate of the point with abscissa $x + k$ for every $x \in X$. This means that the graph repeats itself in intervals of k units along the x-axis. The graphs of the trigonometric functions given in the next section are good illustrations of this type of behavior.

The sine function has period 2π. To see this, we note first that from (5.14) with $n = 1$ we have $\sin(t + 2\pi) = \sin t$. According to (5.15), it is sufficient to prove that there is no smaller positive number k such that $\sin(t + k) = \sin t$. We shall give an indirect proof. Thus suppose there is a positive number k less than 2π such that $\sin(t + k) = \sin t$ for all $t \in \mathbf{R}$. Letting $t = 0$, we obtain $\sin k = \sin 0 = 0$. Therefore, since $0 < k < 2\pi$, $P(k)$ has coordinates $(-1, 0)$, whence $k = \pi$, and we may write $\sin(t + \pi) = \sin t$ for all $t \in \mathbf{R}$. In particular, if $t = \pi/2$, then $\sin(3\pi/2) = \sin \pi/2$, or $-1 = 1$, an absurdity. Therefore $\sin t$ has period 2π.

Similar discussions can be carried on for the other functions. The variation of the cosine function in $[0, 2\pi]$ can be determined by observing the behavior of the abscissa x of $P(t)$ as t varies from 0 to 2π. The miniature table of values given below indicates that the cosine function decreases

t	0	$\pi/2$	π	$3\pi/2$	2π
$\cos t$	1	0	-1	0	1

from 1 to 0 in the interval $[0, \pi/2]$, decreases from 0 to -1 in $[\pi/2, \pi]$, increases from -1 to 0 in $[\pi, 3\pi/2]$, and increases from 0 to 1 in $[3\pi/2, 2\pi]$. This pattern is then repeated in successive intervals of length 2π. We leave it as an exercise to prove that the cosine function has period 2π.

By employing (5.5) and (5.6) it is easy to show that the secant and cosecant functions are periodic of period 2π. We shall discuss the variation of these functions further in Section 4, where their behavior is shown rather strikingly by means of graphs.

If $C(x, y)$ is any point on U, then the point $C'(-x, -y)$ is diametrically opposite C, as is illustrated in Fig. 5.8. Since

$$\tan t = y/x = -y/-x,$$

it follows that the tangent function has the same value at C' as at C. Let t be a real number such that $P(t) = C$, where, as usual, $P(t)$ is the point

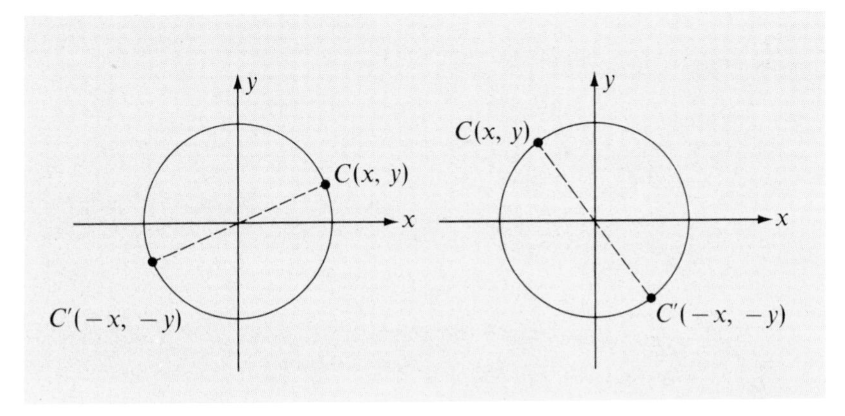

Figure 5.8

which the wrapping function associates with t. Since the arc length $\widehat{CC'}$ (measured in the counterclockwise direction) is π, we have $P(t + \pi) = C'$ and hence

$$\tan (t + \pi) = \tan t.$$

It can be shown that there is no positive real number k smaller than π such that $\tan (t + k) = \tan t$, and hence the tangent function is periodic with period π. The variation of $\tan t$ will be discussed further in Section 4.

Let us return to the problem of finding values of the trigonometric functions. Since the sine function has period 2π, it is enough to know the values of $\sin t$ for $0 \leq t \leq 2\pi$, because these same values are repeated in intervals of length 2π. The same is true for the other trigonometric functions. As a matter of fact, the values of any trigonometric function can be determined if its values in the t-interval $[0, \pi/2]$ are known. In order to prove this, suppose $t \in \mathbf{R}$ and let $P(t)$ be the point that the wrapping function associates with t. The shortest distance t' between $P(t)$ and the

x-axis, measured along U, will be called the *reference number* associated with t. Figure 5.9 illustrates arcs of length t' for positions of $P(t)$ in various quadrants.

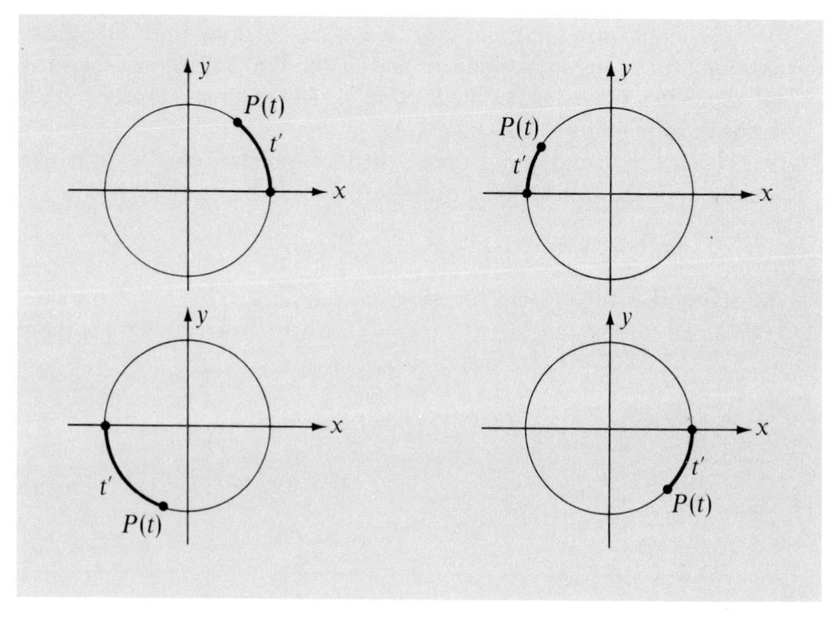

Figure 5.9

EXAMPLE 1. Approximate the reference number t' if t equals (a) 2.5, (b) 4, (c) 15.

Solution: (a) Using $\pi \doteq 3.1416$ and $\pi/2 \doteq 1.5708$, we see that $\pi/2 < 2.5 < \pi$. Hence $P(2.5)$ lies in quadrant II and the reference number t' for 2.5 is $\pi - 2.5$. Approximately, $t' \doteq 3.1416 - 2.5 = 0.6416$.

(b) Since $P(4)$ is in quadrant III, $t' = 4 - \pi \doteq 0.8584$.

(c) Since $2\pi \doteq 6.2832$, the point $P(15)$ is found by making two complete counterclockwise revolutions around U plus a partial revolution of length d. Thus $15 \doteq 2(6.2832) + d$ or $d \doteq 15 - 12.5664 = 2.4336$. Since the point $P(2.4336)$ is in quadrant II, we have $t' \doteq \pi - 2.4336 \doteq 0.7080$.

From the previous discussion we see that if $P(t)$ is not on a coordinate axis, then $0 < t' < \pi/2$. Suppose (x, y) are the rectangular coordinates of $P(t)$ and write $P(t) = P(x, y)$. Let A be the point with coordinates $(1, 0)$ and let $P'(x', y')$ be the point on U in quadrant I such that $\widehat{AP'} = t'$. Illustrations in which $P(x, y)$ lies in quadrants II, III, or IV are given in Fig. 5.10. We see that in all cases

$$x' = |x| \quad \text{and} \quad y' = |y|.$$

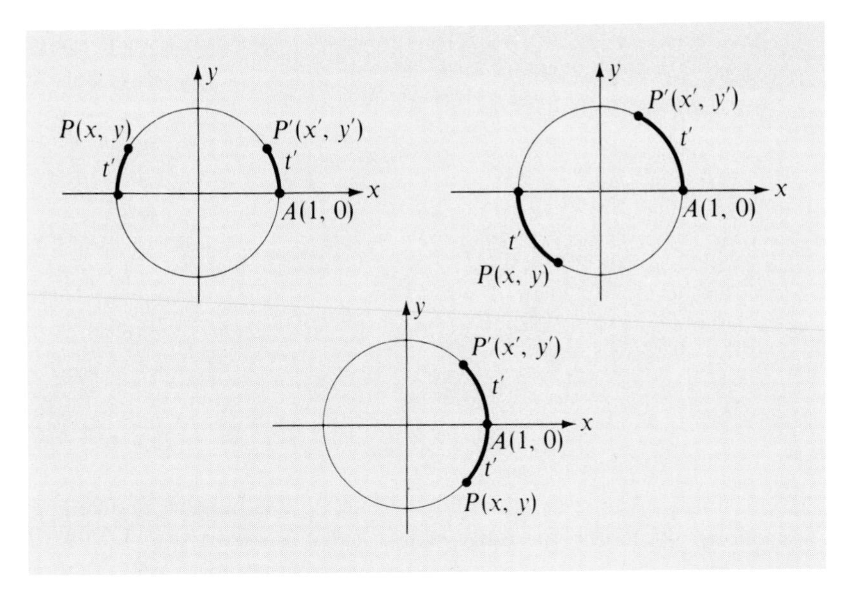

Figure 5.10

Hence we may write

$$|\cos t| = |x| = x' = \cos t',$$
$$|\sin t| = |y| = y' = \sin t'.$$

It is easy to show that the absolute value of *every* trigonometric function at t equals its value at t'. For example,

$$|\tan t| = |y/x| = |y|/|x| = y'/x' = \tan t'.$$

This leads to the following rule:

(5.16) *To find the value of a trigonometric function at a number t, determine its value for the reference number t' associated with t and prefix the appropriate sign.*

EXAMPLE 2. Find $\sin(7\pi/4)$ and $\sec(-7\pi/6)$.

Solution: Since $7\pi/4 = 2\pi - \pi/4$ and $-7\pi/6 = -\pi - \pi/6$, the reference numbers are $\pi/4$ and $\pi/6$, respectively (see Fig. 5.11). Hence by (5.16) and Example 1 on page 143,

$$\sin(7\pi/4) = -\sin(\pi/4) = -\sqrt{2}/2,$$
$$\sec(-7\pi/6) = -\sec(\pi/6) = -2\sqrt{3}/3.$$

By employing more advanced techniques than those available in this book, it is possible to compute, to any degree of accuracy, all the values of the trigonometric functions in the t-interval $[0, \pi/2]$. Table 3 gives

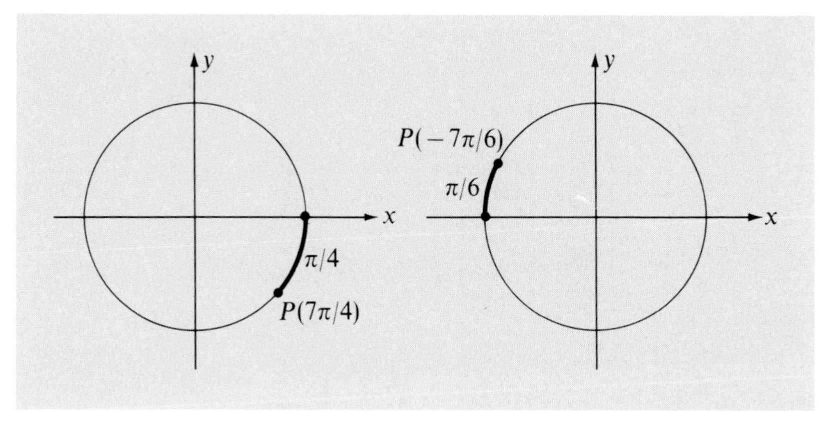

Figure 5.11

approximations to such values for the sine, cosine, tangent, and cotangent functions. Note that the range of t in Table 3 is from 0 to 1.5708. The latter number is a four-decimal-place approximation to $\pi/2$. Although tables for the secant and cosecant are available, we shall not include them in this book. If values of these functions are required, they can be computed by using $\sec t = 1/\cos t$ and $\csc t = 1/\sin t$. Table 3 also includes columns labeled *degrees*, the significance and use of which will be discussed in a later section. As will be seen, the inclusion of the degree columns is the reason that t varies at intervals of approximately 0.0029. For the time being, the reader should ignore the degree columns when using Table 3.

To find the values of the trigonometric functions at a real number t, where $0 \leq t \leq 0.7854$, the labels at the *top* of the columns in Table 3 should be used. For example, we have $\cos (0.1338) \doteq 0.9911$, $\sin (0.4654) \doteq 0.4488$, $\tan (0.5789) \doteq 0.6536$, and so on. On the other hand, if $0.7854 \leq t \leq 1.5708$, then the labels at the *bottom* of the columns should be employed. For example, $\sin (1.2363) \doteq 0.9446$, $\tan (1.5213) \doteq 20.206$, and $\cos (0.8639) \doteq 0.6494$. The reason that the table can be arranged in this way follows from the fact [proved in Chapter Six, (6.4)] that $\sin t = \cos (\pi/2 - t)$ and $\cot t = \tan (\pi/2 - t)$ for all $t \in \mathbf{R}$. In particular, since $\pi/2 \doteq 1.5708$, we obtain

$$\sin (0.1047) \doteq \cos (1.5708 - 0.1047),$$

or

$$\sin (0.1047) \doteq \cos (1.4661).$$

Likewise, the value of the sine function at 1.4661 is the same as the value of the cosine function at 0.1047. Similar remarks are true for the tangent and cotangent functions.

If it is necessary to find trigonometric functional values when t lies *between* numbers given in the table, the method of linear interpolation, as was used for logarithms, may be employed. Similarly, given a functional value, say $\sin t = 0.6371$, one may, by referring to the body of Table 3 and using linear interpolation if necessary, obtain an approximation to t.

EXAMPLE 3. Find approximations for (a) tan (2.3824), (b) cos (0.4).

Solution: (a) Since $P(2.3824)$ is in quadrant II, the reference number t' is $\pi - 2.3824$, or $t' \doteq 3.1416 - 2.3824 = 0.7592$. Using (5.16) and Table 3, we have

$$\begin{aligned} \tan(2.3824) &\doteq -\tan(0.7592) \\ &\doteq -0.9490. \end{aligned}$$

(b) To find cos (0.4) we locate the number 0.4000 between successive values of t in Table 3 and interpolate as follows:

$$0.0029\left\{\begin{matrix} 0.0015\left\{\begin{matrix} \cos(0.3985) \doteq 0.9216 \\ \cos(0.4000) \doteq \ ? \end{matrix}\right\}d \\ \cos(0.4014) \doteq 0.9205 \end{matrix}\right\}0.0011$$

$$\frac{0.0015}{0.0029} = \frac{d}{0.0011},$$

or

$$d = \tfrac{15}{19}(0.0011) \doteq 0.0006.$$

Hence

$$\cos(0.4000) \doteq 0.9216 - 0.0006 = 0.9210.$$

Note that since the cosine function is decreasing in the given interval we must subtract d from 0.9216.

EXAMPLE 4. Approximate the smallest positive real number t such that $\sin t = 0.6635$.

Solution: We locate 0.6635 between successive entries in the sine column of Table 3 and interpolate as follows:

$$0.0029\left\{\begin{matrix} d\left\{\begin{matrix} \sin(0.7243) \doteq 0.6626 \\ \sin t = 0.6635 \end{matrix}\right\}0.0009 \\ \sin(0.7272) \doteq 0.6648 \end{matrix}\right\}0.0022$$

$$\frac{d}{0.0029} = \frac{0.0009}{0.0022},$$

or

$$d = \tfrac{9}{22}(0.0029) \doteq 0.0012.$$

Hence

$$t \doteq 0.7243 + 0.0012 = 0.7255.$$

EXERCISES

1. Prove that the cosine function has period 2π.
2. Prove that the tangent function has period π.

In each of Exercises 3–10 find the reference number t' if t has the indicated value.

3. (a) $7\pi/6$, (b) $5\pi/6$, (c) $-\pi/6$.
4. (a) $-4\pi/3$, (b) $4\pi/3$, (c) $5\pi/3$.
5. (a) $3\pi/4$, (b) $-5\pi/4$, (c) $9\pi/4$.
6. (a) $11\pi/4$, (b) $-\pi/3$, (c) $-5\pi/6$.
7. 1.9. 8. -4.8. 9. 5. 10. 20.

11–14. Use (5.16) and the table on p. 143 to find the values of the sine, cosine, and tangent functions at the numbers t given in Exercises 3–6.

In Exercises 15–22 use Table 3 to approximate the indicated numbers.

15. $\sin(0.5934)$. 16. $\sin(1.4254)$.
17. $\cos(1.1956)$. 18. $\cos(0.7796)$.
19. $\tan(1.8733)$. 20. $\cot(2.9002)$.
21. $\cot(7.1355)$. 22. $\tan(-0.2763)$.

In Exercises 23–26 use interpolation to approximate the given numbers.

23. $\sin(0.53)$. 24. $\cot(1.4)$.
25. $\cos(4)$. 26. $\tan(10)$.

In each of Exercises 27–30 use interpolation to approximate the smallest positive real number t for which the given equality is true.

27. $\cos t = 0.8392$. 28. $\sin t = 0.1174$.
29. $\tan t = 0.3947$. 30. $\cot t = 0.7150$.

4 GRAPHS OF THE TRIGONOMETRIC FUNCTIONS

In our previous work with graphs we used the symbols x and y as labels for the coordinate axes. In the present chapter, since x has been used primarily for the abscissa of a point on the unit circle U, we shall, at the outset, use the symbol t for the horizontal axis. In the t, y coordinate system the graph of the sine function is therefore the same as the graph of the equation $y = \sin t$. Later, when there is less chance for confusion, the reader will be asked to graph the equation $y = \sin x$ in the x, y coordinate system. In the latter case it is important to remember that the symbol x is used in place of t and hence is not to be regarded as the x of definition (5.4).

It is not difficult to sketch the graphs of the trigonometric functions. For example since $|\sin t| \leq 1$ for all $t \in \mathbf{R}$, the graph of the sine function lies between the horizontal lines having y-intercepts 1 and -1. Moreover, since the sine function is periodic, with period 2π, it is sufficient to determine the graph in the interval $[0, 2\pi]$ on the t-axis, for knowing this, the same pattern is repeated in intervals of 2π over the entire t-axis. We have already discussed in some detail the variation of $\sin t$ in the interval $[0, 2\pi]$ (see p. 147). Some of the functional values are given in the table below. If we plot these points, draw a smooth curve through them, and

t	0	$\pi/4$	$\pi/2$	$3\pi/4$	π	$5\pi/4$	$3\pi/2$	$7\pi/4$	2π
$\sin t$	0	$\sqrt{2}/2 \doteq 0.7$	1	$\sqrt{2}/2$	0	$-\sqrt{2}/2$	-1	$-\sqrt{2}/2$	0

extend this configuration to the right and left in periodic fashion, we obtain the portion of the graph shown in Fig. 5.12. Needless to say, the

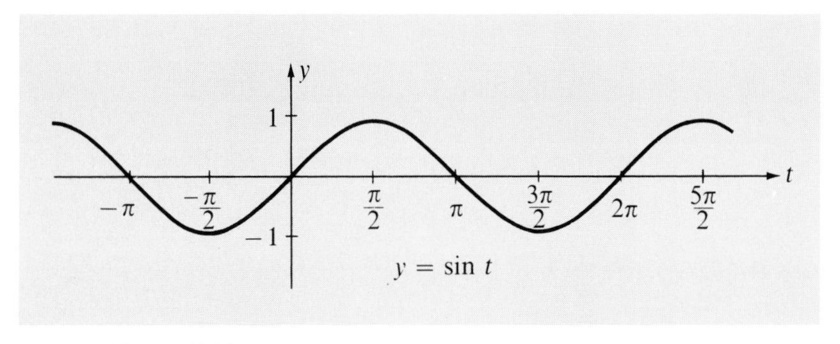

Figure 5.12

graph does not terminate, but continues indefinitely in both directions. If greater accuracy is desired, additional points could be plotted using, for example, $\sin \pi/6 = 1/2$, $\sin \pi/3 = \sqrt{3}/2 \doteq 0.8$, and so on. Also, Table 3 could be used to obtain many additional points on the graph. Because of its appearance, we speak of the part of the graph in the interval $[0, 2\pi]$ as a *sine wave*.

Similarly, it can be shown that the graph of the cosine function has the form shown in Fig. 5.13. We leave the verification of this as an exercise.

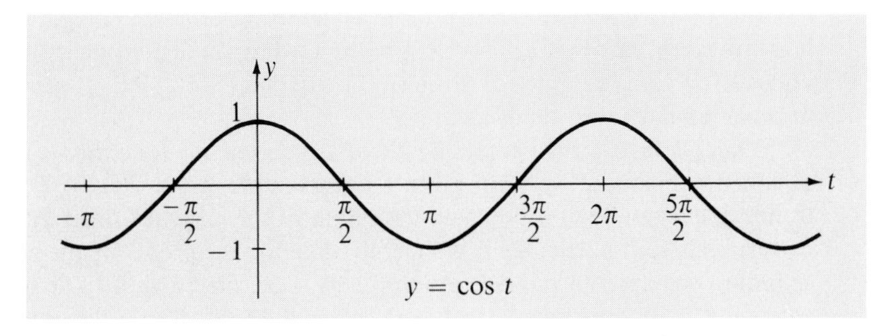

Figure 5.13

A partial table of values for the tangent function is given below. The

t	$-\pi/3$	$-\pi/4$	$-\pi/6$	0	$\pi/6$	$\pi/4$	$\pi/3$
$\tan t$	$-\sqrt{3} \doteq -1.7$	-1	$-\sqrt{3}/3 \doteq -0.6$	0	$\sqrt{3}/3 \doteq 0.6$	1	$\sqrt{3} \doteq 1.7$

corresponding points are plotted in Fig. 5.14. The values of tan t near $t = \pi/2$ demand special consideration. As t increases through positive

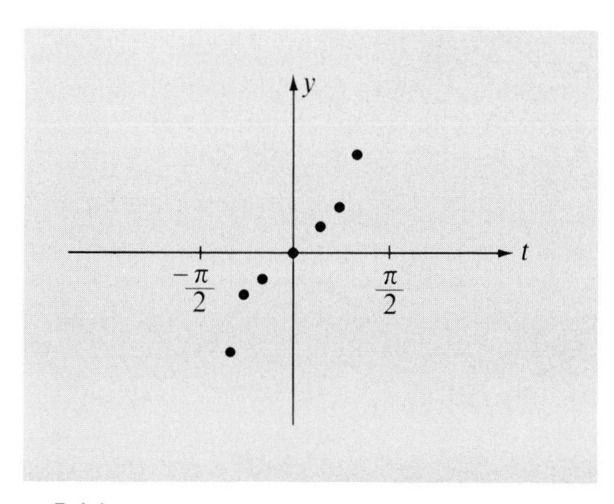

Figure 5.14

values toward $\pi/2$, $P(t)$ in (5.4) approaches the point (0, 1). Hence the abscissa x of $P(t)$ gets close to 0 and, since $\tan t = y/x = \sqrt{1 - x^2}/x$ for $0 \leq t < \pi/2$, it follows that when $t \doteq \pi/2$ and $t < \pi/2$, tan t is a large positive number. Indeed, tan t can be made arbitrarily large by choosing t sufficiently close to $\pi/2$. As on p. 95 the terminology "tan t *increases without bound*" or "tan t *becomes positively infinite*" as t approaches $\pi/2$ is used to describe this situation.

As t approaches $-\pi/2$ through values larger than $-\pi/2$, tan t *decreases without bound*— that is, tan t *becomes negatively infinite*. This is illustrated in Fig. 5.15. The vertical lines that are indicated by dashes are not part of the graph but merely serve as guide lines for sketching. These lines are *vertical asymptotes* for the graph. Owing to the periodicity of the tangent function, the pattern given in the interval $(-\pi/2, \pi/2)$ is repeated in other similar intervals of length π.

The graphs of the remaining three trigonometric functions can be obtained from those we have given. For example, since $\csc t = 1/\sin t$, to find the ordinate of a point on the graph of the cosecant function that corresponds to a particular t, we merely take the reciprocal of the corresponding ordinate on the sine graph. This is possible except for $t = n\pi$, where $n \in \mathbf{Z}$, for in this case $\sin t = 0$. As an aid to sketching the graph of the cosecant function, we sketch, with dashes, the graph of the sine func-

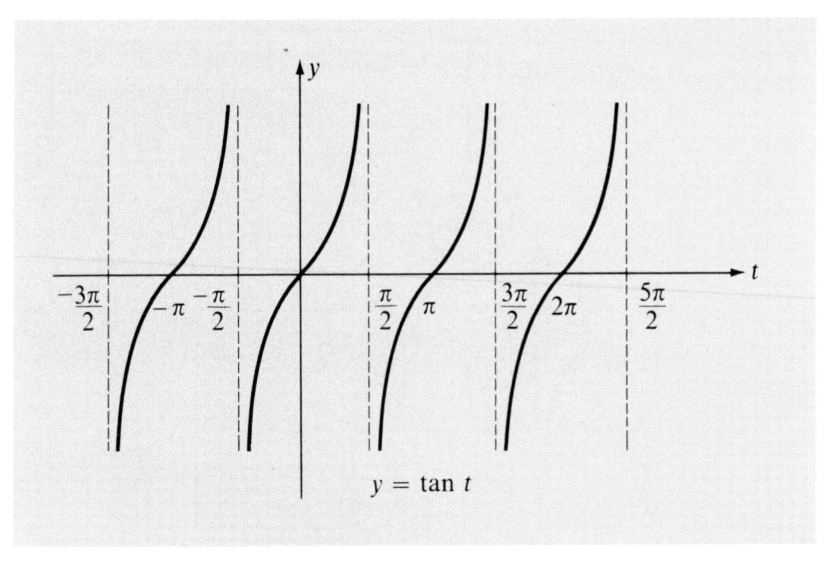

Figure 5.15

tion and then take the reciprocal of ordinates to obtain points on the cosecant graph. The graph is sketched in Fig. 5.16. Notice the manner in which the cosecant function increases or decreases without bound as t approaches $n\pi$, $n \in \mathbf{Z}$. The graph has vertical asymptotes as indicated in the figure.

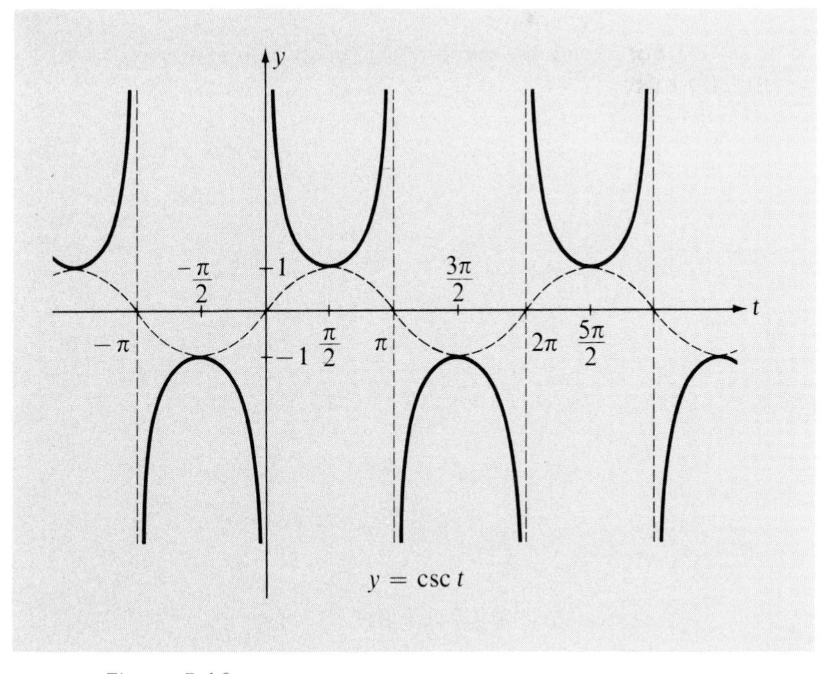

Figure 5.16

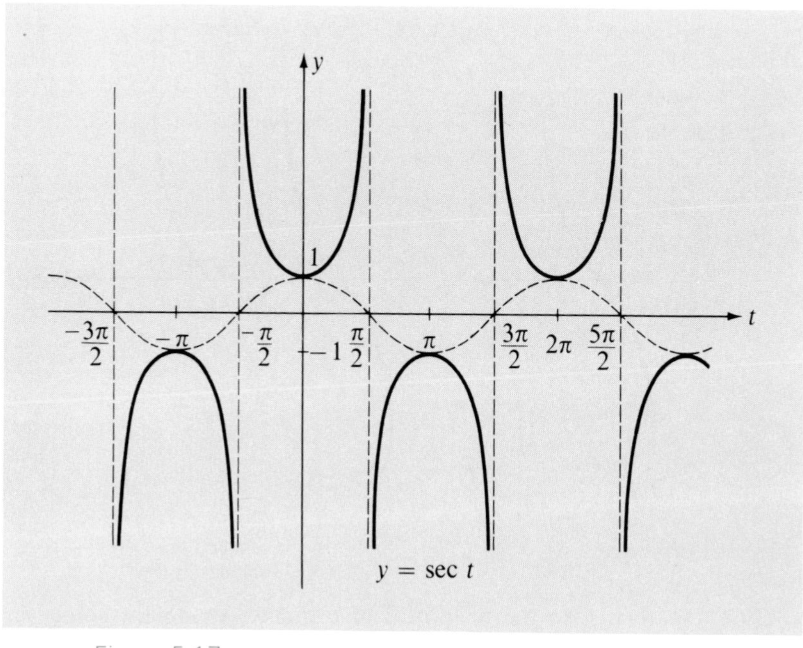

Figure 5.17

The graphs of the secant and cotangent functions may be obtained in a similar manner. These are shown in Figs. 5.17 and 5.18. We leave their verifications as exercises.

EXAMPLE 1. Sketch the graph of the function f defined by $f(t) = 2 \sin t$, for all $t \in \mathbf{R}$.

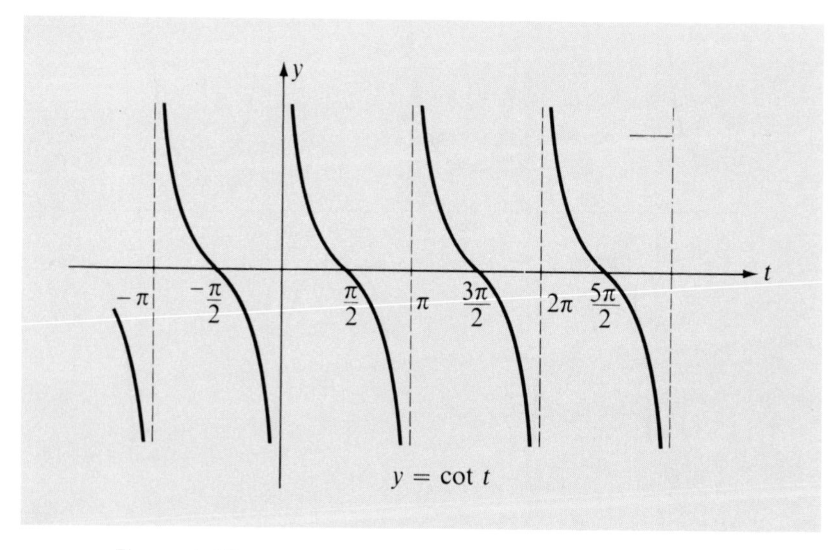

Figure 5.18

Solution: Although the graph could be obtained by plotting points, note that for each t_1 the ordinate $f(t_1)$ is always twice that of the corresponding ordinate on the sine graph. A simple graphical technique is to sketch, with dashes, the graph of $y = \sin t$ and then double each ordinate to find points on the graph of $y = f(t)$, as is illustrated in Fig. 5.19.

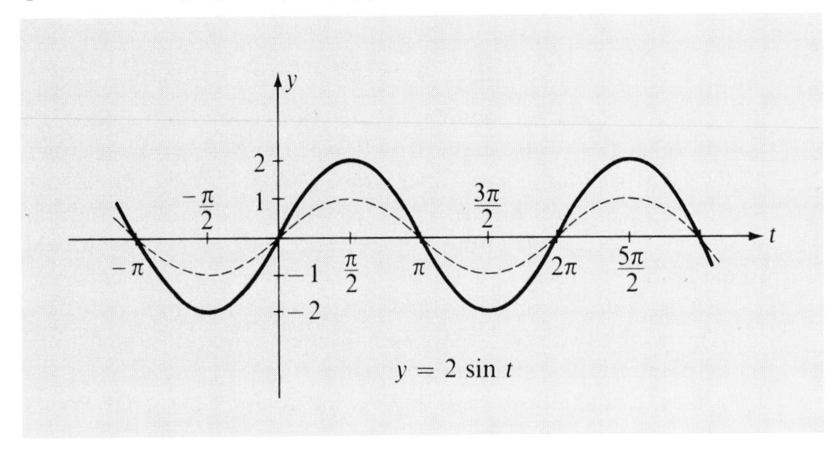

Figure 5.19

In the next chapter we shall discuss graphs that are based on the graphs of the trigonometric functions. For now we prefer to stress the six basic graphs and minor variations such as that given in Example 1. The reader should give much thought to the manner in which the graphs of this section have been obtained and be able to reproduce them rapidly whenever it is necessary.

EXERCISES

1. (a) Verify the graphs of the cosine and secant functions.
   (b) Describe the intervals between -2π and 2π in which the secant function increases.
2. (a) Verify the graph of the cotangent function.
   (b) In what intervals does the cotangent function increase?

In Exercises 3–6 use the method illustrated in Example 1 to sketch the graphs of the functions f which are defined as indicated.

3. $f(t) = 3 \cos t$.
4. $f(t) = 4 \sin t$.
5. $f(t) = \frac{1}{2} \sin t$.
6. $f(t) = 2 \tan t$.
7. Sketch the graph of the equation $y = \sin(-t)$ and describe how it is related to the graph of $y = \sin t$.
8. In what way are the graphs of the equations $y = \cos(-t)$ and $y = \cos t$ related?

In Exercises 9 and 10 sketch the graphs of the given equations after plotting a sufficient number of points.

9. $y = \sin(2t)$.

10. $y = \sin(\frac{1}{2}t)$.

5 ANGLES AND THEIR MEASUREMENT

The definitions of the trigonometric functions can also be based upon the notion of angles. This more traditional approach is quite common in applications of mathematics and hence should not be obscured by our sophisticated version in terms of the wrapping function. Indeed, for a thorough appreciation of the trigonometric functions, it is probably best to blend the two ideas.

In geometry an angle is often thought of as the geometric configuration formed by two rays — that is, half-lines — l_1 and l_2, having the same initial point O. If A and B are points on l_1 and l_2, respectively, as in Fig. 5.20, then, by definition, *angle AOB* is the union of the points on the rays.

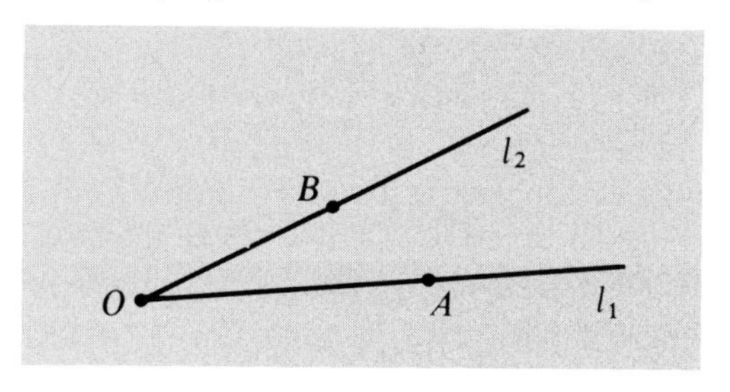

Figure 5.20

The same is true for finite line segments with a common endpoint. For trigonometric purposes it is more convenient to regard an angle as generated by starting with a fixed ray l_1 with endpoint O and rotating it about O, in a plane, to a position specified by a ray l_2. As above, if A and B are points on l_1 and l_2, respectively, the resulting geometric figure is referred to as angle AOB. We call l_1 the *initial side*, l_2 the *terminal side*, and O the *vertex* of the angle. The amount or direction of rotation is not restricted in any way. Thus we might let l_1 make several revolutions in either direction about O before coming to the position l_2.

If a rectangular coordinate system is introduced, then the *standard position* of an angle is obtained by taking the vertex at the origin and letting l_1 coincide with the positive x-axis. If l_1 is rotated in a counterclockwise direction to position l_2, then the angle is considered *positive*, whereas if l_1 is rotated in a clockwise direction, the angle is *negative*. We often denote angles by lower-case Greek letters and specify the direction of rotation by means of a circular arc or spiral with an arrow attached.

Fig. 5.21 contains sketches of two positive angles α and β and a negative angle γ. If the terminal side of an angle is in a certain quadrant, we speak of the *angle* as being in that quadrant. Thus, in Fig. 5.21, α is in quadrant II, β is in quadrant I, and γ is in quadrant III. If the terminal side coincides with a coordinate axis, then the angle is referred to as a *quadrantal angle*. It is important to observe that there are many different angles in standard position which have the same terminal side. Any two such angles are called *coterminal*.

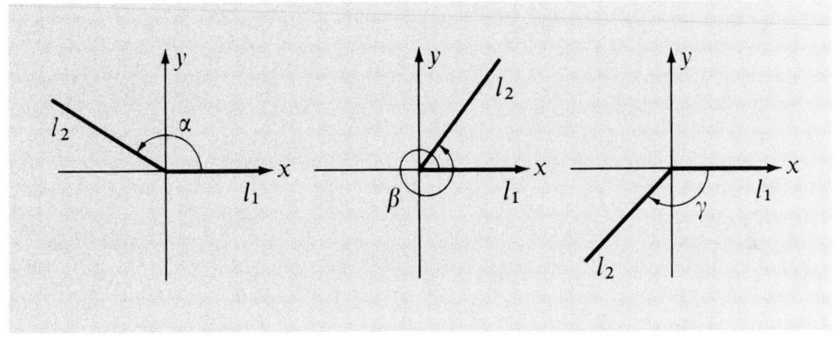

Figure 5.21

We shall now consider the problem of assigning a *measure* to a given angle. Let U be a unit circle with the center at the origin O and let θ be an angle in standard position. As above, we regard θ as generated by rotating the positive x-axis about O. As the axis rotates to its terminal position, its point of intersection with U travels a certain distance t before arriving at its final position P, as is illustrated in Fig. 5.22. If t is considered positive for a counterclockwise rotation and negative for a clockwise rotation, then P is precisely the point which the wrapping function associates with the real number t. A natural way of assigning a measure to θ

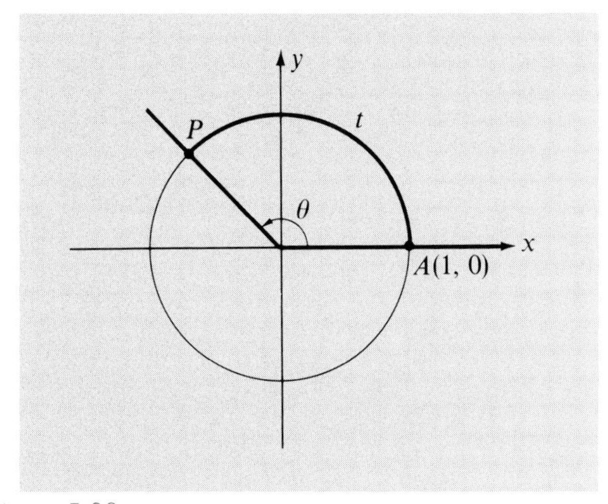

Figure 5.22

is to use the number t. When this is done, we shall say that θ *is an angle of t radians*. Notationally, it is customary to write $\theta = t$ instead of $\theta = t$ radians. In particular, if $\theta = 1$, then θ is an angle that subtends an arc of unit length on the unit circle U. The notation $\theta = -4.3$ means that θ is the angle generated by a clockwise rotation in which the point of intersection P of the terminal side of θ with the unit circle U travels 4.3 units. Several angles, measured in radians, are sketched in Fig. 5.23.

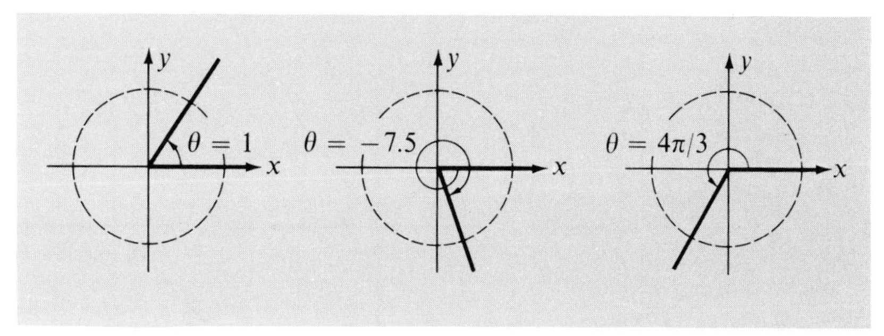

Figure 5.23

The radian measure of an angle can be found by using a circle of any radius. Thus suppose that θ is a central angle of a circle of radius r and that θ subtends an arc of length s, where $0 \leq s < 2\pi r$. Let us show that the radian measure of θ is given by the formula

$$\theta = s/r. \tag{5.17}$$

To prove (5.17), let us place θ in standard position on a rectangular coordinate system and superimpose a unit circle U, as is shown in Fig. 5.24.

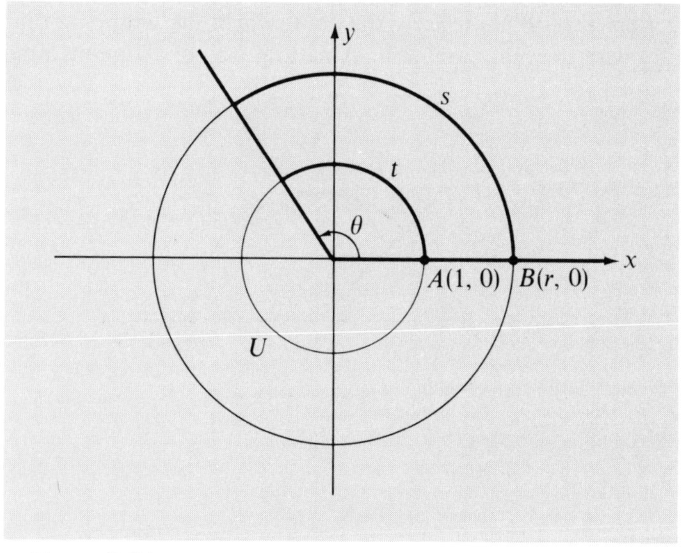

Figure 5.24

If t is the length of arc subtended by θ on U, then $\theta = t$. From geometry, the ratio of the arcs is the same as the ratio of the radii — that is, $s/t = r/1$. Substituting θ for t leads to (5.17). The latter formula may also be written

(5.18) $$s = r\theta.$$

This gives us a formula for finding the length of arc subtended by a central angle of radian measure θ on a circle of radius r.

Another useful formula can be obtained by using the result from plane geometry which states that if θ and θ_1 are radian measures for central angles of a circle of radius r, and if A and A_1 are the areas of the sectors determined by θ and θ_1, respectively, then

$$\frac{A}{A_1} = \frac{\theta}{\theta_1}.$$

In particular, if we let $\theta_1 = 2\pi$, then $A_1 = \pi r^2$ and, substituting in the previous equation, we obtain

$$\frac{A}{\pi r^2} = \frac{\theta}{2\pi}.$$

Multiplying both sides by πr^2 gives us

(5.19) $$A = \tfrac{1}{2}r^2\theta,$$

which may be used to find the area of a sector of a circle of radius r determined by a central angle of radian measure θ.

EXAMPLE 1. A central angle θ subtends an arc 10 inches long on a circle of radius 4 inches. Find the radian measure of θ and the area A of the circular sector determined by θ.

Solution: Substituting in (5.17) gives us the radian measure $\theta = 10/4 = 2.5$. Applying (5.19) we obtain, for the area, $A = \tfrac{1}{2}(4)^2(2.5) = 20$ square inches.

Another unit of measurement for angles is the *degree*. If the angle is placed in standard position on a rectangular coordinate system, then an angle of 1 degree is, by definition, the measure of the angle formed by 1/360 of a complete revolution in the counterclockwise direction. The symbol "$\circ$" is used to denote the number of degrees in the measure of an angle. In Fig. 5.25 several angles, measured in degrees, are shown in standard position on a rectangular coordinate system. It is customary to refer to an angle of measure 90° as a *right angle*. An angle is *acute* if its degree measure is between 0° and 90°. If its measure is between 90° and 180° an angle is *obtuse*.

If smaller measurements than those afforded by the degree and radian are required, one can employ tenths, hundredths, or thousandths of radians or degrees. Another method when degrees are used, is to

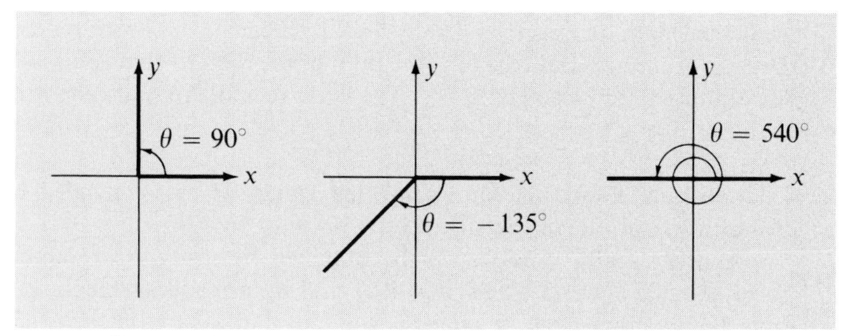

Figure 5.25

divide each degree into 60 equal parts called *minutes* (denoted by "′") and each minute into 60 equal parts called *seconds* (denoted by "″"). Thus $1'$ is $1/60$ of $1°$ and $1''$ is $1/60$ of $1'$. A notation such as $\theta = 73° \, 56' \, 18''$ refers to an angle θ of measure 73 degrees, 56 minutes, and 18 seconds.

It is not difficult to transform angular measure from one system to another. If we consider the angle θ generated by one-half of a complete counterclockwise rotation, then $\theta = 180°$. On the other hand, the radian measure of θ is π. This gives us the basic relation

$$180° = \pi \text{ radians.}$$

Equivalent formulas are

$$1° = \pi/180 \text{ radians} \quad \text{and} \quad 1 \text{ radian} = (180/\pi)°. \tag{5.20}$$

By long division we obtain the fact that $1° \doteq 0.01745$ radians, $1' \doteq 0.00029$ radians, $1'' \doteq 0.00000485$ radians, and $1 \text{ radian} \doteq 57.296°$.

Example 2. (a) Find the radian measure of θ if $\theta = 150°$. (b) Find the degree measure of θ if $\theta = 7\pi/4$.

Solution: (a) By (5.20) there are $\pi/180$ radians in each degree and hence the number of radians in $150°$ can be found by multiplying 150 by $\pi/180$. This gives us $5\pi/6$ radians.

(b) By (5.20) the number of degrees in 1 radian is $180/\pi$. Therefore to find the number of degrees in $7\pi/4$ radians, we multiply by $180/\pi$. This gives us $315°$.

Example 3. If the measure of an angle θ is 3 radians, find the (approximate) measure of θ in terms of degrees, minutes, and seconds.

Solution: Since $1 \text{ radian} \doteq 57.296°$, we have

$$3 \text{ radians} \doteq 171.888° = 171° + 0.888°.$$

Since there are $60'$ in each degree, the number of minutes in $0.888°$ is $60(0.888)$, or $53.28'$. Hence

$$3 \text{ radians} \doteq 171° \, 53.28'.$$

Finally, $0.28' = (0.28)60'' \doteq 17''$. Therefore

$$3 \text{ radians} \doteq 171^\circ\, 53'\, 17''.$$

EXERCISES

In each of Exercises 1–10 sketch, in standard position, the angle with the indicated measure and find the measure of two positive angles and two negative angles which are coterminal with the given angle.

1. 225°.
2. 330°.
3. -60°.
4. 600°.
5. 630°.
6. -180°.
7. $3\pi/4$ radians.
8. $-7\pi/6$ radians.
9. $13\pi/2$ radians.
10. 11π radians.

In each of Exercises 11–14 find the radian measure that corresponds to the given degree measure.

11. (a) 30°, (b) -450°, (c) 240°.
12. (a) 120°, (b) -315°, (c) 495°.
13. (a) -765°, (b) 75°, (c) 100°.
14. (a) 22.5°, (b) 20°, (c) -80°.

In each of Exercises 15–18 find the degree measure that corresponds to the given radian measure.

15. (a) $5\pi/6$, (b) -3π, (c) $7\pi/2$.
16. (a) $-3\pi/2$, (b) $5\pi/3$, (c) $-\pi/4$.
17. (a) $8\pi/3$, (b) $-3\pi/4$, (c) $\pi/5$.
18. (a) $11\pi/6$, (b) -5π, (c) $9\pi/2$.
19. If $\theta = 4$, find the approximate measure of θ in terms of degrees, minutes, and seconds.
20. Same as Exercise 19 if $\theta = 5$.
21. A central angle θ subtends an arc 5 inches long on a circle of radius 3 inches.
    (a) Find an approximation to the measure of θ in radians and in degrees.
    (b) Find the area of the circular sector determined by θ.
22. Same as Exercise 21 if θ subtends an arc 2 feet long on a circle of radius 16 inches.
23. (a) Find an approximation to the length of arc subtended by a central angle of 40° on a circle of radius 6 inches.
    (b) Find the area of the circular sector determined by the angle in part (a).
24. Same as Exercise 23 if the central angle is 1.75 radians and the radius of the circle is 3 yards.

6 TRIGONOMETRIC FUNCTIONS OF ANGLES

In certain applications it is convenient to change the domain of the trigonometric functions from a subset of **R** to the set of angles. This is very easy to do. If θ denotes an angle, we merely have to agree on the values $\sin\theta$, $\cos\theta$, and so on. The usual way of assigning values is to use the radian measure of θ. Specifically, we have the following definition.

(5.21) Definition of Trigonometric Functions of Angles

If θ is an angle and if the radian measure of θ is t, then the value of each trigonometric function at θ is its value at the real number t.

From (5.21) we see that $\sin\theta = \sin t$, $\cos\theta = \cos t$, and so on, where t is the radian measure of θ. For convenience we shall use the terminology *trigonometric functions* regardless of whether angles or real numbers are employed for the domain. So that the unit of angular measure will be clear, we shall use the degree symbol and write $\sin 65°$, $\tan 150°$, etc., when the angle is measured in degrees, whereas numerals without any symbol attached, such as $\cos 3$, $\csc \pi/6$, etc., will indicate that radian measure should be used. This will not lead to a conflict with our previous work where, for example, $\cos 3$ meant the value of the cosine function at the real number 3, since by (5.21) the cosine of 3 radians is identical with the cosine of the real number 3.

In Section 3 it was pointed out that all the values of the trigonometric functions can be found if the values are known in the t-interval $[0, \pi/2]$. Since an angle of $\pi/2$ radians is the same as an angle of $90°$, it follows that a table of functional values for the domain $0°$–$90°$ is sufficient for finding all values of the trigonometric functions. It is convenient to introduce the following analogue of the reference number defined on page 150. If θ is an angle in standard position and θ is not a quadrantal angle, then the *reference angle associated with* θ is the acute angle θ' that the terminal side of θ makes with the x-axis. If θ lies in quadrant I, then $\theta = \theta'$. The situations in which θ lies in quadrants II, III, or IV are illustrated in Fig. 5.26. We have not included the angle θ in the sketches since there

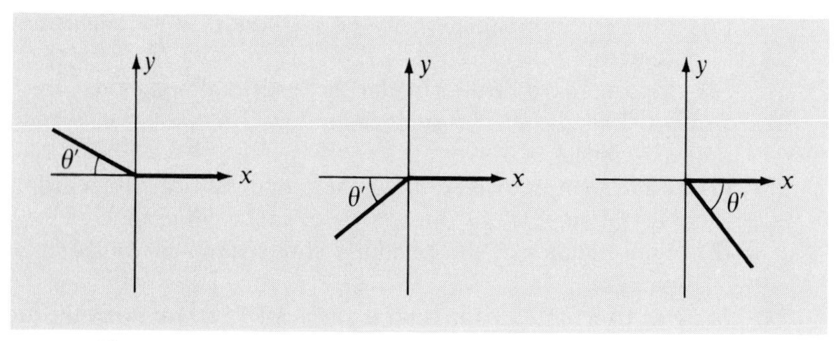

Figure 5.26

are an infinite number of angles with a given terminal side. It is important to note that we always have $0 < \theta' < 90°$. If a unit circle is introduced as illustrated in Fig. 5.27 and if t' is the reference number associated with

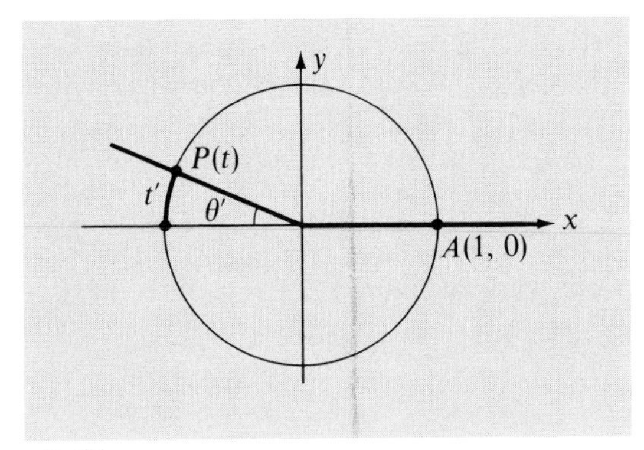

Figure 5.27

$P(t)$, then evidently $\sin \theta' = \sin t'$, $\cos \theta' = \cos t'$, and so forth. The following rule is a consequence of our discussion in Section 3 (cf. (5.16)).

(5.22) *To find the value of a trigonometric function at an angle θ, find its value at the reference angle θ' associated with θ and prefix the appropriate sign.*

EXAMPLE 1. Find each of the following:

(a) $\sin 150°$, (b) $\tan 315°$, (c) $\sec(-240°)$.

Solution: The angles and their reference angles are shown in Fig. 5.28. By (5.22) and (5.21) we have the following:

$$\begin{aligned}
&\text{(a)}\ \sin 150° &&= \sin 30° &&= \sin \pi/6 &&= 1/2,\\
&\text{(b)}\ \tan 315° &&= -\tan 45° &&= -\tan \pi/4 &&= -1,\\
&\text{(c)}\ \sec(-240°) &&= -\sec 60° &&= -\sec \pi/3 &&= -2.
\end{aligned}$$

Table 3 is arranged so that functional values corresponding to angles that are expressed in degree measure may be found directly. Angular

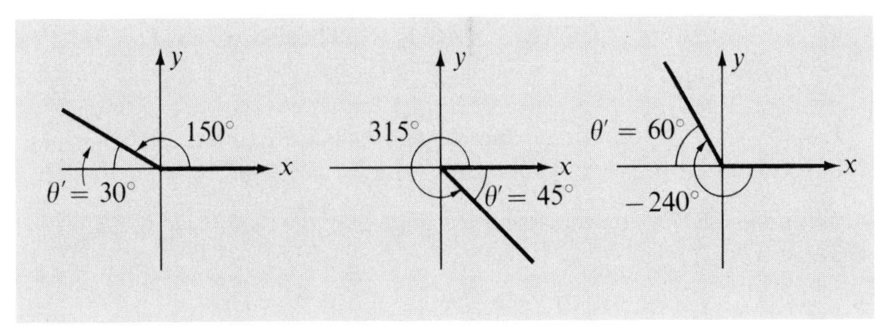

Figure 5.28

measures are given at 10′ intervals from 0° to 90° and interpolation may be used to approximate values that correspond to angles between those listed. When Table 3 is used, we shall round off answers to the nearest minute. As in our previous work, if the angular measure is between 0° and 45°, labels at the top of the columns are used, whereas if the measure is between 45° and 90°, labels at the bottom are employed.

EXAMPLE 2. Approximate tan 155° 44′.

Solution: Since the angle is in quadrant II, the reference angle is $180° - 155°\,44' = 24°\,16'$ and we have, by (5.22), $\tan 155°\,44' = -\tan 24°\,16'$. We arrange our work as follows:

$$10'\left\{\begin{array}{l} 6'\left\{\begin{array}{l}\tan 24°\,10' \doteq 0.4487\\ \tan 24°\,16' = \ ?\end{array}\right\}d \\ \quad\ \tan 24°\,20' \doteq 0.4522\end{array}\right\}0.0035.$$

Then

$$\frac{d}{0.0035} = \frac{6}{10}, \quad \text{or} \quad d = \frac{6}{10}(0.0035) \doteq 0.0021.$$

Therefore $\tan 24°\,16' \doteq 0.4487 + 0.0021 = 0.4508$ and $\tan 155°\,44' \doteq -0.4508$.

EXAMPLE 3. Approximate $\cos(-117°\,47')$.

Solution: The angle is in quadrant III and the reference angle is 62° 13′. Hence, by (5.22), $\cos(-117°\,47') = -\cos 62°\,13'$. Interpolating, we have

$$10'\left\{\begin{array}{l} 3'\left\{\begin{array}{l}\cos 62°\,10' \doteq 0.4669\\ \cos 62°\,13' \doteq \ ?\end{array}\right\}d \\ \quad\ \cos 62°\,20' \doteq 0.4643\end{array}\right\}0.0026$$

and

$$\frac{d}{0.0026} = \frac{3}{10}, \quad \text{or} \quad d \doteq 0.0008.$$

Since the cosine function is decreasing in this interval, we have $\cos 62°\,13' \doteq 0.4669 - 0.0008 = 0.4661$. Therefore $\cos(-117°\,47') \doteq -0.4661$.

EXAMPLE 4. Approximate the degree measure of all angles θ which lie in the interval [0°, 360°] such that $\sin\theta = -0.7963$.

Solution: Let θ' be the reference angle, so that $\sin\theta' = 0.7963$. From Table 3

$$10'\left\{\begin{array}{l} d\left\{\begin{array}{l}\sin 52°\,40' \doteq 0.7951\\ \sin\theta' \quad\ \ = 0.7963\end{array}\right\}0.0012 \\ \quad\ \sin 52°\,50' \doteq 0.7969\end{array}\right\}0.0018.$$

Then

$$\frac{d}{10} = \frac{0.0012}{0.0018}, \quad \text{or} \quad d \doteq 7'.$$

Hence $\theta' \doteq 52^\circ\, 47'$. Since $\sin\theta$ is negative, θ lies in quadrant III or IV, and since the reference angle is $52^\circ\, 47'$ we have $\theta \doteq 180^\circ + 52^\circ\, 47'$ or $\theta \doteq 360^\circ - 52^\circ\, 47'$. Therefore $\theta \doteq 232^\circ\, 47'$ or $\theta \doteq 307^\circ\, 13'$.

The values of the trigonometric functions at an angle θ may be determined by means of an arbitrary point on the terminal side of θ. To prove this, let θ be an angle in standard position and let $P(x, y)$ be any point on the terminal side of θ, where $d(O, P) = r > 0$. Figure 5.29

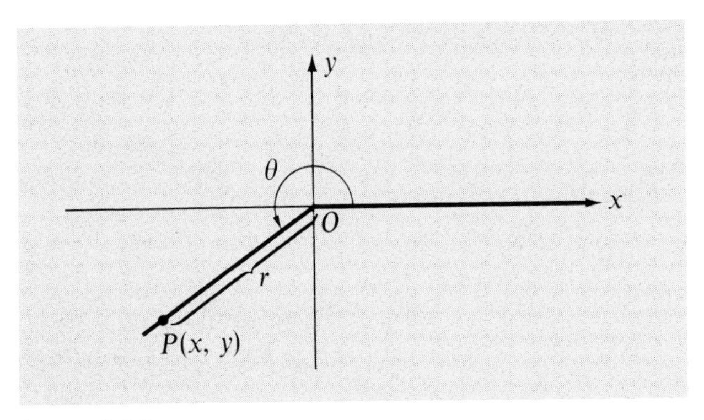

Figure 5.29

illustrates the case in which the terminal side lies in quadrant III; however, our discussion applies to any angle. The point $P(x, y)$ is not necessarily a point assigned by the wrapping function since r may be different from 1. Now suppose $P'(x', y')$ is the point on the terminal side of θ such that $d(O, P') = 1$. Therefore P' is on the unit circle U and consequently, if t is the radian measure of θ, by (5.21) and (5.4) we have

$$\sin\theta = \sin t = y',$$
$$\cos\theta = \cos t = x',$$

and so on. As in Fig. 5.30, we construct lines through P' and P parallel to

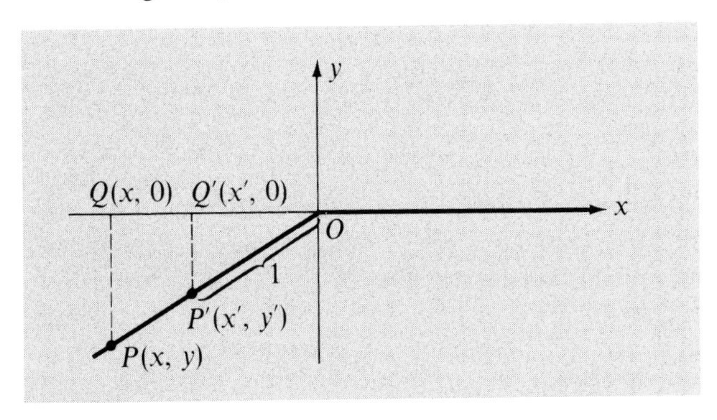

Figure 5.30

the y-axis and intersecting the x-axis at $Q'(x', 0)$ and $Q(x, 0)$, respectively. Since triangles $OP'Q'$ and OPQ are similar, we have

$$\frac{d(Q', P')}{d(O, P')} = \frac{d(Q, P)}{d(O, P)},$$

or

$$\frac{|y'|}{1} = \frac{|y|}{r}.$$

Since y and y' always have the same sign, this gives us

$$y' = y/r$$

and hence

$$\sin\theta = y/r.$$

In similar fashion we obtain

$$\cos\theta = x/r.$$

If we now use (5.5)–(5.9), we obtain the following theorem.

(5.23) Theorem

Let θ be an angle in standard position on a rectangular coordinate system and let $P(x, y)$ be any point other than O on the terminal side of θ. If $d(O, P) = r$, then

$$\begin{aligned} &\sin\theta = y/r, && \csc\theta = r/y \quad (\text{if } y \neq 0),\\ &\cos\theta = x/r, && \sec\theta = r/x \quad (\text{if } x \neq 0),\\ &\tan\theta = y/x \quad (\text{if } x \neq 0), && \cot\theta = x/y \quad (\text{if } y \neq 0). \end{aligned}$$

It can be shown, by using similar triangles, that the formulas given in (5.23) are independent of the point $P(x, y)$ that is chosen on the terminal side of θ. Of course, if $r = 1$, then (5.23) reduces to (5.4). We might remark in passing that in some books on trigonometry, (5.23) is used to *define* the trigonometric functions. Since our approach has been non-angular, it appears here as a theorem.

Theorem (5.23) is extremely important for certain applications of trigonometry. In the next section we shall use it to solve problems concerned with right triangles. Another reason for its importance is that it can be used to obtain values of trigonometric functions without resorting to the rather cumbersome process involving arc length on a unit circle. Indeed, by (5.23) it is sufficient to find *one* point (other than O) on the terminal side of an angle which is in standard position on a rectangular coordinate system.

EXAMPLE 5. If θ is an angle in standard position on a rectangular coordinate system and if the point $P(-15, 8)$ is on the terminal side of θ, find the values of the trigonometric functions of θ.

Solution: By (1.15), the distance r from the origin O to *any* point $P(x, y)$ is $r = \sqrt{(x - 0)^2 + (y - 0)^2} = \sqrt{x^2 + y^2}$. Hence, for $P(-15, 8)$, we have

$$r = \sqrt{(-15)^2 + 8^2} = \sqrt{225 + 64} = \sqrt{289} = 17.$$

Applying (5.23) with $x = -15$, $y = 8$, and $r = 17$, we obtain $\sin\theta = 8/17$, $\cos\theta = -15/17$, $\tan\theta = -8/15$, $\csc\theta = 17/8$, $\sec\theta = -17/15$, and $\cot\theta = -15/8$.

EXAMPLE 6. Find the values of the trigonometric functions of θ if the terminal side of θ lies on the line $4y = 3x$ and θ is in quadrant III.

Solution: We begin by choosing any point, say $P(-4, -3)$, on the terminal side of θ. Since $d(O, P) = 5$, we have $\sin\theta = -3/5$, $\cos\theta = -4/5$, $\tan\theta = -3/(-4) = 3/4$, $\csc\theta = -5/3$, $\sec\theta = -5/4$, and $\cot\theta = 4/3$.

In our proof of (5.23) we considered θ as a nonquadrantal angle; however, the formulas we derived are also true if the terminal side of θ lies on either the x- or y-axis. This is illustrated by the next example.

EXAMPLE 7. Find the values of the trigonometric functions of θ if $\theta = 270°$.

Solution: The terminal side of θ coincides with the negative y-axis. To use (5.23) we may choose any point on the terminal side of θ and hence, for simplicity, we consider $P(0, -1)$. In this case $r = 1$ and by (5.23), $\sin\theta = -1/1 = -1$, $\cos\theta = 0/1 = 0$, $\csc\theta = 1/(-1) = -1$, and $\cot\theta = 0/(-1) = 0$. The tangent and secant functions are undefined, since the meaningless expressions $\tan\theta = -1/0$ and $\sec\theta = 1/0$ occur when we substitute in the appropriate formulas.

EXERCISES

In each of Exercises 1 and 2 find the reference angle θ' if θ has the indicated measure.

1. (a) $230°$, (b) $-165°$, (c) $423°$, (d) $117° 22'$, (e) $-359°$, (f) $1500°$, (g) $342° 57' 13''$.
2. (a) $285°$, (b) $138°$, (c) $-77°$, (d) $529°$, (e) $162° 47'$, (f) $2345°$, (g) $159° 12' 43''$.

In Exercises 3–6 use Table 3 to find the given numbers.

3. (a) $\sin 37° 40'$, (b) $\tan 12° 20'$, (c) $\cos 72° 10'$.
4. (a) $\cot 40° 10'$, (b) $\sin 11° 30'$, (c) $\tan 80° 20'$.
5. (a) $\cot(-235°)$, (b) $\sin 325° 40'$, (c) $\cos 800° 50'$.
6. (a) $\cos 100°$, (b) $\tan(-70° 40')$, (c) $\sin 1225°$.

In Exercises 7–12 approximate the numbers by using interpolation.

7. $\sin 23° 37'$.
8. $\cos 37° 23'$.
9. $\cos 51° 53'$.
10. $\tan 69° 7'$.
11. $\tan 349° 28'$.
12. $\cot 253° 34'$.

In Exercises 13–20 use interpolation in Table 3 to approximate, to the nearest minute, the degree measure of all angles θ which lie in the interval $[0°, 360°]$.

13. $\sin \theta = 0.4759$.
14. $\cos \theta = 0.1088$.
15. $\tan \theta = 1.1822$.
16. $\sin \theta = 0.4444$.
17. $\cot \theta = -2.2390$.
18. $\tan \theta = -0.1180$.
19. $\cos \theta = -0.7302$.
20. $\cos \theta = -0.8441$.

In Exercises 21–28 use (5.23) to find the values of the six trigonometric functions of θ if θ is in standard position and satisfies the given condition.

21. The point $P(-2, 1)$ is on the terminal side of θ.
22. The point $P(-4, -5)$ is on the terminal side of θ.
23. The terminal side of θ lies on the line with equation $y = -3x$ in quadrant IV.
24. The terminal side of θ lies on the line with equation $2y + 5x = 0$ in quadrant II.
25. $\theta = 90°$.
26. $\theta = 180°$.
27. $\theta = 135°$.
28. $\theta = 315°$.
29. Prove geometrically that the formulas given in (5.23) are independent of the point $P(x, y)$ that is chosen on the terminal side of θ.
30. If (5.23) were used for the definition of the trigonometric functions, how could one prove that $\sin^2 \theta + \cos^2 \theta = 1$ for all angles θ?

7 RIGHT TRIANGLE TRIGONOMETRY

A triangle is called a *right triangle* if one of its angles is a right angle. If θ is an acute angle, then it can be regarded as an angle of a right triangle and we may refer to the lengths of the *hypotenuse*, the *opposite side*, and the *adjacent side* in the usual way. For convenience we shall use *hyp*, *opp*, and *adj*, respectively, to denote these numbers. Introducing a rectangular coordinate system as in Fig. 5.31, we see that the lengths of the adjacent side and the opposite side for θ are the abscissa and ordinate,

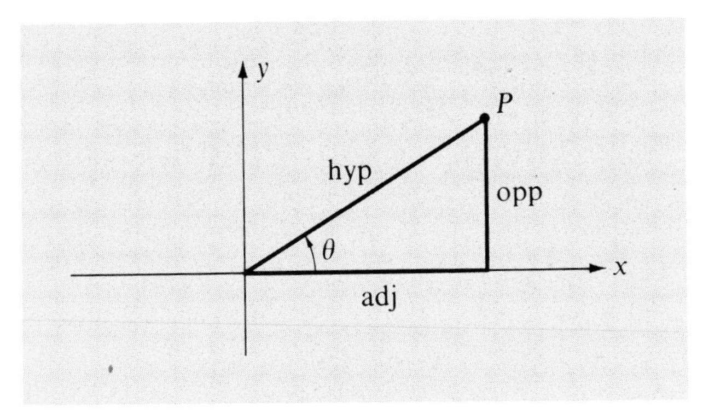

Figure 5.31

respectively, of a point P on the terminal side of θ and, by (5.23), we may write

$$
\begin{aligned}
\sin\theta &= \frac{\text{opp}}{\text{hyp}}, & \csc\theta &= \frac{\text{hyp}}{\text{opp}},\\
\cos\theta &= \frac{\text{adj}}{\text{hyp}}, & \sec\theta &= \frac{\text{hyp}}{\text{adj}},\\
\tan\theta &= \frac{\text{opp}}{\text{adj}}, & \cot\theta &= \frac{\text{adj}}{\text{opp}}.
\end{aligned} \tag{5.24}
$$

The formulas given in (5.24) are very useful in work with triangles and should be memorized.

EXAMPLE 1. Use (5.24) to find the values of $\sin\theta$, $\cos\theta$, and $\tan\theta$ when

$$\text{(a)}\quad \theta = 60^\circ, \quad \text{(b)}\quad \theta = 30^\circ, \quad \text{(c)}\quad \theta = 45^\circ.$$

Solution: Consider an equilateral triangle having sides equal to 2. The median from one vertex to the opposite side bisects the angle at that vertex and we have the first drawing shown in Fig. 5.32. Using (5.24) to calculate functional values, we obtain

$$\text{(a)}\quad \sin 60^\circ = \sqrt{3}/2,\ \cos 60^\circ = 1/2, \quad \tan 60^\circ = \sqrt{3}/1 = \sqrt{3};$$

$$\text{(b)}\quad \sin 30^\circ = 1/2, \quad \cos 30^\circ = \sqrt{3}/2,\ \tan 30^\circ = 1/\sqrt{3} = \sqrt{3}/3.$$

To find the functional values at 45°, consider an isosceles right triangle whose two equal sides have length 1, as in the second drawing in Fig. 5.32. Again using (5.24), we obtain

$$\text{(c)}\quad \sin 45^\circ = 1/\sqrt{2} = \sqrt{2}/2 = \cos 45^\circ, \quad \tan 45^\circ = 1.$$

In our work with triangles in this section and later, in Chapter Six, we shall often use the following notation. The vertices of the triangle will be denoted by A, B, and C. The angles of the triangle at A, B, and C

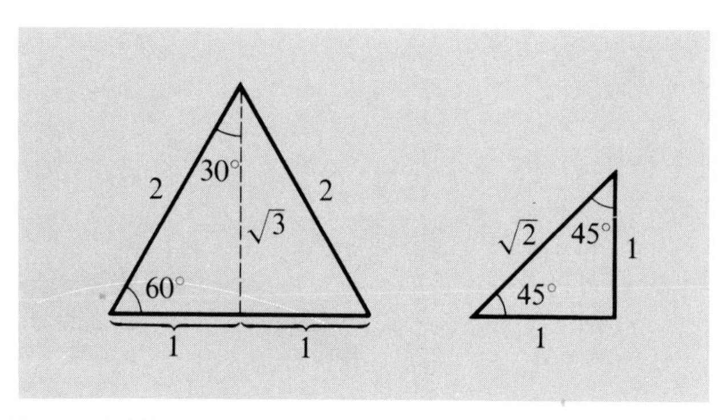

Figure 5.32

will be denoted by α, β, and γ, respectively, and the lengths of the sides opposite these angles by a, b, and c, respectively. The triangle itself will often be referred to as *triangle ABC*. To *solve* a triangle means to find all of its parts — that is, the lengths of the three sides and the measures of the three angles. If the triangle is a right triangle and if one of the acute angles and a side are known, or if two sides are given, then (5.24) may be used to find the remaining parts.

EXAMPLE 2. If, in triangle ABC, $\gamma = 90°$, $\alpha = 34°$, and $b = 10.5$, approximate the remaining parts of the triangle.

Solution: The triangle is sketched in Fig. 5.33. Since the sum of the angles is 180°, we immediately have $\beta = 56°$. Using (5.24), we may write

$$\tan 34° = \frac{a}{10.5}, \quad \text{or} \quad a = (10.5) \tan 34°.$$

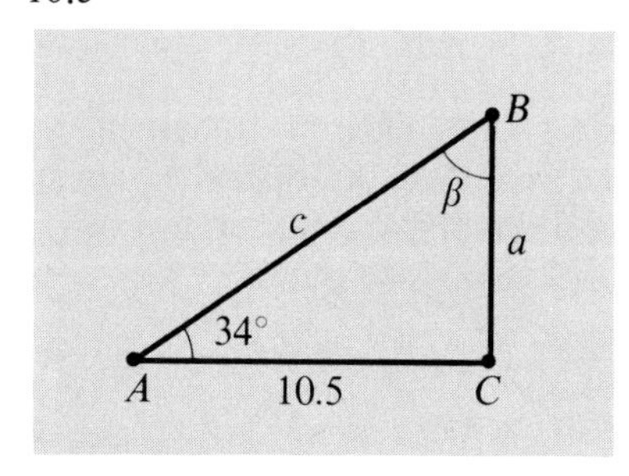

Figure 5.33

Substituting from Table 3, we obtain

$$a \doteq (10.5)(0.6745) = 7.08225.$$

Similarly,

$$\cos 34° = \frac{10.5}{c}, \quad \text{or} \quad c = \frac{10.5}{\cos 34°}.$$

Again using Table 3, we obtain

$$c \doteq (10.5)/(0.8290) \doteq 12.67.$$

We shall round off these answers to $a \doteq 7.1$ and $c \doteq 12.7$. If a table of values for the secant function were available, we could have avoided the division in the calculation of c by writing $c = (10.5) \sec 34^\circ$.

When working with triangles, we shall, as illustrated in Example 2, round off answers. There are several reasons for this. In applications, lengths of sides of triangles and measures of angles are usually found by some mechanical device and hence are only approximations to the exact values. Because of this, the number 10.5 is assumed to have been rounded off to the nearest tenth. One cannot expect more accuracy in the calculated values for the remaining sides, and consequently they also should be rounded off to the nearest tenth.

In some problems a large number of digits, such as 13,647.29, may be given for a number. If Table 3 is used, a number of this type should be rounded off to four significant figures and written as 13,650 before beginning any calculations. The situation here is similar to that described on page 125 with regard to the use of the table of logarithms. Since the values of the trigonometric functions given in Table 3 have been rounded off to four significant figures, we cannot expect more than four-figure accuracy in our computations.

As a final remark on approximations, answers should also be rounded off when Table 3 is used to find angles. In general, we shall use the following rules: if the sides of a triangle are known to four significant figures, then measures of angles calculated from Table 3 should be rounded off to the nearest minute; if the sides are known to three significant figures, then calculated measures of angles should be rounded off to the nearest multiple of ten minutes; and if the sides are known to only two significant figures, then calculations should be rounded off to the nearest degree. In order to justify these rules, we would have to make a much deeper analysis of problems involving approximate data.

EXAMPLE 3. If, in triangle ABC, $\gamma = 90^\circ$, $a = 12.3$, and $b = 31.6$, find the remaining parts.

Solution: Applying (5.24) to Fig. 5.34 we may write

$$\tan \alpha = \frac{12.3}{31.6} \doteq 0.3892.$$

By the rule just stated, α should be rounded off to the nearest multiple of $10'$. Referring to Table 3, we see that $\alpha \doteq 21^\circ 20'$. Consequently

$$\beta \doteq 90^\circ - 21^\circ 20' = 68^\circ 40'.$$

From $\cos \alpha = (31.6)/c$ we have $c = (31.6)/\cos 21^\circ 20' \doteq (31.6)/(0.9315)$. Dividing and rounding off, we obtain $c \doteq 33.9$.

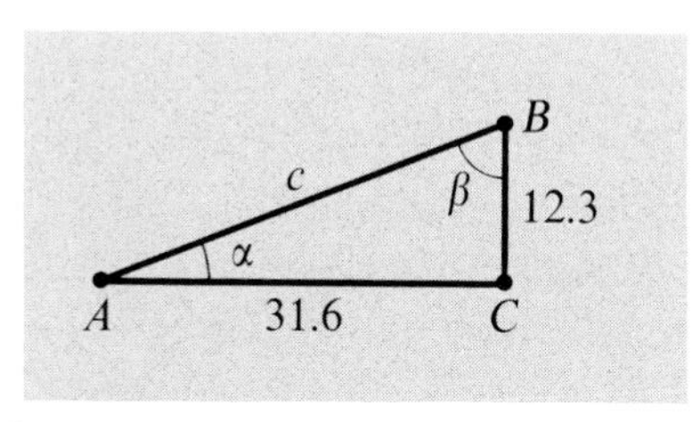

Figure 5.34

In certain triangle problems it is convenient to use logarithms. Such problems could be worked by first using Table 3 to find values of the trigonometric functions and then employing Table 1 to find the logarithms. The work can be shortened, however, by using Table 4, which contains the logarithms of $\sin\theta$, $\cos\theta$, $\tan\theta$, and $\cot\theta$ for values of θ from $0°$ to $90°$ at $10'$ intervals. If θ is between $0°$ and $90°$ then $0 < \sin\theta < 1$ and $0 < \cos\theta < 1$ and consequently the characteristic of $\log\sin\theta$ or $\log\cos\theta$ is negative. The same is true for $\log\tan\theta$ when $0 < \theta < 45°$. In order to conserve space, Table 4 is constructed so that -10 should be added to each entry. For example,

$$\log\sin 38°\,40' = 9.7957 - 10, \quad \log\cot 86°\,20' = 8.8067 - 10,$$

and so on.

If we wish to solve the problem given in Example 2 by means of logarithms, we have, from $a = (10.5)\tan 34°$,

$$\log a = \log 10.5 + \log\tan 34°.$$

Using Tables 1 and 4, we obtain

$$\begin{aligned}\log a &\doteq 1.0212 + [9.8290 - 10]\\ &= 0.8502,\end{aligned}$$

and, from Table 1, $a \doteq 7.1$.

Similarly, the measure of angle α of Example 3 may be found from $\tan\alpha = 12.3/31.6$ by writing

$$\begin{aligned}\log\tan\alpha &= \log 12.3 - \log 31.6\\ &\doteq 1.0899 - 1.4997\\ &= (11.0899 - 10) - 1.4997\\ &= 9.5902 - 10.\end{aligned}$$

Consulting Table 4, we find $\alpha \doteq 21°\,20'$.

Needless to say, if an angle between one of those in Table 4 is given, then an approximation to the logarithm of a functional value may be determined by interpolation.

Right triangles are useful in solving various types of applied problems. The following examples give two illustrations; others will be found in the exercises.

EXAMPLE 4. From a point on level ground 135 feet from the foot of a tower, the angle of elevation of the top of the tower is $57^\circ\,20'$. Find the height of the tower.

Solution: The *angle of elevation* is the angle which the line of sight makes with the horizontal. Referring to Fig. 5.35, we may write

$$\tan 57^\circ\,20' = d/135, \quad \text{or} \quad d = (135)\tan 57^\circ\,20'.$$

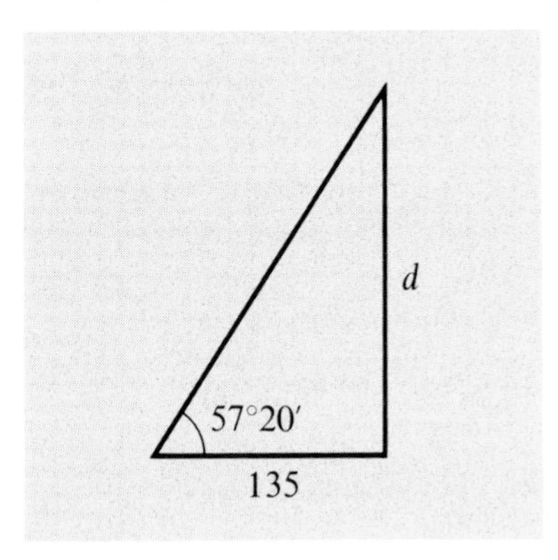

Figure 5.35

Substituting from Table 3 gives us

$$d = (135)(1.5597) \doteq 211.$$

On the other hand, if logarithms are used to obtain d, we have

$$\begin{aligned}\log d &= \log 135 + \log \tan 57^\circ\,20' \\ &\doteq 2.1303 + (10.1930 - 10) \\ &= 2.3233.\end{aligned}$$

From Table 1 we obtain $d \doteq 211$.

EXAMPLE 5. From the top of a building which overlooks an ocean, a man sees a boat sailing directly toward him. If the man is 100 feet above sea level and if the angle of depression of the boat changes from 25° to 40° during the period of observation, find the approximate distance the boat travels during this time.

Solution: The *angle of depression* is the angle between the line of sight and the horizontal. Let A and B be the positions of the boat which correspond to the 25° and 40° angles, respectively. Suppose the man is at point D and let C be the point 100 feet directly below him. Let d

denote the distance the boat travels and let g denote the distance from B to C. This gives us the drawing in Fig. 5.36, where α and β denote

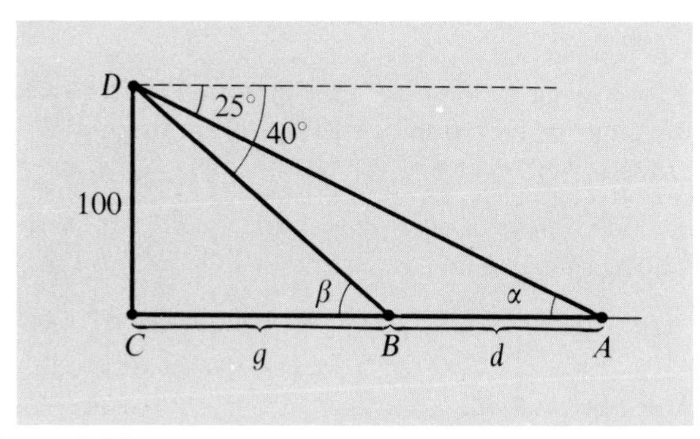

Figure 5.36

angles DAC and DBC, respectively. Therefore $\alpha = 25°$ and $\beta = 40°$ (Why?). From triangle BCD we have $\cot \beta = g/100$, or $g = 100 \cot \beta$. From triangle DAC we have $\cot \alpha = (d + g)/100$, or $d + g = 100 \cot \alpha$. Substituting $100 \cot \beta$ for g in the latter equation and simplifying gives us

$$\begin{aligned} d &= 100(\cot \alpha - \cot \beta) \\ &= 100(\cot 25° - \cot 40°) \\ &\doteq 100(2.1445 - 1.1918) \\ &= 100(0.9527). \end{aligned}$$

Hence $d \doteq 95$ feet.

In certain navigation and surveying problems the *direction*, or *bearing*, from a point P to a point Q is often specified by stating the acute angle which the half-line from P through Q varies to the east or west from the north-south line. Figure 5.37(a) illustrates four such lines. The north-south and east-west lines are labeled NS and WE, respectively. The

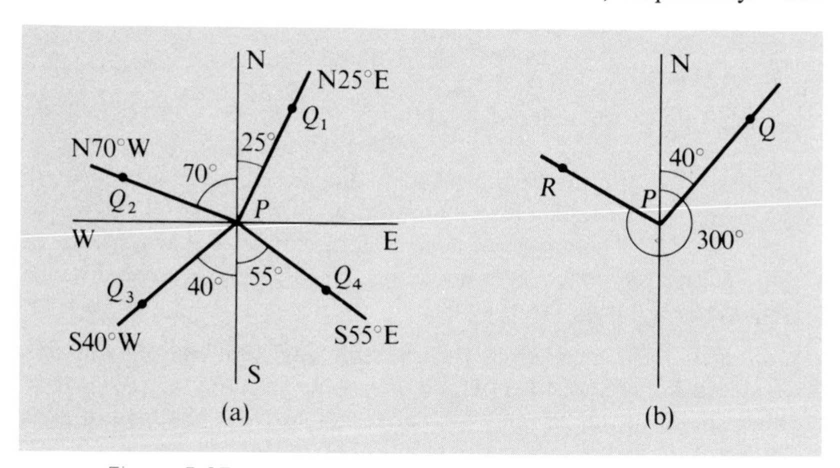

Figure 5.37

bearing from P to Q_1 is 25° east of north and is denoted by N25°E. We also refer to the *direction* N25°E, meaning the direction from P to Q_1. The bearings from P to Q_2, Q_3, and Q_4 are represented in a similar manner in the figure.

In air navigation, directions and bearings are specified by measuring from the north in a clockwise direction. In this particular situation a positive measure is assigned to the angle instead of the negative measure to which we are accustomed for clockwise rotations. Thus, referring to Fig. 5.37(b) we see that the direction of PQ is 40° whereas the direction of PR is 300°.

We shall use the above notations in several of the following exercises.

EXERCISES

In Exercises 1–10, given the indicated parts of triangle ABC with $\gamma = 90°$, approximate the remaining parts.

1. $\beta = 42°$, $a = 19$.
2. $\alpha = 27°$, $b = 34$.
3. $\alpha = 19°\ 20'$, $a = 7.3$.
4. $\beta = 73°\ 40'$, $a = 29.7$.
5. $\beta = 68°\ 24'$, $b = 132$.
6. $\alpha = 33°\ 33'$, $a = 333$.
7. $a = 14$, $b = 39$.
8. $a = 27$, $b = 8$.
9. $a = 569$, $b = 341$.
10. $a = 37.61$, $b = 62.95$.
11. Approximate the angle of elevation of the sun if a man 6 feet tall casts a shadow 10 feet long on level ground.
12. From a point 234 feet above level ground an observer measures the angle of depression of an object on the ground as 57° 20′. Approximately how far is the object from the point on the ground directly beneath the observer?
13. The string on a kite is taut and makes an angle of 27° 40′ with the horizontal. Find the approximate height of the kite above level ground if 170 feet of string are out and the end of the string is held 4 feet above the ground.
14. Generalize Exercise 13 to the case where the angle is α, the number of feet of string out is d, and the end of the string is held c feet above the ground. Express the height h of the kite in terms of α, d, and c.
15. From a point P on level ground the angle of elevation of the top of a tower is 36° 10′. From a point 78 feet closer to the tower and on the same line with P and the base of the tower, the angle of elevation of the top is 48° 20′. Approximate the height of the tower.
16. A ladder 24 feet long leans against the side of a building. If the angle between the ladder and the building is 18° 10′, approximately how far is the bottom of the ladder from the building? If the distance from the bottom of the ladder to the building is increased 2 feet, approximately how far does the top of the ladder move down the building?

17. From a point A 20 feet above level ground, the angle of elevation of the top of a building is $41^\circ 50'$ and the angle of depression of the base of the building is $19^\circ 10'$. Approximate the height of the building.
18. Generalize Exercise 17 to the case where point A is d feet above ground and the angles of elevation and depression are α and β, respectively. Express the height h of the building in terms of d, α, and β.
19. As a balloon rises vertically, its angle of elevation from a point P on the level ground 350 feet from the point Q directly underneath the balloon changes from $49^\circ 10'$ to $67^\circ 40'$. Approximately how far does the balloon rise during this period?
20. Generalize Exercise 19 to the case where the distance from P to Q is d feet and the angle of elevation changes from α to β.
21. A ship leaves port at 1:00 P.M. and sails in the direction N34°W at a rate of 24 miles per hour. Another ship leaves port at 1:30 P.M. and sails in the direction N56°E at a rate of 18 miles per hour. Approximately how far apart are the ships at 3:00 P.M.?
22. An airplane, flying at a speed of 360 miles per hour, flies from a point A in the direction 137° for 30 minutes and then flies in the direction 227° for 45 minutes. Approximately how far is the airplane from A?
23. From an observation point A a forest ranger sights a fire in the direction S48° 20′W. From a point B 5 miles due west of A another ranger sights the same fire in the direction S54° 10′E. Approximate the distance of the fire from A.

8 REVIEW EXERCISES

Oral

Define or discuss each of the following.

1. The wrapping function.
2. The trigonometric functions.
3. Graphs of the trigonometric functions.
4. The fundamental identities.
5. Periodic function.
6. Angles.
7. Standard position of an angle.
8. Terminal side of an angle.
9. Positive and negative angles.
10. Radian measure of an angle.
11. The relationship between radian measure and degree measure of an angle.
12. Trigonometric functions of angles.

Written

In Exercises 1–3, $P(t)$ denotes the point on the unit circle U which the wrapping function associates with the real number t.

1. Find the rectangular coordinates of $P(3\pi)$, $P(3\pi/2)$, $P(-7\pi/2)$, $P(7\pi/6)$, $P(2\pi/3)$, and $P(5\pi/4)$.

2. If $P(t)$ has coordinates $(8/17, -15/17)$ find the rectangular coordinates of $P(t + \pi)$, $P(t - \pi)$, $P(-t)$, and $P(2\pi - t)$.

3. Find the quadrant which contains $P(t)$ if:

   (a) $\cos t < 0$ and $\sin t > 0$.
   (b) $\tan t < 0$ and $\cos t < 0$.
   (c) $\sec t > 0$ and $\cot t < 0$.

4. Find the values of all six trigonometric functions at t if:

   (a) $\tan t = 3/4$ and $\sin t < 0$.
   (b) $\sec t = -13/5$ and $\cot t < 0$.
   (c) $\sin t = 1/3$ and $\cos t < 0$.

5. Use fundamental identities to transform the first expression into the second:

   (a) $\cos t \tan t$, $\sin t$.
   (b) $(\cot^2 t)/(\csc t - 1)$, $\csc t + 1$.
   (c) $(1 - \sin^2 x)/\cos^2 x$, 1.

6. Sketch the graph of the function f defined by:
   (a) $f(t) = 3 \sin t$. (b) $f(t) = \frac{1}{2} \cos t$.
   (c) $f(t) = -\tan t$.

7. (a) Find the reference number t' if t equals:

   $$13\pi/4; \quad -7\pi/6; \quad -4\pi/3.$$

   (b) Find the reference angle θ' if θ has measure:

   $$220°; \quad 157°\ 12'; \quad 1111°.$$

8. (a) Find the radian measures that correspond to the following degree measures:

   $$60°; \quad -150°; \quad 270°; \quad 225°; \quad 85°.$$

   (b) Find the degree measures that correspond to the following radian measures:

   $$3\pi/4; \quad 2\pi/3; \quad -3\pi/2; \quad 7\pi/6; \quad \pi/5.$$

9. Find, without the aid of tables, the value of:
   (a) $\sin 225°$. (b) $\tan \pi/3$. (c) $\cos(-\pi/6)$.
   (d) $\sec \pi$. (e) $\cot 330°$. (f) $\csc(-210°)$.

10. Use interpolation in Table 3 to approximate:
    (a) $\cos 21°\ 33'$. (b) $\tan 74°\ 52'$. (c) $\sin 248°\ 19'$.

11. Use interpolation to approximate, to the nearest minute, the degree measure of all angles θ which are in the interval $[0°, 360°)$ if:
    (a) $\cos \theta = 0.6318$. (b) $\sin \theta = -0.8412$.
    (c) $\tan \theta = 0.9635$.

12. Find the values of the six trigonometric functions of θ if θ is in standard position and satisfies the stated condition:
    (a) The point $P(8, -15)$ is on the terminal side of θ.
    (b) The terminal side of θ is in quadrant IV on the line having equation $y = -2x$.
    (c) $\theta = 270°$.

13. Given the following parts of triangle ABC with $\gamma = 90°$, approximate the remaining parts:
    (a) $\alpha = 60°, \quad b = 10$.
    (b) $\beta = 52°\ 10', \quad a = 32.0$.
    (c) $a = 64, \quad b = 41$.

14. If the side of a rectangular pentagon is 16 inches, approximate the radius of the circumscribed circle.

Supplementary Questions

1. Show that the cosecant function has a minimum value but no maximum value on the open interval $(0, \pi)$. Does it have a maximum or minimum value on $(0, 2\pi)$? Explain. Find all discontinuities of the cosecant function.
2. Show that a constant function is periodic but has no period. Show that the function f defined in Supplementary Question 3 of Chapter Three is periodic. Does f have a period?
3. The hyperbolic sine and cosine functions were defined in Supplementary Question 4 of Chapter Four. The *hyperbolic tangent, cotangent, secant,* and *cosecant* functions are defined by

$$\tanh x = \frac{\sinh x}{\cosh x}, \qquad \coth x = \frac{1}{\tanh x}$$

$$\operatorname{sech} x = \frac{1}{\cosh x}, \qquad \operatorname{csch} x = \frac{1}{\sinh x}.$$

Verify the following identities:

$$\tanh^2 x = 1 - \operatorname{sech}^2 x$$

$$\coth^2 x = 1 + \operatorname{csch}^2 x$$

$$\tanh(x + y) = \frac{\tanh x + \tanh y}{1 + \tanh x \tanh y}$$

$$|\sinh \tfrac{1}{2}x| = \sqrt{\frac{\cosh x - 1}{2}}$$

$$|\cosh \tfrac{1}{2}x| = \sqrt{\frac{\cosh x + 1}{2}}$$

$$|\tanh \tfrac{1}{2}x| = \sqrt{\frac{\cosh x - 1}{\cosh x + 1}}.$$

4. For certain applications it is convenient to employ the *versine, coversine,* and *exsecant functions,* denoted by vers, covers, and exsec, respectively, and defined by

$$\operatorname{vers} x = 1 - \cos x$$

$$\operatorname{covers} x = 1 - \sin x$$

$$\operatorname{exsec} x = \sec x - 1.$$

Sketch the graphs of these three functions and verify the following identities:

$$\text{vers } x = 2 \sin^2 \frac{x}{2}$$

$$\text{covers } x = \frac{\cos^2 x}{1 + \sin x}$$

$$\text{exsec } x = \frac{2 \tan^2 (x/2)}{1 - \tan^2 (x/2)}.$$

chapter six Analytic Trigonometry

In this chapter we examine various algebraic and geometric aspects of trigonometry. In Sections 1 through 6 the emphasis is on identities and equations. This is followed by work on graphs that are related to graphs of the trigonometric functions. In Section 9 we consider inverse functions for the trigonometric functions. Methods for solving oblique triangles are discussed in the last two sections.

1 TRIGONOMETRIC IDENTITIES

Any mathematical expression that contains symbols such as $\sin x$, $\cos \beta$, $\tan v$, and so on, where the letters x, β, v represent variables, will be referred to as a *trigonometric expression.* Trigonometric expressions may be very simple or very complicated. Some examples are

$$x + \sin x, \qquad \frac{\cos (3y + 1)}{x^2 + \tan^2 (z - y^2)}, \qquad \frac{\sec (\cot \theta)}{|\theta| \log \sin \theta}.$$

As usual, we tacitly assume that the domain of the variables is the set of real numbers (or angles) for which the expressions are meaningful. Many trigonometric expressions can be simplified or changed in form by employing the fundamental identities (5.5)–(5.12). This was illustrated in Example 3 on page 145. The next example contains a similar illustration.

EXAMPLE 1. Simplify the expression $(\sec \theta + \tan \theta)(1 - \sin \theta)$.

Solution: The reader should supply reasons for the following steps:

$$\begin{aligned}(\sec\theta + \tan\theta)(1 - \sin\theta) &= \left(\frac{1}{\cos\theta} + \frac{\sin\theta}{\cos\theta}\right)(1 - \sin\theta)\\ &= \left(\frac{1 + \sin\theta}{\cos\theta}\right)(1 - \sin\theta)\\ &= \frac{1 - \sin^2\theta}{\cos\theta}\\ &= \frac{\cos^2\theta}{\cos\theta}\\ &= \cos\theta.\end{aligned}$$

There are other ways to simplify the expression in Example 1. We could begin by multiplying the two factors and then proceed to simplify and combine terms. The method we employed — of changing all expressions to expressions which involve only sines and cosines — is often worthwhile. However, this technique does not always lead to the shortest possible simplification. As the reader becomes familiar with trigonometric expressions, he should gain facility in rapid simplifications.

A *trigonometric equation* is any statement of the form $p = q$, where at least one of the indicated expressions is trigonometric. Exactly as in our discussion of algebraic equations on p. 10, we define *solution* and *solution set* of a trigonometric equation. *Conditional equations* and *identities* have the same meaning as before. In our work with trigonometric equations we may, of course, express solution sets either in terms of a set of real numbers or in terms of a set of angles.

We encountered several trigonometric identities in Chapter Five. Some of these, for example the fundamental identities, are basic and should be memorized; others are introduced merely to supply practice in manipulating trigonometric expressions and hence should not be memorized. We shall take the latter point of view in this section. Thus the identities to be investigated are unimportant in their own right. The important thing is the manipulative practice that is gained. The ability to carry out trigonometric manipulations is essential for problems which the student will encounter in advanced mathematics courses.

We shall usually use the phrase "verify the identity $p = q$" instead of "prove that the equation $p = q$ is an identity." When verifying an identity, we shall not, by means of (5.4), return to the coordinates (x, y) of a point on the unit circle U. Although many of the identities could be proved in this way, it is not the type of practice we desire. Rather, we shall use the fundamental identities and algebraic manipulations to change the form of trigonometric expressions in a manner similar to our work in Example 1. The preferred method of showing that an equation $p = q$ is an identity is to transform one side into the other. This method is illustrated in the next three examples. The reader should supply reasons for all steps in the solutions.

EXAMPLE 2. Verify the identity

$$\frac{\tan t + \cos t}{\sin t} = \sec t + \cot t.$$

Solution: We shall transform the left side into the right side. Thus

$$\begin{aligned}\frac{\tan t + \cos t}{\sin t} &= \frac{\tan t}{\sin t} + \frac{\cos t}{\sin t}\\ &= \frac{(\sin t/\cos t)}{\sin t} + \cot t\\ &= \frac{1}{\cos t} + \cot t\\ &= \sec t + \cot t.\end{aligned}$$

EXAMPLE 3. Verify the identity $\sec \alpha - \cos \alpha = \sin \alpha \tan \alpha$.

Solution:

$$\begin{aligned}\sec \alpha - \cos \alpha &= \frac{1}{\cos \alpha} - \cos \alpha\\ &= \frac{1 - \cos^2 \alpha}{\cos \alpha}\\ &= \frac{\sin^2 \alpha}{\cos \alpha}\\ &= \sin \alpha \left(\frac{\sin \alpha}{\cos \alpha}\right)\\ &= \sin \alpha \tan \alpha.\end{aligned}$$

EXAMPLE 4. Verify the identity

$$\frac{\cos x}{1 - \sin x} = \frac{1 + \sin x}{\cos x}.$$

Solution: We begin by multiplying the numerator and denominator of the fraction on the left by $1 + \sin x$. Thus

$$\begin{aligned}\frac{\cos x}{1 - \sin x} &= \frac{\cos x}{1 - \sin x} \cdot \frac{1 + \sin x}{1 + \sin x}\\ &= \frac{\cos x(1 + \sin x)}{1 - \sin^2 x}\\ &= \frac{\cos x(1 + \sin x)}{\cos^2 x}\\ &= \frac{1 + \sin x}{\cos x}.\end{aligned}$$

Another technique for showing that an equation $p = q$ is an identity is to begin by transforming the left side p into another expression s, making sure that each step is *reversible*, in the sense that it is possible to transform s back into p by reversing the procedure which has been used. In this event the equation $p = s$ is an identity. Next, as a *separate* exercise, it is shown that the right side q can also be transformed to the expression s by means of reversible steps, and hence that $q = s$ is an identity. It then follows that $p = q$ is an identity. This method is illustrated in the next example.

EXAMPLE 5. Verify the identity

$$(\tan\theta - \sec\theta)^2 = \frac{1 - \sin\theta}{1 + \sin\theta}.$$

Solution: We shall verify the identity by showing that each side of the equality can be transformed into the same expression. For the left side we may write

$$\begin{aligned}(\tan\theta - \sec\theta)^2 &= \tan^2\theta - 2\tan\theta\sec\theta + \sec^2\theta \\ &= \frac{\sin^2\theta}{\cos^2\theta} - \frac{2\sin\theta}{\cos^2\theta} + \frac{1}{\cos^2\theta} \\ &= \frac{\sin^2\theta - 2\sin\theta + 1}{\cos^2\theta}.\end{aligned}$$

The right-hand side of the original equation may be changed by multiplying numerator and denominator by $1 - \sin\theta$. Thus

$$\begin{aligned}\frac{1 - \sin\theta}{1 + \sin\theta} &= \frac{1 - \sin\theta}{1 + \sin\theta}\cdot\frac{1 - \sin\theta}{1 - \sin\theta} \\ &= \frac{1 - 2\sin\theta + \sin^2\theta}{1 - \sin^2\theta} \\ &= \frac{1 - 2\sin\theta + \sin^2\theta}{\cos^2\theta},\end{aligned}$$

which is the same expression as that obtained above for $(\tan\theta - \sec\theta)^2$.

EXERCISES

Verify the following identities.

1. $\cos\theta\csc\theta = \cot\theta$.
2. $\tan\theta\cos\theta = \sin\theta$.
3. $\csc\theta - \sin\theta = \cot\theta\cos\theta$.
4. $\cos\theta(\tan\theta + \cot\theta) = \csc\theta$.
5. $\sin t(\csc t - \sin t) = \cos^2 t$.
6. $\cot t + \tan t = \csc t\sec t$.

7. $\dfrac{\sec^2 u - 1}{\sec^2 u} = \sin^2 u.$

8. $(\tan u + \cot u)(\cos u + \sin u) = \sec u + \csc u.$

9. $(\cos^2 x - 1)(\tan^2 x + 1) = 1 - \sec^2 x.$

10. $\dfrac{1 + \cos^2 y}{\sin^2 y} = 2 \csc^2 y - 1.$

11. $\sec^2 \theta \csc^2 \theta = \sec^2 \theta + \csc^2 \theta.$

12. $\dfrac{\sec x - \cos x}{\tan x} = \dfrac{\tan x}{\sec x}.$

13. $\dfrac{1 + \cos t}{\sin t} + \dfrac{\sin t}{1 + \cos t} = 2 \csc t.$

14. $\tan^2 \alpha - \sin^2 \alpha = \tan^2 \alpha \sin^2 \alpha.$

15. $\dfrac{1 + \tan^2 v}{\tan^2 v} = \csc^2 v.$

16. $\dfrac{\cos x \cot x}{\cot x - \cos x} = \dfrac{\cot x + \cos x}{\cos x \cot x}.$

17. $(\sec u - \tan u)(\csc u + 1) = \cot u.$

18. $\dfrac{\cot \theta - \tan \theta}{\sin \theta + \cos \theta} = \csc \theta - \sec \theta.$

19. $\dfrac{\cot \alpha - 1}{1 - \tan \alpha} = \cot \alpha.$

20. $\dfrac{1 + \sec \beta}{\tan \beta + \sin \beta} = \csc \beta.$

21. $\cot^4 t - \csc^4 t = -\cot^2 t - \csc^2 t.$

22. $\cos^4 \theta + \sin^2 \theta = \sin^4 \theta + \cos^2 \theta.$

23. $\dfrac{\cos \beta}{1 - \sin \beta} = \sec \beta + \tan \beta.$

24. $\dfrac{1}{\csc y - \cot y} = \csc y + \cot y.$

25. $\dfrac{\tan^2 x}{\sec x + 1} = \dfrac{1 - \cos x}{\cos x}.$

26. $\dfrac{\cot x}{\csc x + 1} = \dfrac{\csc x - 1}{\cot x}.$

27. $\dfrac{\cot u - 1}{\cot u + 1} = \dfrac{1 - \tan u}{1 + \tan u}.$

28. $\dfrac{1 + \sec x}{\sin x + \tan x} = \csc x.$

29. $\sec^2 \gamma + \tan^2 \gamma = (1 - \sin^4 \gamma)(\sec^4 \gamma).$

30. $\dfrac{\sin t}{1 - \cos t} = \csc t + \cot t.$

31. $(\sin^2 \theta + \cos^2 \theta)^3 = 1.$

32. $\left(\dfrac{\sin^2 x}{\tan^4 x}\right)^3 \left(\dfrac{\csc^3 x}{\cot^6 x}\right)^2 = 1.$

33. $\dfrac{\cos^3 x - \sin^3 x}{\cos x - \sin x} = 1 + \sin x \cos x.$

34. $\dfrac{\sin\theta + \cos\theta}{\tan^2\theta - 1} = \dfrac{\cos^2\theta}{\sin\theta - \cos\theta}.$

35. $(\csc t - \cot t)^4(\csc t + \cot t)^4 = 1.$

36. $(a\cos t - b\sin t)^2 + (a\sin t + b\cos t)^2 = a^2 + b^2.$

37. $\sin^6 v + \cos^6 v = 1 - 3\sin^2 v\cos^2 v.$

38. $\dfrac{\sin\alpha\cos\beta + \cos\alpha\sin\beta}{\cos\alpha\cos\beta - \sin\alpha\sin\beta} = \dfrac{\tan\alpha + \tan\beta}{1 - \tan\alpha\tan\beta}.$

39. $\sqrt{\dfrac{1 - \cos t}{1 + \cos t}} = \dfrac{1 - \cos t}{|\sin t|}.$

40. $-\log|\sec\theta - \tan\theta| = \log|\sec\theta + \tan\theta|.$

2 CONDITIONAL EQUATIONS

If a trigonometric equation $p = q$ is not an identity, then methods similar to those used for algebraic equations may be employed for finding solution sets. The main difference here is that we usually solve for $\sin x$, $\cos\theta$, and so forth, and then find x and θ. Some methods of solution are illustrated in the following examples.

EXAMPLE 1. Solve the equation $2\sin^2 t - \cos t - 1 = 0$.

Solution: A chain of equations, each of which is equivalent to the given equation, is the following:

$$2(1 - \cos^2 t) - \cos t - 1 = 0,$$

$$2\cos^2 t + \cos t - 1 = 0,$$

$$(2\cos t - 1)(\cos t + 1) = 0.$$

The solution set of the last equation is the union of the solution sets of the equations

$$\cos t = 1/2 \quad \text{and} \quad \cos t = -1.$$

It is sufficient to find the solutions which lie in the interval $[0, 2\pi)$, for once these are known, all solutions may be found by adding multiples of 2π.

If $\cos t = 1/2$, then the reference number (or reference angle) is $\pi/3$ (or $60°$). Since $\cos t$ is positive, $P(t)$ (or the angle denoted by t) lies in quadrant I or quadrant IV. Hence in the interval $[0, 2\pi)$ we have $t = \pi/3$

or $t = 2\pi - \pi/3 = 5\pi/3$. If $\cos t = -1$, then $t = \pi$. It follows that the totality of solutions for the given equation consists of all the numbers

$$\pi/3 + 2\pi n, \quad 5\pi/3 + 2\pi n, \quad \text{and} \quad \pi + 2\pi n,$$

where $n \in \mathbf{Z}$. If we wish to write the solutions in terms of degrees, we have

$$60^\circ + 360^\circ n, \quad 300^\circ + 360^\circ n, \quad \text{and} \quad 180^\circ + 360^\circ n,$$

where $n \in \mathbf{Z}$.

EXAMPLE 2. Find the solutions of the equation

$$4 \sin^2 x \tan x - \tan x = 0$$

which are in the interval $[0, 2\pi)$.

Solution: Factoring, we have

$$\tan x(4 \sin^2 x - 1) = 0.$$

The solution set is therefore the union of the solution sets of

$$\tan x = 0 \quad \text{and} \quad \sin^2 x = 1/4$$

or of

$$\tan x = 0, \quad \sin x = 1/2, \quad \text{and} \quad \sin x = -1/2.$$

The equation $\tan x = 0$ has solutions 0 and π in the given interval. The equation $\sin x = 1/2$ gives us the numbers in $[0, \pi)$ with reference number $\pi/6$, namely $\pi/6$ and $\pi - \pi/6 = 5\pi/6$. The equation $\sin x = -1/2$ gives us the number in $[\pi, 2\pi)$ with reference number $\pi/6$. These are $\pi + \pi/6 = 7\pi/6$ and $2\pi - \pi/6 = 11\pi/6$. Hence the solutions of the given equation which lie in the interval $[0, 2\pi)$ are

$$\{0, \pi, \pi/6, 5\pi/6, 7\pi/6, 11\pi/6\}.$$

EXAMPLE 3. Find, in degree measure, approximations to the solutions of the equation $5 \sin \theta \tan \theta - 10 \tan \theta + 3 \sin \theta - 6 = 0$ which are in the interval $[0^\circ, 360^\circ)$.

Solution: The equation may be factored by grouping terms as follows:

$$5 \tan \theta(\sin \theta - 2) + 3(\sin \theta - 2) = 0,$$
$$(5 \tan \theta + 3)(\sin \theta - 2) = 0.$$

The solution set of $\sin \theta = 2$ is $\varnothing$ (Why?). Consequently the solution set of the given equation is the same as the solution set of

$$\tan \theta = -3/5 = -0.6000.$$

We obtain approximations to θ by interpolating in Table 3. First we approximate the reference angle θ':

$$10'\left\{\begin{array}{l} d\left\{\begin{array}{ll}\tan 30°\ 50' & \doteq 0.5969 \\ \tan\theta' & = 0.6000\end{array}\right\}0.0031 \\ \quad\tan 31°\ 00' \doteq 0.6009\end{array}\right\}0.0040.$$

We have

$$d/10 = 31/40, \quad \text{or} \quad d = 31/4 \doteq 8'.$$

Consequently $\theta' \doteq 30°\ 58'$. Since θ lies in quadrant II or quadrant IV, this gives us

$$\theta \doteq 180° - 30°\ 58' = 149°\ 2',$$

or

$$\theta \doteq 360° - 30°\ 58' = 329°\ 2'.$$

EXAMPLE 4. Find the solution set of the equation $\csc^4 2u - 4 = 0$.

Solution: Factoring the equation gives us

$$(\csc^2 2u - 2)(\csc^2 2u + 2) = 0.$$

Since the solution set of $\csc^2 2u + 2 = 0$ is $\varnothing$ (Why?), it follows that the solution set of the original equation is the union of the solution sets of

$$\csc 2u = \sqrt{2} \quad \text{and} \quad \csc 2u = -\sqrt{2}.$$

If $\csc 2u = \sqrt{2}$, then $2u = \pi/4 + n2\pi$ or $2u = 3\pi/4 + n2\pi$, where $n \in \mathbf{Z}$. Therefore the solution set of $\csc 2u = \sqrt{2}$ consists of all numbers of the form

$$\pi/8 + n\pi \quad \text{or} \quad 3\pi/8 + n\pi,$$

where $n \in \mathbf{Z}$.

Similarly, from $\csc 2u = -\sqrt{2}$ we obtain $2u = 5\pi/4 + n2\pi$ or $2u = 7\pi/4 + n2\pi$, where $n \in \mathbf{Z}$. This implies that

$$u = 5\pi/8 + n\pi \quad \text{or} \quad u = 7\pi/8 + n\pi,$$

where $n \in \mathbf{Z}$.

Thus the solution set of the given equation is

$$\{\pi/8 + n(\pi/4) \mid n \in \mathbf{Z}\}.$$

EXERCISES

In Exercises 1–8 find the solution sets of the given equations.

1. $3 \sin t + 2 = \sin t + 3$.
2. $2 \sin\theta + \sqrt{3} = 0$.
3. $(2 \cos\theta + 1)(2 \sin\theta + \sqrt{2}) = 0$.

4. $(\sin u - 1)(\cos u - 1) = 0.$
5. $\tan^2 x - 3 = 0.$
6. $4 \cos^2 t - 3 = 0.$
7. $\tan 2x(\sec 2x - 2) = 0.$
8. $\cot^2 \theta + \cot \theta = 0.$

In Exercises 9–30 find, in degree measure, the solutions of the given equations which are in the interval $[0°, 360°)$.

9. $\sin x \tan^2 x = \sin x.$
10. $\csc 2\theta \sec 2\theta = 2 \csc 2\theta.$
11. $2 \sin^2 t + \sin t - 1 = 0.$
12. $\cos t + 2 \sin^2 t = 1.$
13. $\cos \theta = \cot \theta.$
14. $\sin t - \cos t = 0.$
15. $2 \sin^3 \theta - \sin \theta = 0.$
16. $4 \sec x - 3 \sec^3 x = 0.$
17. $\sin^2 x + \sin x - 6 = 0.$
18. $\sin \theta \cos \theta - 4 \cos \theta + 3 \sin \theta - 12 = 0.$
19. $1 - \sqrt{3} \sin t = \cos t.$
20. $\cos u + 1 = \sin u.$
21. $\cos \theta + \sin \theta - 1 = 0.$
22. $2 \tan t - \sec^2 t = 0.$
23. $\tan x + \sec x = 1.$
24. $2 \sin \alpha + \cot \alpha = \csc \alpha.$
25. $\sin u \tan u + \cos u = \sec u.$
26. $\tan \theta + \cot \theta = \sec \theta \csc \theta.$
27. $2 \sin^3 t + \sin^2 t - 2 \sin t - 1 = 0.$
28. $\sec^5 \theta - 4 \sec \theta = 0.$
29. $2 \sec x \tan x + \tan x + 2 \sec x + 1 = 0.$
30. $2 \sec u \sin u - 4 \sin u - \sec u + 2 = 0.$

In Exercises 31–34 use Table 3 to approximate, to the nearest multiple of ten minutes, the solutions of the given equations which are in the interval $[0°, 360°)$.

31. $12 \sin^2 t - 5 \sin t - 2 = 0.$
32. $5 \cos^2 u + 3 \cos u - 2 = 0.$
33. $\sin^2 \theta - 4 \sin \theta + 1 = 0.$
34. $\tan^2 x + 3 \tan x - 2 = 0.$

3 THE ADDITION FORMULAS

In this section and the next two sections, we shall establish several identities for the trigonometric functions that are extremely important for advanced work in mathematics.

Let t_1 and t_2 be any real numbers and let $P(t_1)$ and $P(t_2)$ be the corresponding points on the unit circle U assigned by the wrapping function. In terms of rectangular coordinates, we shall denote these points by $P_1(x_1, y_1)$ and $P_2(x_2, y_2)$, respectively. Let us consider the real number $t_1 - t_2$ and denote the rectangular coordinates of $P(t_1 - t_2)$ by $P_3(x_3, y_3)$. Then by (5.4) we have

$$\text{(6.1)} \quad \begin{aligned} \cos t_1 &= x_1, & \cos t_2 &= x_2, & \cos (t_1 - t_2) &= x_3, \\ \sin t_1 &= y_1, & \sin t_2 &= y_2, & \sin (t_1 - t_2) &= y_3. \end{aligned}$$

Our goal is to obtain a formula for $\cos (t_1 - t_2)$ in terms of functional values of t_1 and t_2. For convenience we shall specialize values so that t_1 and t_2 are in the interval $[0, 2\pi]$ and $0 \leq t_1 - t_2 < t_2$. The latter inequality implies that the length t_1 of $\widehat{AP_1}$ is greater than or equal to the length t_2 of $\widehat{AP_2}$ and that the length $t_1 - t_2$ of $\widehat{AP_3}$ is less than the length of $\widehat{AP_2}$. Figure 6.1 illustrates one arrangement of points P_1 and P_2

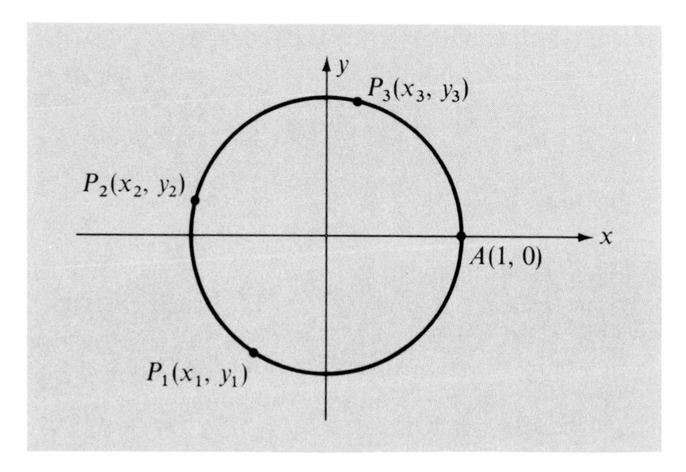

Figure 6.1

under these conditions. It is possible to extend our argument to cover all values of t. Since the arc length of $\widehat{P_2P_1}$ and $\widehat{AP_3}$ are both $t_1 - t_2$, the line segments P_2P_1 and AP_3 also have the same length, that is

$$d(A, P_3) = d(P_1, P_2)$$

or, by the distance formula,

$$\sqrt{(x_3 - 1)^2 + (y_3 - 0)^2} = \sqrt{(x_2 - x_1)^2 + (y_2 - y_1)^2}.$$

Squaring both sides and the indicated terms underneath the radical gives us

$$x_3^2 - 2x_3 + 1 + y_3^2 = x_2^2 - 2x_1x_2 + x_1^2 + y_2^2 - 2y_1y_2 + y_1^2.$$

Since P_1, P_2, and P_3 lie on U, we may substitute 1 for $x_1^2 + y_1^2$, $x_2^2 + y_2^2$, and $x_3^2 + y_3^2$ in the last equation. Doing this and simplifying, we obtain

$$2 - 2x_3 = 2 - 2x_1x_2 - 2y_1y_2,$$

which reduces to

$$x_3 = x_1x_2 + y_1y_2.$$

Substituting from (6.1) gives us the desired formula, namely

$$\cos(t_1 - t_2) = \cos t_1 \cos t_2 + \sin t_1 \sin t_2.$$

In order to make the formula less cumbersome, we shall eliminate subscripts by changing the symbols for the variables from t_1 and t_2 to u and v, respectively. Our identity then takes on the form

$$\cos(u - v) = \cos u \cos v + \sin u \sin v. \tag{6.2}$$

Note that, in general, $\cos(u - v) \neq \cos u - \cos v$.

EXAMPLE 1. Find the exact value of cos 15°.

Solution:

$$\begin{aligned}\cos 15^\circ = \cos(60^\circ - 45^\circ) &= \cos 60^\circ \cos 45^\circ + \sin 60^\circ \sin 45^\circ \\ &= (1/2)(\sqrt{2}/2) + (\sqrt{3}/2)(\sqrt{2}/2) \\ &= \frac{\sqrt{2} + \sqrt{6}}{4}.\end{aligned}$$

It is easy to obtain a formula for $\cos(u + v)$. We simply write $u + v = u - (-v)$ and employ (6.2). Thus

$$\begin{aligned}\cos(u + v) &= \cos[u - (-v)] \\ &= \cos u \cos(-v) + \sin u \sin(-v).\end{aligned}$$

By (5.13), $\cos(-v) = \cos v$ and $\sin(-v) = -\sin v$, for all $v \in \mathbf{R}$. This gives us the following identity:

$$\cos(u + v) = \cos u \cos v - \sin u \sin v. \tag{6.3}$$

EXAMPLE 2. Find $\cos 7\pi/12$ by using $\pi/3$ and $\pi/4$.

Solution:

$$\begin{aligned}\cos 7\pi/12 = \cos(\pi/3 + \pi/4) &= \cos \pi/3 \cos \pi/4 - \sin \pi/3 \sin \pi/4 \\ &= (1/2)(\sqrt{2}/2) - (\sqrt{3}/2)(\sqrt{2}/2) \\ &= \frac{\sqrt{2} - \sqrt{6}}{4}.\end{aligned}$$

Identities similar to (6.2) and (6.3) may be proved for the sine function. First we establish the following identities, which are of some interest in themselves:

$$\begin{aligned}\cos(\pi/2 - u) &= \sin u, \\ \sin(\pi/2 - u) &= \cos u, \\ \tan(\pi/2 - u) &= \cot u.\end{aligned} \tag{6.4}$$

Similar identities are true for the other trigonometric functions (see Exercise 28).

Using (6.2), we may write

$$\begin{aligned}\cos(\pi/2 - u) &= \cos \pi/2 \cos u + \sin \pi/2 \sin u \\ &= 0 \cdot \cos u + 1 \cdot \sin u \\ &= \sin u,\end{aligned}$$

which gives us the first identity of (6.4). Substituting $\pi/2 - v$ for u in this identity, we get

$$\cos[\pi/2 - (\pi/2 - v)] = \sin(\pi/2 - v),$$

or

$$\cos v = \sin(\pi/2 - v),$$

for all $v \in \mathbf{R}$. This leads to the second identity of (6.4).

The third formula in (6.4) can be obtained as follows:

$$\tan(\pi/2 - u) = \frac{\sin(\pi/2 - u)}{\cos(\pi/2 - u)} = \frac{\cos u}{\sin u} = \cot u.$$

If θ denotes an angle expressed in degree measure, then formulas (6.4) may be written in the form

(6.5)
$$\begin{aligned} \sin(90^\circ - \theta) &= \cos\theta, \\ \cos(90^\circ - \theta) &= \sin\theta, \\ \tan(90^\circ - \theta) &= \cot\theta. \end{aligned}$$

The angles θ and $90^\circ - \theta$ are *complementary*, since their sum is 90°. It is customary to refer to the sine and cosine functions as *cofunctions* of one another. Similarly, the tangent and cotangent functions are cofunctions, as are the secant and cosecant. Consequently (6.5) is a partial description of the fact that *any functional value of θ equals the cofunction of the complementary angle* $90^\circ - \theta$.

The following relationships for the sine and tangent now follow readily:

(6.6)
$$\sin(u + v) = \sin u \cos v + \cos u \sin v,$$

(6.7)
$$\sin(u - v) = \sin u \cos v - \cos u \sin v,$$

(6.8)
$$\tan(u + v) = \frac{\tan u + \tan v}{1 - \tan u \tan v},$$

(6.9)
$$\tan(u - v) = \frac{\tan u - \tan v}{1 + \tan u \tan v}.$$

We shall prove (6.6) and (6.8) and leave the others as exercises. Thus

$$\begin{aligned} \sin(u + v) &= \cos[\pi/2 - (u + v)] && (6.4) \\ &= \cos[(\pi/2 - u) - v] && \text{(Why?)} \\ &= \cos(\pi/2 - u)\cos v + \sin(\pi/2 - u)\sin v && (6.2) \\ &= \sin u \cos v + \cos u \sin v. && (6.4) \end{aligned}$$

To prove (6.8), we begin as follows:

$$\tan(u + v) = \frac{\sin(u + v)}{\cos(u + v)}$$

$$= \frac{\sin u \cos v + \cos u \sin v}{\cos u \cos v - \sin u \sin v}.$$

Next, dividing numerator and denominator by $\cos u \cos v$ (assuming, of course, that $\cos u \cos v \neq 0$), we obtain

$$\tan (u + v) = \frac{\left(\frac{\sin u}{\cos u}\right)\left(\frac{\cos v}{\cos v}\right) + \left(\frac{\cos u}{\cos u}\right)\left(\frac{\sin v}{\cos v}\right)}{\left(\frac{\cos u}{\cos u}\right)\left(\frac{\cos v}{\cos v}\right) - \left(\frac{\sin u}{\cos u}\right)\left(\frac{\sin v}{\cos v}\right)},$$

which may be written in the form (6.8). The formulas derived in this section are known as the *addition formulas.*

EXAMPLE 3. Given $\sin \alpha = 4/5$, where α is an angle in quadrant I, and $\cos \beta = -12/13$, where β is in quadrant II, find $\sin (\alpha + \beta)$, $\tan (\alpha + \beta)$, and the quadrant in which $\alpha + \beta$ lies.

Solution: From Fig. 6.2 and (5.23) we see that $\cos \alpha = 3/5$, $\tan \alpha = 4/3$,

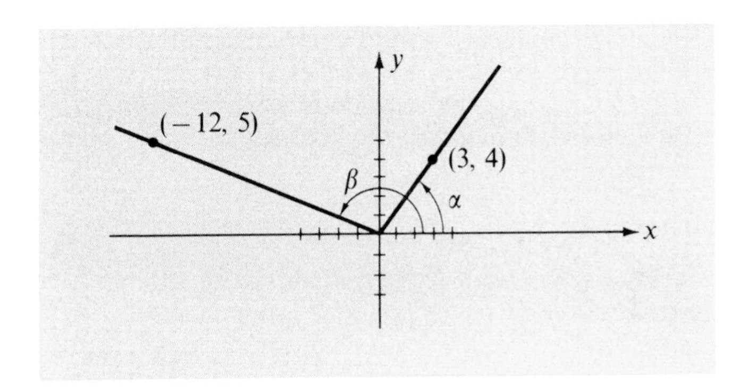

Figure 6.2

$\sin \beta = 5/13$, and $\tan \beta = -5/12$. Hence

$$\begin{aligned}\sin (\alpha + \beta) &= \sin \alpha \cos \beta + \cos \alpha \sin \beta \\ &= (4/5)(-12/13) + (3/5)(5/13) \\ &= -33/65.\end{aligned}$$

$$\begin{aligned}\tan (\alpha + \beta) &= \frac{\tan \alpha + \tan \beta}{1 - \tan \alpha \tan \beta} \\ &= \frac{4/3 + (-5/12)}{1 - (4/3)(-5/12)} \\ &= 33/56.\end{aligned}$$

Since $\sin (\alpha + \beta)$ is negative and $\tan (\alpha + \beta)$ is positive, it follows that $\alpha + \beta$ lies in quadrant III.

EXERCISES

Use (6.4) and (6.5) to write the expressions in Exercises 1 and 2 in terms of cofunctions of complementary angles.

1. (a) $\sin 27° 14'$, (b) $\cos \pi/6$, (c) $\tan 67° 10'$.
2. (a) $\tan 49° 23' 37''$, (b) $\sin 3\pi/8$, (c) $\cos 18° 47'$.

Find the exact functional values in Exercises 3–6.

3. (a) $\sin 45° + \sin 30°$, (b) $\sin 75°$.
4. (a) $\cos 150° + \cos 45°$, (b) $\cos 195°$.
5. (a) $\tan 330° - \tan 45°$, (b) $\tan 285°$.
6. (a) $\sin 3\pi/4 - \sin \pi/6$, (b) $\sin 7\pi/12$.
7. If α and β are acute angles such that $\cos \alpha = 3/5$ and $\tan \beta = 15/8$, calculate $\cos(\alpha + \beta)$, $\sin(\alpha + \beta)$, and find the quadrant in which $\alpha + \beta$ lies.
8. If $\sin \alpha = -2/5$ and $\sec \beta = 4/3$, where α is a third-quadrant angle and β is a fourth-quadrant angle, find $\sin(\alpha + \beta)$, $\tan(\alpha + \beta)$, and the quadrant in which $\alpha + \beta$ lies.
9. If $\sec \alpha = -25/24$ and $\cot \beta = 4/3$, where α is in the second quadrant and β is in the third quadrant, find $\sin(\alpha + \beta)$, $\cos(\alpha + \beta)$, $\tan(\alpha + \beta)$, $\sin(\alpha - \beta)$, $\cos(\alpha - \beta)$, and $\tan(\alpha - \beta)$.
10. If $P(t_1)$ and $P(t_2)$ are in quadrant III such that $\cos t_1 = -1/3$ and $\cos t_2 = -2/3$, find $\tan(t_1 - t_2)$, $\cos(t_1 - t_2)$, and the quadrant in which $P(t_1 - t_2)$ lies.

Verify the identities in Exercises 11–22.

11. $\sin(t + \pi/4) = (\sqrt{2}/2)(\sin t + \cos t)$.
12. $\cos(45° + \theta) = (\sqrt{2}/2)(\cos \theta - \sin \theta)$.
13. $\cos(x + 3\pi/2) = \sin x$.
14. $\tan(\pi/2 + t) = -\cot t$.
15. $\tan(\theta + \pi/4) = \dfrac{1 + \tan \theta}{1 - \tan \theta}$.
16. $\cot(t - \pi/3) = \dfrac{\sqrt{3}\tan t + 1}{\tan t - \sqrt{3}}$.
17. $\sin(u + v) \cdot \sin(u - v) = \sin^2 u - \sin^2 v$.
18. $\cos(u + v) \cdot \cos(u - v) = \cos^2 u - \sin^2 v$.
19. $\cos(u + v) + \cos(u - v) = 2 \cos u \cos v$.
20. $\sin(u + v) + \sin(u - v) = 2 \sin u \cos v$.
21. $\sin 2u = 2 \sin u \cos u$. [*Hint:* $2u = u + u$.]
22. $\cos 2u = \cos^2 u - \sin^2 u$.
23. Express $\sin(u + v + w)$ in terms of functions of u, v, and w. [*Hint:* Write $\sin(u + v + w) = \sin[(u + v) + w]$ and use (6.6) two times.]
24. Same as Exercise 23 for $\tan(u + v + w)$.

25. Use (6.8) and (5.7) to derive the formula

$$\cot(u + v) = \frac{\cot u \cot v - 1}{\cot u + \cot v}.$$

26. Verify the identity

$$\frac{\sin(u + v)}{\sin(u - v)} = \frac{\tan u + \tan v}{\tan u - \tan v}.$$

27. Prove (6.7) and (6.9).
28. Prove the following analogues of (6.4):
    (a) $\sec(\pi/2 - u) = \csc u$. (b) $\csc(\pi/2 - u) = \sec u$.
    (c) $\cot(\pi/2 - u) = \tan u$.

4 MULTIPLE ANGLE FORMULAS

In the previous section we were interested primarily in formulas which involved functional values of $u \pm v$. In this section we shall obtain formulas for values of $n \cdot u$, where n represents certain integers or rational numbers. The formulas are referred to as *multiple angle formulas.* Let us list some of these formulas for reference before proceeding to the proofs. As usual, the following equations are true for all values of u for which the expressions on both sides are meaningful:

(6.10)
$$\sin 2u = 2 \sin u \cos u,$$

(6.11)
$$\begin{aligned} \cos 2u &= \cos^2 u - \sin^2 u \\ &= 1 - 2\sin^2 u \\ &= 2\cos^2 u - 1, \end{aligned}$$

(6.12)
$$\tan 2u = \frac{2 \tan u}{1 - \tan^2 u}.$$

These formulas are easily proved by choosing $u = v$ in the appropriate addition formulas of Section 3. Thus, using (6.6), we have

$$\begin{aligned} \sin 2u &= \sin(u + u) \\ &= \sin u \cos u + \cos u \sin u \\ &= 2 \sin u \cos u. \end{aligned}$$

Similarly, from (6.3) we obtain

$$\cos 2u = \cos(u + u) = \cos^2 u - \sin^2 u.$$

The other forms in (6.11) may be obtained from the first by using the fundamental identity (5.10). Thus

$$\begin{aligned}\cos 2u &= \cos^2 u - \sin^2 u\\ &= (1 - \sin^2 u) - \sin^2 u\\ &= 1 - 2\sin^2 u.\end{aligned}$$

Similarly, if we substitute for $\sin^2 u$ instead of $\cos^2 u$, we obtain

$$\begin{aligned}\cos 2u &= \cos^2 u - (1 - \cos^2 u)\\ &= 2\cos^2 u - 1.\end{aligned}$$

The final formula (6.12) is obtained by taking $u = v$ in (6.8).

When u is an angle, the formulas in (6.10)–(6.12) are often called the *double angle formulas.*

EXAMPLE 1. Express $\cos 3\theta$ in terms of $\cos \theta$.

Solution:

$$\begin{aligned}\cos 3\theta &= \cos(2\theta + \theta)\\ &= \cos 2\theta \cos\theta - \sin 2\theta \sin\theta\\ &= (2\cos^2\theta - 1)\cos\theta - (2\sin\theta\cos\theta)\sin\theta\\ &= 2\cos^3\theta - \cos\theta - 2\sin^2\theta\cos\theta\\ &= 2\cos^3\theta - \cos\theta - 2(1 - \cos^2\theta)\cos\theta\\ &= 4\cos^3\theta - 3\cos\theta.\end{aligned}$$

If we solve the second and third equations of (6.11) for $\sin^2 u$ and $\cos^2 u$, respectively, the following identities are obtained:

$$\text{(6.13)}\qquad \sin^2 u = \frac{1 - \cos 2u}{2},\qquad \cos^2 u = \frac{1 + \cos 2u}{2}.$$

Since $\tan^2 u = \sin^2 u/\cos^2 u$, (6.13) leads to

$$\text{(6.14)}\qquad \tan^2 u = \frac{1 - \cos 2u}{1 + \cos 2u}.$$

Identities (6.13) and (6.14) may be used to reduce exponents as shown in the next example.

EXAMPLE 2. Express $\cos^4 t$ in terms of functional values with exponent 1.

Solution:

$$\begin{aligned}
\cos^4 t &= (\cos^2 t)^2 && \text{(Why?)} \\
&= \left(\frac{1 + \cos 2t}{2}\right)^2 && (6.13) \\
&= \tfrac{1}{4}(1 + 2\cos 2t + \cos^2 2t) && \text{(squaring)} \\
&= \tfrac{1}{4}\left(1 + 2\cos 2t + \frac{1 + \cos 4t}{2}\right) && (6.13) \\
&= \tfrac{3}{8} + \tfrac{1}{2}\cos 2t + \tfrac{1}{8}\cos 4t && \text{(simplifying).}
\end{aligned}$$

The following alternate forms for (6.13) and (6.14) are obtained by substituting $v/2$ for u:

$$\sin^2 \frac{v}{2} = \frac{1 - \cos v}{2},$$

(6.15) $$\cos^2 \frac{v}{2} = \frac{1 + \cos v}{2},$$

$$\tan^2 \frac{v}{2} = \frac{1 - \cos v}{1 + \cos v}.$$

If we take the square root of both sides of the equations in (6.15) and use the fact that $\sqrt{a^2} = |a|$ for all real numbers a, the following identities result:

$$\left|\sin \frac{v}{2}\right| = \sqrt{\frac{1 - \cos v}{2}},$$

(6.16) $$\left|\cos \frac{v}{2}\right| = \sqrt{\frac{1 + \cos v}{2}},$$

$$\left|\tan \frac{v}{2}\right| = \sqrt{\frac{1 - \cos v}{1 + \cos v}}.$$

The absolute value signs may be eliminated if more information is known about $v/2$. Thus if it is known that the point $P(v/2)$ assigned by the wrapping function is in either quadrant I or II (or, equivalently, if the angle determined by $v/2$ lies in one of these quadrants), then $\sin v/2$ is positive and we may write

$$\sin \frac{v}{2} = \sqrt{\frac{1 - \cos v}{2}}.$$

On the other hand, if $v/2$ leads to a point (or angle) in either quadrant III or IV, then we must write

$$\sin \frac{v}{2} = -\sqrt{\frac{1 - \cos v}{2}}.$$

Similar remarks are true for the other formulas. An alternate form for $\tan v/2$ can be obtained. From (6.16) we may write

$$\left|\tan\frac{v}{2}\right| = \sqrt{\frac{1-\cos v}{1+\cos v}\cdot\frac{1-\cos v}{1-\cos v}}$$

$$= \sqrt{\frac{(1-\cos v)^2}{\sin^2 v}}$$

$$= \frac{1-\cos v}{|\sin v|},$$

where the absolute value sign is unnecessary in the numerator since $1 - \cos v$ is never negative. It can be shown that $\tan v/2$ and $\sin v$ always have the same sign. For example, if $0 < v < \pi$, then $0 < v/2 < \pi/2$, and hence $\sin v$ and $\tan v/2$ are both positive. If $\pi < v < 2\pi$, then $\pi/2 < v/2 < \pi$, and hence $\sin v$ and $\tan v/2$ are both negative. It is possible to generalize these remarks to all values of v for which the expressions $\tan v/2$ and $(1 - \cos v)/|\sin v|$ have meaning. Because of this we may write

(6.17) $$\tan\frac{v}{2} = \frac{1-\cos v}{\sin v},$$

for all $v \in \mathbf{R}$ such that $\sin v \neq 0$.

If in (6.17) we multiply numerator and denominator on the right-hand side by $1 + \cos v$ and simplify, we obtain

(6.18) $$\tan\frac{v}{2} = \frac{\sin v}{1+\cos v}.$$

If v is an angle, formulas (6.15)–(6.18) are called the *half-angle formulas.*

EXAMPLE 3. Find the exact value of $\sin 22.5°$ and $\cos 22.5°$.

Solution: From (6.16) and the fact that $22.5°$ is in quadrant I, we have

$$\sin 22.5° = \sin\frac{45°}{2} = \sqrt{\frac{1-\cos 45°}{2}}$$

$$= \sqrt{\frac{1-\sqrt{2}/2}{2}}$$

$$= \frac{\sqrt{2-\sqrt{2}}}{2},$$

$$\cos 22.5° = \sqrt{\frac{1+\cos 45°}{2}}$$

$$= \frac{\sqrt{2+\sqrt{2}}}{2}.$$

EXAMPLE 4. If $\tan \alpha = -4/3$, where α is in quadrant IV, find $\tan \alpha/2$.

Solution: We see from Fig. 6.3 and (5.23) that $\sin \alpha = -4/5$ and

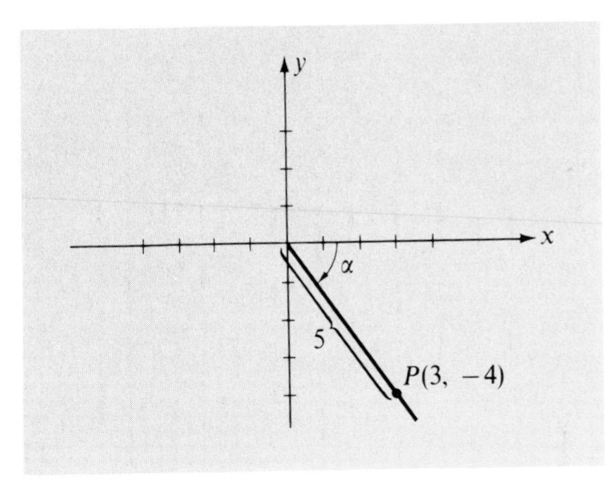

Figure 6.3

$\cos \alpha = 3/5$. Hence from (6.17) we have

$$\tan \frac{\alpha}{2} = \frac{1 - \cos \alpha}{\sin \alpha} = \frac{1 - 3/5}{-4/5} = -\frac{1}{2}$$

EXERCISES

In Exercises 1–4 find the exact values of $\sin 2\theta$, $\cos 2\theta$, and $\tan 2\theta$ subject to the given conditions.

1. $\sin \theta = 4/5$ and θ acute.
2. $\tan \theta = 1/3$ and $180° < \theta < 270°$.
3. $\sec \theta = -5/2$ and $90° < \theta < 180°$.
4. $\cos \theta = 8/17$ and $270° < \theta < 360°$.

In Exercises 5–8 find the exact values of $\sin \theta/2$, $\cos \theta/2$, and $\tan \theta/2$ subject to the given conditions.

5. $\csc \theta = 3/2$ and θ acute.
6. $\sin \theta = -2/5$ and $-90° < \theta < 0°$.
7. $\cot \theta = 1$ and $-180° < \theta < -90°$.
8. $\sec \theta = -4$ and $180° < \theta < 270°$.
9. Find the exact value of (a) $\cos 67°\,30'$, (b) $\sin 3\pi/8$.
10. Find the exact value of (a) $\tan 15°$, (b) $\tan 7°\,30'$.

Verify the identities given in Exercises 11–22.

11. $(\sin t + \cos t)^2 = 1 + \sin 2t$.

12. $\csc 2u = \frac{1}{2} \sec u \csc u$.
13. $\sin 3u = \sin u(3 - 4 \sin^2 u)$.
14. $\sin 4t = 4 \cos t \sin t(1 - 2 \sin^2 t)$.
15. $\cos 4\theta = 8 \cos^4 \theta - 8 \cos^2 \theta + 1$.
16. $\cos 6t = 32 \cos^6 t - 48 \cos^4 t + 18 \cos^2 t - 1$.
17. $\sin^4 t = \frac{3}{8} - \frac{1}{2} \cos 2t + \frac{1}{8} \cos 4t$.
18. $\cos^4 x - \sin^4 x = \cos 2x$.
19. $\sec 2\theta = \dfrac{\sec^2 \theta}{2 - \sec^2 \theta}$.
20. $\cot 2u = \dfrac{\cot^2 u - 1}{2 \cot u}$.
21. $\tan \frac{1}{2}\theta = \csc \theta - \cot \theta$.
22. $\cot^2 \frac{1}{2}u = \dfrac{\sec u + 1}{\sec u - 1}$.

In Exercises 23–30 find, in degree measure, all solutions of the given equations which are in the interval $[0°, 360°)$.

23. $\sin 2t + \sin t = 0$.
24. $\cos t - \sin 2t = 0$.
25. $\cos u + \cos 2u = 0$.
26. $\cos 2\theta - \tan \theta = 1$.
27. $\tan 2x = \tan x$.
28. $\tan 2t - 2 \cos t = 0$.
29. $\sin \frac{1}{2}u + \cos u = 1$.
30. $2 - \cos^2 x = 4 \sin^2 \frac{1}{2}x$.
31. If $a > 0$, $b > 0$, and $0 < u < \pi/2$ prove that

$$a \sin u + b \cos u = \sqrt{a^2 + b^2} \sin (u + v),$$

where $0 < v < \pi/2$, $\sin v = b/\sqrt{a^2 + b^2}$, and $\cos v = a/\sqrt{a^2 + b^2}$.
32. Use Exercise 31 to express $8 \sin u - 15 \cos u$ in the form $c \sin (u + v)$.

5 SUM AND PRODUCT FORMULAS

The addition formulas of Section 3 can be used to obtain several other useful identities. If the expressions on the left- and right-hand sides of (6.6) and (6.7) are added, the term involving $\cos u \sin v$ drops out and we obtain

$$\sin (u + v) + \sin (u - v) = 2 \sin u \cos v. \tag{6.19}$$

In like manner,

$$\sin (u + v) - \sin (u - v) = 2 \cos u \sin v. \tag{6.20}$$

Using (6.2) and (6.3) in similar fashion gives us

$$\cos (u + v) + \cos (u - v) = 2 \cos u \cos v, \tag{6.21}$$

$$\cos (u - v) - \cos (u + v) = 2 \sin u \sin v. \tag{6.22}$$

The above identities are sometimes called the *product formulas*. They can be used to express a product as a sum, as is shown in the following example.

EXAMPLE 1. Express $\sin 4\theta \cos 3\theta$ as a sum.

Solution: Using (6.19) with $u = 4\theta$ and $v = 3\theta$, we have

$$2 \sin 4\theta \cos 3\theta = \sin (4\theta + 3\theta) + \sin (4\theta - 3\theta)$$

or

$$\sin 4\theta \cos 3\theta = \tfrac{1}{2} \sin 7\theta + \tfrac{1}{2} \sin \theta.$$

An equivalent formula could have been found by using (6.20).

Identities (6.19)–(6.22) may also be used to express a sum as a product. In order to obtain a form which can be applied more easily, let us change the notation as follows. Let

$$u + v = a \quad \text{and} \quad u - v = b. \tag{6.23}$$

Consequently $(u + v) + (u - v) = a + b$, which simplifies to

$$u = (a + b)/2. \tag{6.24}$$

Similarly, from $(u + v) - (u - v) = a - b$ we obtain

$$v = (a - b)/2. \tag{6.25}$$

If we now substitute the expressions from (6.23)–(6.25) into the expressions given in (6.19)–(6.22), we obtain

$$\begin{aligned}
\sin a + \sin b &= 2 \cos \frac{a - b}{2} \sin \frac{a + b}{2}, \\
\sin a - \sin b &= 2 \cos \frac{a + b}{2} \sin \frac{a - b}{2}, \\
\cos a + \cos b &= 2 \cos \frac{a + b}{2} \cos \frac{a - b}{2}, \\
\cos b - \cos a &= 2 \sin \frac{a + b}{2} \sin \frac{a - b}{2}.
\end{aligned} \tag{6.26}$$

The identities (6.26) are sometimes referred to as the *sum* or *factoring formulas*.

EXAMPLE 2. Express $\sin 5x - \sin 3x$ as a product.

Solution: Using the second identity in (6.26) with $a = 5x$ and $b = 3x$, we have

$$\begin{aligned}
\sin 5x - \sin 3x &= 2 \cos \frac{5x + 3x}{2} \sin \frac{5x - 3x}{2} \\
&= 2 \cos 4x \sin x.
\end{aligned}$$

EXAMPLE 3. Verify the identity

$$\frac{\sin 3t + \sin 5t}{\cos 3t - \cos 5t} = \cot t.$$

Solution: Using the first and fourth identities given in (6.26), we may write

$$\frac{\sin 3t + \sin 5t}{\cos 3t - \cos 5t} = \frac{2 \cos \dfrac{3t - 5t}{2} \sin \dfrac{3t + 5t}{2}}{2 \sin \dfrac{5t + 3t}{2} \sin \dfrac{5t - 3t}{2}}$$

$$= \frac{2 \cos(-t) \sin 4t}{2 \sin 4t \sin t}$$

$$= \frac{\cos(-t)}{\sin t}$$

$$= \cot t,$$

where the last step follows from (5.13) and (5.9).

EXAMPLE 4. Find the solution set of the equation

$$\cos t - \sin 2t - \cos 3t = 0.$$

Solution: From (6.26) we can write

$$\cos t - \cos 3t = 2 \sin \frac{3t + t}{2} \sin \frac{3t - t}{2}$$

$$= 2 \sin 2t \sin t.$$

Hence the given equation is equivalent to

$$2 \sin 2t \sin t - \sin 2t = 0,$$

or

$$\sin 2t(2 \sin t - 1) = 0.$$

Therefore the solution set is the union of the solutions sets of

$$\sin 2t = 0 \quad \text{and} \quad \sin t = 1/2.$$

The first equation gives us $2t = n\pi$, or $t = n \cdot (\pi/2)$, where $n \in \mathbf{Z}$. The second equation gives us the solutions $t = \pi/6 + 2n\pi$ and $t = 5\pi/6 + 2n\pi$, where $n \in \mathbf{Z}$.

The addition formulas may also be employed to derive the so-called *reduction formulas*. The latter formulas can be used to write expressions such as

$$\sin\left(\theta + n \cdot \frac{\pi}{2}\right) \quad \text{and} \quad \cos\left(\theta + n \cdot \frac{\pi}{2}\right),$$

where n is any integer, in terms of only $\sin \theta$ or $\cos \theta$. Similar formulas are true for the other trigonometric functions. For example, (6.4) illustrates the case in which $n = 1$ and $\theta = -u$. We shall not derive general reduction formulas, but merely illustrate, by means of examples, several special cases.

EXAMPLE 5. Express $\sin(\theta - 3\pi/2)$ and $\cos(\theta + \pi)$ in terms of a value of θ.

Solution: From (6.7) and (6.3)

$$\begin{aligned}\sin(\theta - 3\pi/2) &= \sin\theta\cos 3\pi/2 - \cos\theta\sin 3\pi/2\\ &= \sin\theta\cdot 0 - \cos\theta\cdot(-1)\\ &= \cos\theta,\end{aligned}$$

$$\begin{aligned}\cos(\theta + \pi) &= \cos\theta\cos\pi - \sin\theta\sin\pi\\ &= \cos\theta\cdot(-1) - \sin\theta\cdot 0\\ &= -\cos\theta.\end{aligned}$$

EXERCISES

In Exercises 1–6 express each product as a sum or difference.

1. $2\sin 5x \sin 2x$.
2. $2\cos 4\theta \sin 6\theta$.
3. $\cos 3t \sin 2t$.
4. $\sin(-3u)\cos 4u$.
5. $3\cos(-5u)\cos(-9u)$.
6. $\cos 3x \cos 5x$.

In Exercises 7–12 write each expression as a product.

7. $\sin 6u + \sin 4u$.
8. $\sin 3x - \sin 5x$.
9. $\cos 5t - \cos 7t$.
10. $\cos 5\theta + \cos 6\theta$.
11. $\sin 2t - \sin 4t$.
12. $\cos 3t - \cos t$.

Verify the identities in Exercises 13–18.

13. $\dfrac{\sin 4t + \sin 6t}{\cos 4t - \cos 6t} = \cot t$.
14. $\dfrac{\sin\theta + \sin 3\theta}{\cos\theta + \cos 3\theta} = \tan 2\theta$.
15. $\dfrac{\sin u + \sin v}{\cos u + \cos v} = \tan\dfrac{u + v}{2}$.
16. $\dfrac{\sin u - \sin v}{\cos u - \cos v} = -\cot\dfrac{u + v}{2}$.
17. $\sin 2x + \sin 4x + \sin 6x = 4\cos x \cos 2x \sin 3x$.
18. $\dfrac{\cos t + \cos 4t + \cos 7t}{\sin t + \sin 4t + \sin 7t} = \cot 4t$.

In Exercises 19 and 20 find the solution sets of the given equations.

19. $\sin 5t + \sin 3t = 0.$

20. $\sin t + \sin 3t = \sin 2t.$

In Exercises 21–26 verify the given reduction formulas by using addition formulas.

21. $\sin(\theta + \pi) = -\sin\theta.$

22. $\sin\left(\theta + \dfrac{3\pi}{2}\right) = -\cos\theta.$

23. $\cos\left(\theta - \dfrac{5\pi}{2}\right) = \sin\theta.$

24. $\cos(\theta - 3\pi) = -\cos\theta.$

25. $\tan(\pi - \theta) = -\tan\theta.$ [*Hint:* $\tan(\pi - \theta) = \sin(\pi - \theta)/\cos(\pi - \theta)$.]

26. $\tan\left(\theta + \dfrac{\pi}{2}\right) = -\cot\theta.$

6 SUMMARY OF FORMULAS

In order to make the principal trigonometric identities and formulas which we have developed readily available for reference, we list them below.

Fundamental Identities

$$\csc u = \frac{1}{\sin u}, \qquad \sec u = \frac{1}{\cos u}, \qquad \cot u = \frac{1}{\tan u},$$

$$\tan u = \frac{\sin u}{\cos u}, \qquad \cot u = \frac{\cos u}{\sin u},$$

$$\sin^2 u + \cos^2 u = 1,$$

$$\tan^2 u + 1 = \sec^2 u,$$

$$\cot^2 u + 1 = \csc^2 u,$$

$$\sin(-u) = -\sin u, \quad \cos(-u) = \cos u, \quad \tan(-u) = -\tan u.$$

Addition Formulas

$$\sin(u \pm v) = \sin u \cos v \pm \cos u \sin v,$$

$$\cos(u \pm v) = \cos u \cos v \mp \sin u \sin v,$$

$$\tan(u \pm v) = \frac{\tan u \pm \tan v}{1 \mp \tan u \tan v}.$$

Double Angle Formulas

$$\sin 2u = 2 \sin u \cos u,$$

$$\cos 2u = \cos^2 u - \sin^2 u = 1 - 2\sin^2 u = 2\cos^2 u - 1,$$

$$\tan 2u = \frac{2 \tan u}{1 - \tan^2 u}.$$

Half-Angle Formulas

$$\left|\sin \frac{u}{2}\right| = \sqrt{\frac{1 - \cos u}{2}},$$

$$\left|\cos \frac{u}{2}\right| = \sqrt{\frac{1 + \cos u}{2}},$$

$$\tan \frac{u}{2} = \frac{1 - \cos u}{\sin u} = \frac{\sin u}{1 + \cos u}$$

Product Formulas

$$2 \sin u \cos v = \sin(u + v) + \sin(u - v),$$

$$2 \cos u \sin v = \sin(u + v) - \sin(u - v),$$

$$2 \cos u \cos v = \cos(u + v) + \cos(u - v),$$

$$2 \sin u \sin v = \cos(u - v) - \cos(u + v).$$

Factoring Formulas

$$\sin u \pm \sin v = 2 \cos \frac{u \mp v}{2} \sin \frac{u \pm v}{2},$$

$$\cos u + \cos v = 2 \cos \frac{u + v}{2} \cos \frac{u - v}{2},$$

$$\cos u - \cos v = 2 \sin \frac{v + u}{2} \sin \frac{v - u}{2}.$$

7 TRIGONOMETRIC GRAPHS

In this section we consider graphs of functions f that are defined by expressions of the form

$$f(x) = a \sin(bx + c), \tag{6.27}$$

where $a, b, c \in \mathbf{R}$, or by similar expressions that involve the other trigonometric functions. Instead of using a t, y coordinate system as in Chapter Five, we shall now use the conventional x, y coordinate system. To sketch the graph of (6.27), one could begin by plotting many points. However, it is generally easier to use what is known about graphs of the trigonometric functions. Let us consider the special case of (6.27) in which $c = 0$, $b = 1$, and $a > 0$. Thus we wish to find the graph of the equation $y = a \sin x$. Since $y = a \sin x$, to find the ordinate of a point with abscissa x_1 we may multiply by a the ordinate of the point on the graph of $y = \sin x$ which has abscissa x_1. Thus if $y = 2 \sin x$, we multiply by 2; if $y = \frac{1}{2} \sin x$, we multiply by $\frac{1}{2}$; and so on. The graph of $y = 2 \sin x$ is shown in Fig. 5.19 and the graph of $y = \frac{1}{2} \sin x$ is sketched in Fig. 6.4, where for comparison we have indicated the graph of $y = \sin x$ with dashes.

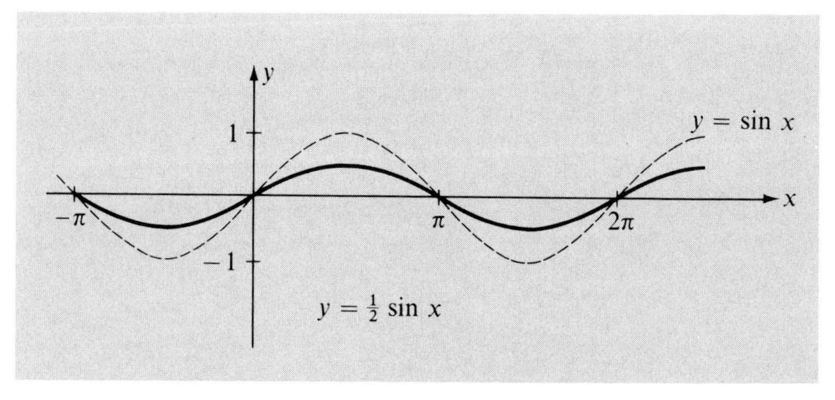

Figure 6.4

If $a < 0$, then the ordinates of points on the graph of $y = a \sin x$ are negatives of the corresponding ordinates of points on the graph of $y = |a| \sin x$. This is illustrated by the graph of the equation $y = -2 \sin x$ sketched in Fig. 6.5, which may be thought of as the *reflection*, through the x-axis, of the graph of $y = 2 \sin x$.

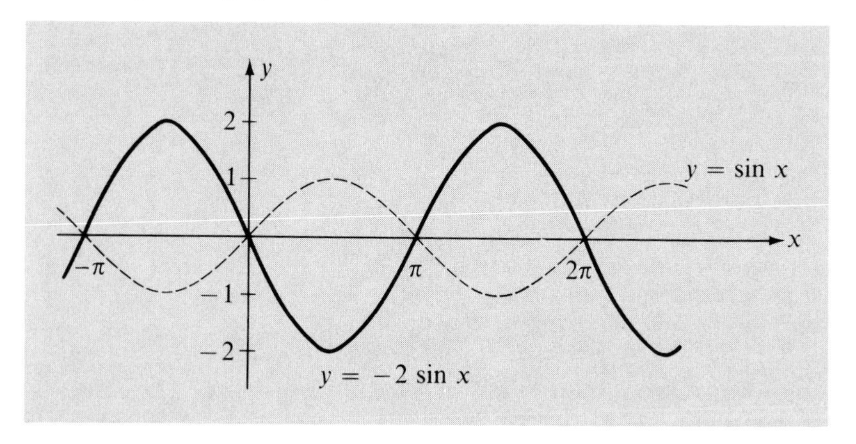

Figure 6.5

The *amplitude* of the function f defined by (6.27) is the maximum ordinate of points on the graph. If $a > 0$, the maximum ordinate occurs when $\sin(bx + c) = 1$, whereas if $a < 0$, we must take $\sin(bx + c) = -1$. In either case the amplitude is $|a|$. For example, if $f(x) = \frac{1}{2}\sin x$, the amplitude is $\frac{1}{2}$; if $f(x) = -3\sin x$, the amplitude is 3; and so on. We shall also speak of the amplitude of a *graph* in the obvious way. More generally, for any periodic function, the amplitude is defined as $|M - m|/2$, where M is the maximum value and m is the minimum value of the function, provided they exist. Thus the amplitude of $y = a\sin x$ is $|a - (-a)|/2 = |2a|/2 = |a|$.

Next, let us consider the graph of a function f defined by $f(x) = \sin bx$, $b > 0$. As bx ranges from 0 to 2π we obtain one complete sine wave of amplitude 1. If b is fixed, then bx ranges from 0 to 2π if and only if x ranges from 0 to $2\pi/b$. This implies that the period of f is $2\pi/b$. To illustrate, if $f(x) = \sin 2x$, then the period of f is $2\pi/2$, or π, and there is one sine wave in the interval $[0, \pi]$. On the other hand, if $f(x) = \sin(\frac{1}{2})x$, then the period is $2\pi/(\frac{1}{2})$, or 4π. This can also be seen by plotting points. In Fig. 6.6 the graphs of the two functions just mentioned are

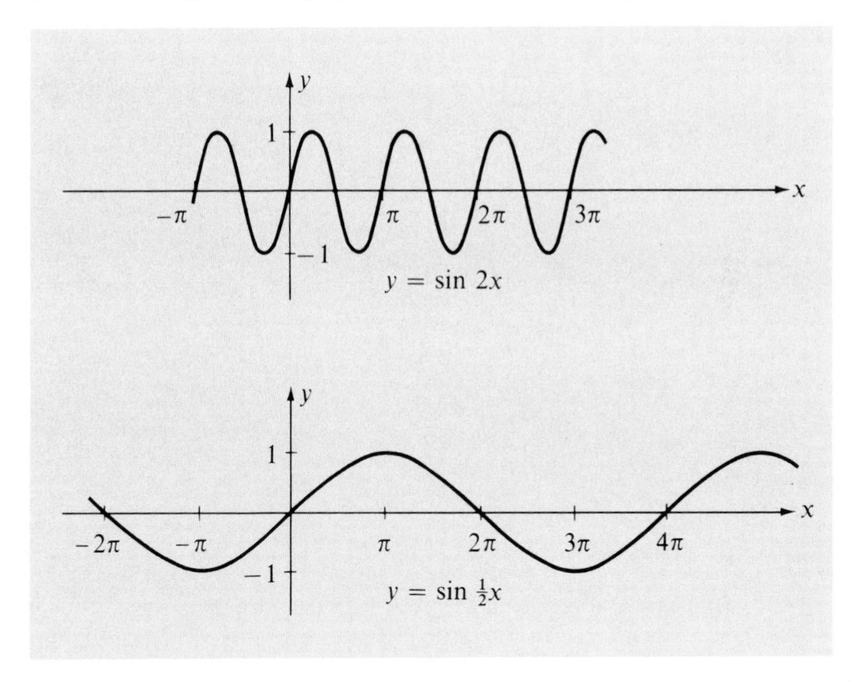

Figure 6.6

sketched, where for clarity different scales on the x- and y-axes have been used. Note in general that if b is large, then $2\pi/b$ is small and the waves are close together — indeed, there are b sine waves in an interval of 2π units. On the other hand, if b is small, $2\pi/b$ is large and the waves are shallow. For example, if $y = \sin(\frac{1}{10})x$, an interval 20π units long is required for one complete wave.

If $b < 0$, we can use the fact that $\sin(-u) = -\sin u$ to obtain the graph. To illustrate, the graph of $y = \sin(-2x)$ is the same as the graph of $y = -\sin 2x$.

By combining the above discussions we can arrive at a technique for sketching the graph of a function f defined by $f(x) = a \sin bx$. The graph has the basic sine wave pattern. However, the amplitude is $|a|$ and the period is $2\pi/|b|$. If $a < 0$ or $b < 0$, we make adjustments on the signs of ordinates, as discussed earlier.

EXAMPLE 1. Sketch the graph of f if $f(x) = 2 \sin(-3x)$.

Solution: We may write $f(x) = -2 \sin 3x$. Thus the amplitude is 2 and the period is $2\pi/3$. The minus sign indicates a reflection through the x-axis. If we mark off an interval of length $2\pi/3$ and sketch a sine wave of amplitude 2 (reflected through the x-axis), the shape of the graph is apparent. The configuration given in the interval $[0, 2\pi/3]$ is carried along periodically to obtain the graph (see Fig. 6.7).

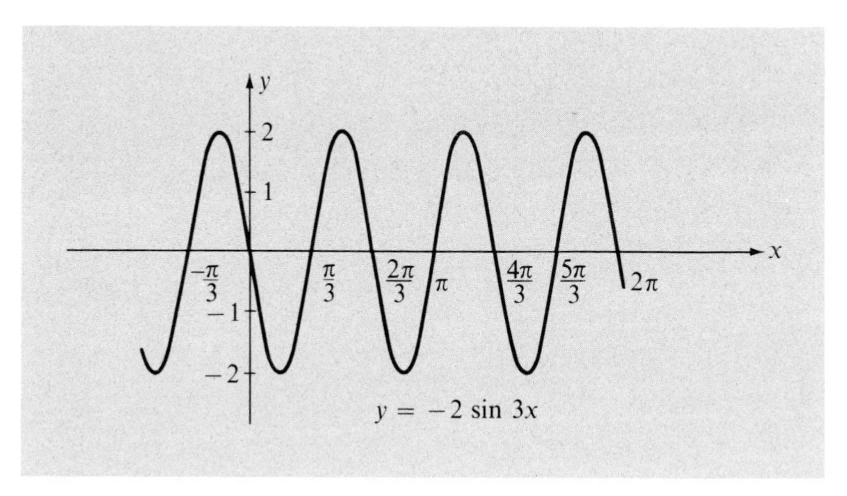

Figure 6.7

Similar discussions can be given if f is defined by $f(x) = a \cos bx$ or by $f(x) = a \tan bx$. In the latter case the period is π/b since the tangent function has period π. Since there is no maximum or minimum ordinate for points on the graph of the tangent function, we do not refer to its amplitude. However, we may still use the process of multiplying tangent ordinates by a in order to obtain points on the graph of $y = a \tan bx$.

Finally, consider the graph of a function f defined by (6.27) — that is,

$$f(x) = a \sin(bx + c).$$

One complete sine wave of amplitude a is obtained as $bx + c$ ranges from 0 to 2π — that is, as bx ranges from $-c$ to $2\pi - c$. In turn, the latter variation is obtained by taking the domain of x from $-c/b$ to $(2\pi - c)/b$. If $-c/b > 0$, this amounts graphically to *shifting* the graph of $y = a \sin bx$ to the right $-c/b$ units. If $-c/b < 0$, the shift is to the left. The number $-c/b$ is sometimes called the *phase shift* associated with the function. Needless to say, similar remarks can be made for the other functions.

EXAMPLE 2. Sketch the graph of the function f defined by $f(x) = 3 \sin (2x - \pi/2)$.

Solution: Although we could use the above discussion, let us work directly so as to avoid memorization. In order to obtain an interval containing one sine wave, we let $2x - \pi/2$ range from 0 to 2π. Since $2x - \pi/2 = 0$ when $x = \pi/4$ and $2x - \pi/2 = 2\pi$ when $x = 5\pi/4$, the desired interval is $[\pi/4, 5\pi/4]$. The amplitude is 3. The graph is sketched in Fig. 6.8. Note that the phase shift is $\pi/4$ and the period is $5\pi/4 - \pi/4 = \pi$.

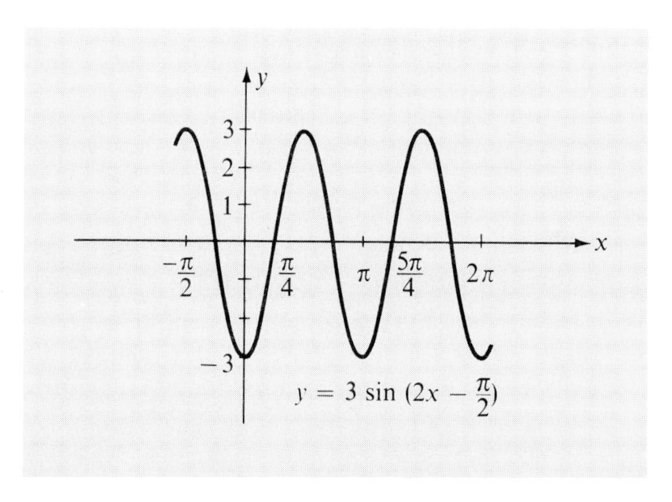

Figure 6.8

EXERCISES

1. Without plotting many points, sketch the graph of each function f defined as follows, determining the amplitude and period in each case:

   (a) $f(x) = 3 \sin x$. (b) $f(x) = \sin 3x$.
   (c) $f(x) = \frac{1}{3} \sin x$. (d) $f(x) = \sin (x/3)$.
   (e) $f(x) = 2 \sin (x/3)$. (f) $f(x) = \frac{1}{3} \sin (-2x)$.

2. Sketch the graphs of the functions involving the cosine which are analogous to those defined in Exercise 1.

In Exercises 3–14 sketch the graphs of the given equations.

3. $y = \frac{1}{2} \sec 3x$.
4. $y = -2 \csc \frac{1}{2}x$.
5. $y = \tan(-x/2)$.
6. $y = \frac{1}{3} \cot 2x$.
7. $y = 2 \cos \frac{1}{2}\pi x$.
8. $y = 4 \sin \pi x$.
9. $y = 2 \cot \frac{1}{4}\pi x$.
10. $y = \frac{1}{2} \tan \frac{1}{2}\pi x$.
11. $y = 3 \cos(2x + \pi/4)$.
12. $y = 2 \sin(3x - \pi/2)$.
13. $y = -4 \sin \frac{1}{2}(3x - 4)$.
14. $y = \frac{1}{2} \cos \frac{1}{2}(x + 1)$.

8 GRAPHS AND THEIR APPLICATIONS

In mathematical applications it is common to encounter functions that are defined in terms of sums and products of expressions, some of which are trigonometric. For example, a function f might be defined by expressions of the form

$$f(x) = \sin 2x + \cos x, \qquad (6.28)$$
$$f(x) = 2^{-x} \sin x,$$

and so on. When working with a sum of two expressions such as given in the first equation of (6.28) it is convenient to use the method of addition of ordinates discussed on p. 41. The next two examples are illustrations of this graphical technique.

EXAMPLE 1. Use the method of addition of ordinates to sketch the graph of the function f defined by $f(x) = \cos x + \sin x$.

Solution: We begin by sketching, with dashes, the graphs of the equations $y = \cos x$ and $y = \sin x$. Then, for various numbers x_1, we add ordinates as indicated by the sketch in Fig. 6.9. After a sufficient number of ordinates are added and a pattern emerges, we draw a smooth curve through

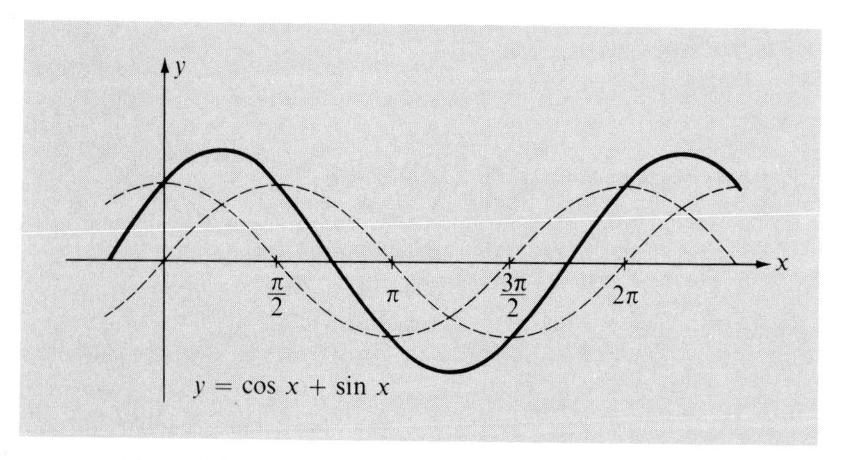

Figure 6.9

the points. As a check, it would be worthwhile to obtain some points on the graph by substituting numbers for x. We shall leave such verifications to the reader. It can be seen from the graph that the function f is periodic with period 2π.

EXAMPLE 2. Sketch the graph of the equation $y = \cos x + \sin 2x$.

Solution: We sketch, with dashes, the graphs of the equations $y = \cos x$ and $y = \sin 2x$ on the same coordinate axes and use the method of addition of ordinates. The graph is shown in Fig. 6.10. Evidently f is periodic with period 2π.

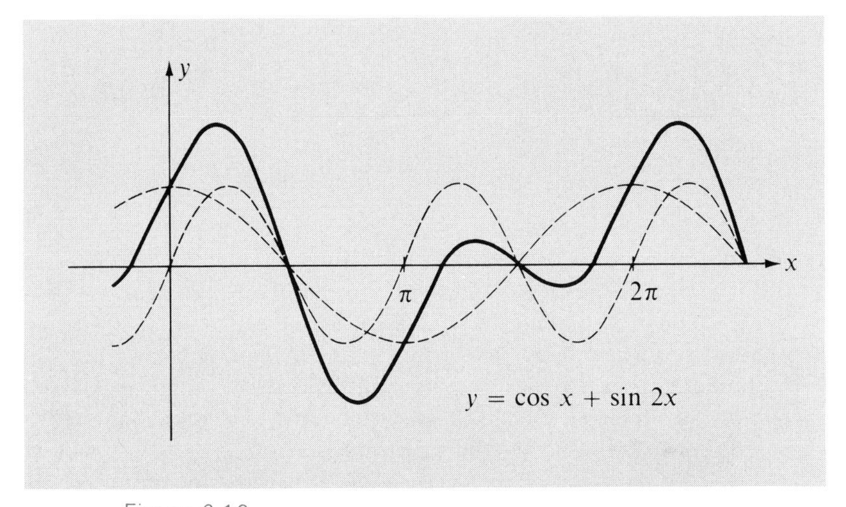

Figure 6.10

Another interesting function f is that defined by the second equation of (6.28). Since $|f(x)| = |2^{-x}|\,|\sin x|$ and since $|\sin x| \leq 1$ and $2^{-x} > 0$, we have $|f(x)| \leq 2^{-x}$ for all $x \in \mathbf{R}$. Consequently, by (1.9),

$$-2^{-x} \leq f(x) \leq 2^{-x}$$

for all $x \in \mathbf{R}$, which implies that the graph of f lies between the graphs of the equations $y = -2^{-x}$ and $y = 2^{-x}$. The graph of f will coincide with one of the latter graphs when $|\sin x| = 1$ — that is, when $x = \pi/2 + n\pi$, where $n \in \mathbf{Z}$. Since $2^{-x} > 0$, the x-intercepts on the graph of f occur at $\sin x = 0$ — that is, at $x = n\pi$, $n \in \mathbf{Z}$. With this information we obtain the sketch shown in Fig. 6.11.

The graph of $y = 2^{-x} \sin x$ is called a *damped sine wave* and 2^{-x} is called the *damping factor*. By using different damping factors, we may obtain other variations of such "compressed" or "expanded" sine waves. The analysis of such graphs is important in electrical theory.

The graphical analysis of functions that are defined in terms of the trigonometric functions is important in the investigation of a great number

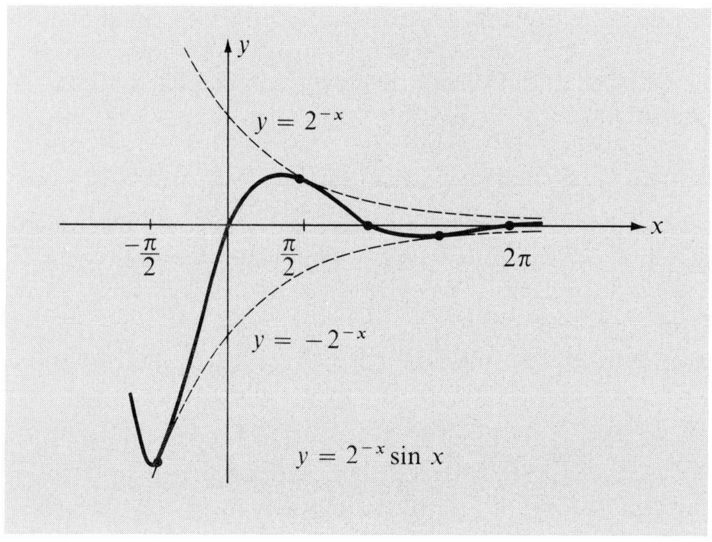

Figure 6.11

of physical phenomena related to oscillatory motion. An illustration of the type of motion we have in mind is given by a particle in a vibrating guitar string or in a spring that has been compressed or elongated and then released to vibrate back and forth. Another illustration is given by the motion of a particle of air brought about by certain sound waves. A large part of the theory needed for studying radio and television signals is based on functions of the type we have described.

The fundamental type of particle displacement inherent in the above illustrations is termed *harmonic motion*. Harmonic motion can be illustrated very simply by considering a point P which moves at a constant rate of speed along the circumference of a circle of radius a with center at the origin O of a rectangular coordinate system. Suppose the initial position of P is $A(a, 0)$ and let θ be the angle generated by the ray OP at the end of t units of time (see Fig. 6.12). The *angular velocity* ω of the ray OP is, by definition, the rate at which the measure of angle AOP changes per unit time. It follows that to state that P moves along the circle at a constant rate is equivalent to stating that the angular velocity is constant. If ω is constant, then we may write $\theta = \omega t$. To illustrate, if $\omega = \pi/6$ radians per second, then when $t = 1$ second, $\theta = \pi/6$ and P is one-third of the way from $A(a, 0)$ to $B(0, a)$. When $t = 2$ seconds, $\theta = 2\pi/6$ and P is two-thirds of the way from $A(a, 0)$ to $B(0, a)$. When $t = 6$ seconds, $\theta = \pi$ and P is at $A'(-a, 0)$, and so on. If the coordinates of P are (x, y), then from (5.23) and the fact that $\theta = \omega t$, we have

$$(6.29)\qquad \begin{aligned} x &= a \cos \omega t, \\ y &= a \sin \omega t. \end{aligned}$$

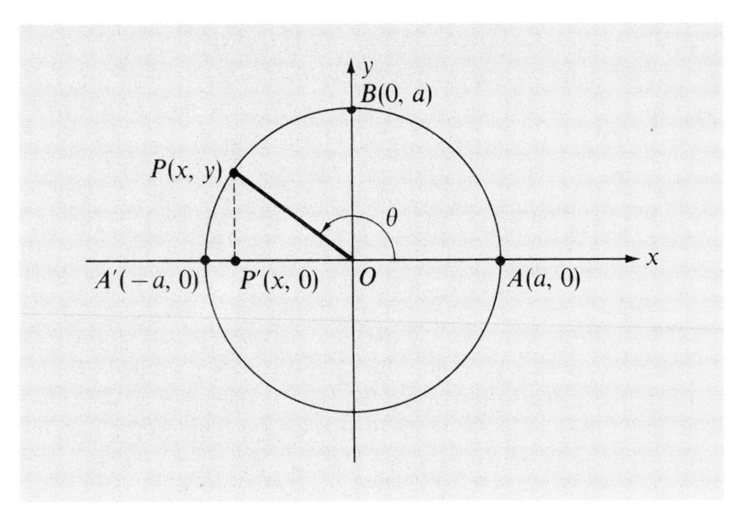

Figure 6.12

As in Fig. 6.12, consider the point $P'(x, 0)$, which is the *projection* of P on the x-axis — that is, the point of intersection of the x-axis with a vertical line through P. The position of P' on the x-axis at time t may be determined from the first equation of (6.29). As P moves around the circle, P' oscillates between A and A'. The motion of P' is given a special name according to the following definition.

(6.30) Definition of Simple Harmonic Motion

If a point moves on a real axis such that its distance d from the origin at time t is given by either $d = a \cos \omega t$ or $d = a \sin \omega t$, where a and ω are real numbers, then the motion of the point is termed *simple harmonic.*

If a point moves according to (6.30), the maximum displacement a of the point from the origin is called the *amplitude of the motion.* The *period* is the time t required for one complete oscillation and can be found by writing $\omega t = 2\pi$, whence $t = 2\pi/\omega$. The *frequency* of the motion is $1/(2\pi/\omega)$, or $\omega/2\pi$, which gives the number of oscillations per unit time.

Let us briefly analyze the motion of P' as given by the first equation of (6.29) in the special case where $a = 1$, $\omega = \pi/6$ radians per second, and t is in seconds. The table below shows the directed distance x from the origin to P' at various times t. Note that in the first second, P' travels

t	0	1	2	3	4	5	6
ωt	0	$\pi/6$	$\pi/3$	$\pi/2$	$2\pi/3$	$5\pi/6$	π
x	1	$\frac{\sqrt{3}}{2} \doteq 0.87$	0.5	0	-0.5	$-\frac{\sqrt{3}}{2} \doteq -0.87$	-1

approximately 0.13 units, in the next second about 0.37 units, in the third second 0.5 units, in the fourth 0.5 units, in the fifth approximately 0.37 units, and so on. We see that P' starts at $A(a, 0)$, moves to the left, gaining speed as it approaches the origin O, after which its speed decreases until it arrives at $A'(-a, 0)$. Then the direction of motion is reversed and it gains speed until it reaches O, after which the speed decreases until it reaches A. Then the direction is again reversed and the pattern is repeated. Note that the period is $2\pi/(\pi/6) = 12$ seconds and the frequency, or number of oscillations per second, is $1/12$.

There are many other illustrations; however, the one we have given is typical of the behavior of a particle which is in simple harmonic motion. Incidentally, if the initial angle AOP is ϕ when $t = 0$, then the displacement of P' is given by $x = a \cos(\omega t + \phi)$. This is precisely the type of variation considered in Section 7.

EXERCISES

Sketch the graphs of the equations in Exercises 1–8 by the method of addition of ordinates.

1. $y = \cos x + 2 \sin x$.
2. $y = \sin x + 2 \cos x$.
3. $y = \sin x + \cos 2x$.
4. $y = \cos x - \sin x$.
5. $y = 3 \sin x - \frac{1}{2} \cos 2x$.
6. $y = 2 \sin x + \sin \frac{1}{2}x$.
7. $y = x + \sin x$.
8. $y = \dfrac{|x + 1|}{2} + \cos 2x$.

In Exercises 9–12 sketch the graph of the function f defined by the given expression.

9. $f(x) = 2^x \sin x$.
10. $f(x) = 2^{-x} \cos x$.
11. $f(x) = x \cos x$.
12. $f(x) = (1 + x^2) \sin x$.
13. A point moves harmonically on a real axis according to the equation $y = 8 \sin 4\pi t$, where t is in seconds. Determine the amplitude, period, frequency, and describe the motion of the point.
14. Same as Exercise 13 if $x = \frac{1}{2} \cos \pi t/6$.

9 THE INVERSE TRIGONOMETRIC FUNCTIONS

Inverse functions were discussed in Section 8 of Chapter One. At that time we pointed out that if f is a one-to-one function with domain X and range Y, then for each element y in Y there is one and only one element x in X such that $f(x) = y$. In this case we can define a function g from Y to X by letting $g(y) = x$. The function g is called the *inverse function* of

f and is often denoted by f^{-1}. In words, f^{-1} *reverses* the correspondence from X to Y given by f— that is

$$f^{-1}(u) = v \quad \text{if and only if} \quad f(v) = u, \tag{.31}$$

for all $u \in Y$ and $v \in X$.

As on p. 46, the inverse function f^{-1} of f is characterized by the equations

$$f^{-1}(f(v)) = v \quad \text{for all } v \in X,$$

(.32) and

$$f(f^{-1}(u)) = u \quad \text{for all } u \in Y.$$

Since the trigonometric functions are not one-to-one, they do not have inverses. However, by restricting the domain it is possible to obtain functions that behave in the same way as the trigonometric functions (over the smaller domain) and which do possess inverse functions.

To begin with, consider only the sine function, whose domain X is all of **R** and range Y is the set of real numbers in the interval $[-1, 1]$. This function is not one-to-one, since, for example, numbers such as $\pi/6$, $5\pi/6$, and $-7\pi/6$ have the same image $\frac{1}{2}$ in Y. It is easy to find a subset X' of **R** with the property that as x ranges through X', $\sin x$ takes on each value between -1 and 1 once and only once. For convenience we shall choose, for X', the interval $[-\pi/2, \pi/2]$. The new function obtained by restricting the domain of the sine function to $[-\pi/2, \pi/2]$ will be referred to as the Sine (pronounced cap-sine) function and its values will be denoted by $\text{Sin } x$, where $x \in X'$. Thus we have

$$\text{Sin } x = \sin x, \quad -\pi/2 \le x \le \pi/2. \tag{.33}$$

Since the Sine function coincides with the sine function on $[-\pi/2, \pi/2]$, their graphs are identical on this interval. If $|x| > \pi/2$, then $\text{Sin } x$ is undefined. Therefore Fig. 6.13 is a sketch of the complete graph of

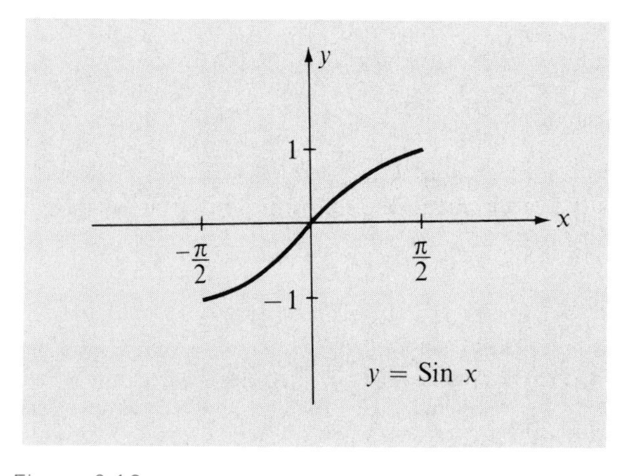

Figure 6.13

$y = \text{Sin } x$. The Sine function, being one-to-one, has an inverse function. Using (6.31), we arrive at the following definition.

(6.34) Definition of the Inverse Sine Function

The *inverse Sine function*, denoted by Sin^{-1}, is defined by

$$\text{Sin}^{-1} u = v \quad \text{if and only if} \quad \text{Sin } v = u,$$

where $-1 \leq u \leq 1$ and $-\pi/2 \leq v \leq \pi/2$.

It is also customary to refer to this function as the *Arcsine function* and use the notation Arcsin u in place of $\text{Sin}^{-1} u$. The expression Arcsin u is used because if $v = \text{Arcsin } u$, then $\text{Sin } v = u$—that is, v is a number (or an *arc*length) whose Sine is u. Since both notations Sin^{-1} and Arcsin are commonly used in mathematics and its applications, we shall employ both of them in our work. Note that by (6.34) we have $-\pi/2 \leq \text{Sin}^{-1} u \leq \pi/2$, or equivalently, $-\pi/2 \leq \text{Arcsin } u \leq \pi/2$.

The graph of $v = \text{Sin}^{-1} u$ (or $v = \text{Arcsin } u$) may be found by graphing $u = \text{Sin } v$. If the u-axis is taken horizontally, this gives us the sketch shown in Fig. 6.14.

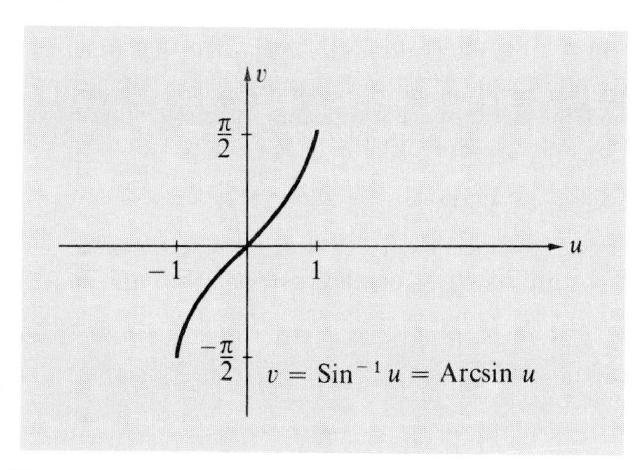

Figure 6.14

Using (6.32) we obtain the following important identities:

$$\text{(6.35)} \quad \begin{aligned} \text{Sin}^{-1} (\text{Sin } v) &= v \\ \text{Sin } (\text{Sin}^{-1} u) &= u \end{aligned}$$

for $-1 \leq u \leq 1$ and $-\pi/2 \leq v \leq \pi/2$. Of course, (6.35) may also be written in the form $\text{Arcsin } (\text{Sin } v) = v$ and $\text{Sin } (\text{Arcsin } u) = u$.

EXAMPLE 1. (a) Find $\text{Sin}^{-1} (\sqrt{2}/2)$ and $\text{Arcsin } (-1/2)$.
(b) Find $\text{Sin}^{-1} (\tan 3\pi/4)$.

Solution: (a) If $v = \text{Sin}^{-1}(\sqrt{2}/2)$, then $\text{Sin } v = \sqrt{2}/2$ and consequently $v = \pi/4$. Note that it is essential to choose v in the interval $[-\pi/2, \pi/2]$. Thus a number such as $3\pi/4$ is incorrect, even though $\sin(3\pi/4) = \sqrt{2}/2$. In like manner, if $v = \text{Arcsin}(-1/2)$, then $\text{Sin } v = -1/2$ and hence $v = -\pi/6$.

(b) If we let $v = \text{Sin}^{-1}(\tan 3\pi/4) = \text{Sin}^{-1}(-1)$, then $\text{Sin } v = -1$ and consequently $v = -\pi/2$.

Similar discussions may be given for the other trigonometric functions. The idea always is to determine a convenient subset of the domain so that a one-to-one correspondence is obtained. The new function will be distinguished from the old by using a capital first letter, and the notation for the inverse function will be similar to that used for the inverse Sine function. We shall illustrate the technique with the cosine and tangent functions.

The Cosine function, denoted by Cos, is defined by

$$\text{Cos } x = \cos x, \quad 0 \le x \le \pi.$$

Note that the Cosine function determines a one-to-one correspondence from the interval $[0, \pi]$ to the interval $[-1, 1]$.

(6.36) Definition of the Inverse Cosine Function

The *inverse Cosine function*, denoted by Cos^{-1}, is defined by

$$\text{Cos}^{-1} u = v \quad \text{if and only if} \quad \text{Cos } v = u,$$

where $-1 \le u \le 1$ and $0 \le v \le \pi$.

We shall also refer to the inverse Cosine function as the *Arccosine function* and write $\text{Arccos } u$ in place of $\text{Cos}^{-1} u$.

From (6.36) we see that if $-1 \le u \le 1$, then $0 \le \text{Cos}^{-1} u \le \pi$, or equivalently, $0 \le \text{Arccos } u \le \pi$. If $0 \le v \le \pi$, then we may write

$$\text{Cos}(\text{Cos}^{-1} u) = \text{Cos}(\text{Arccos } u) = u,$$

and

$$\text{Cos}^{-1}(\text{Cos } v) = \text{Arccos}(\text{Cos } v) = v.$$

The graphs of the Arccosine and Cosine functions are sketched in Fig. 6.15.

EXAMPLE 2. Approximate $\text{Cos}^{-1}(-0.7951)$.

Solution: If $v = \text{Cos}^{-1}(-0.7951)$, then $\text{Cos } v = -0.7951$. If v' is the reference number for v, then $\text{Cos } v' = 0.7951$ and, from Table 3, $v' \doteq 0.6516$. Since $\pi/2 < v < \pi$ (Why?), we obtain $v \doteq 3.1416 - 0.6516$, that is, $\text{Cos}^{-1}(-0.7951) \doteq 2.4900$.

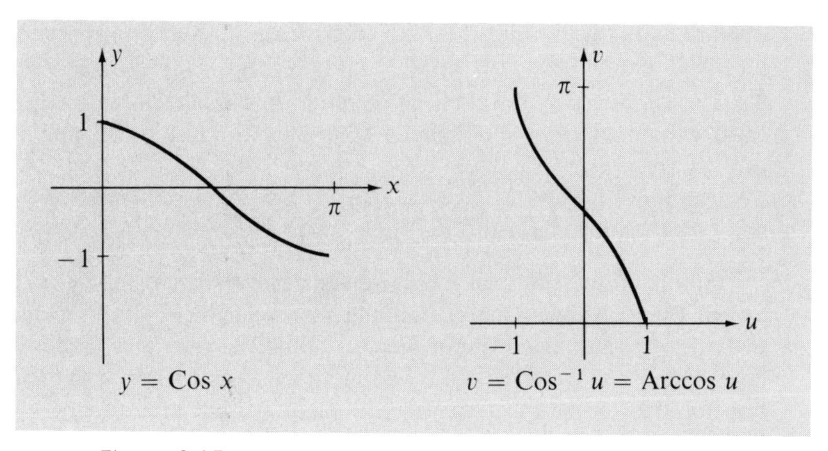

Figure 6.15

The Tangent function, denoted by Tan, is defined by

$$\text{Tan } x = \tan x, \quad -\pi/2 < x < \pi/2.$$

The restriction on x implies that this function is one-to-one.

(6.37) Definition of the Inverse Tangent Function

The *inverse Tangent* or *Arctangent function*, denoted by Tan^{-1} or Arctan, is defined by

$$\text{Tan}^{-1} u = \text{Arctan } u = v \quad \text{if and only if} \quad \text{Tan } v = u,$$

where $u \in \mathbf{R}$ and $-\pi/2 < v < \pi/2$.

Note that the domain of the Arctangent function is all of **R**. The graphs of the Tangent and inverse Tangent functions are sketched in Fig. 6.16.

An analogous procedure may be used for the other trigonometric functions. The functions we have defined are generally referred to as the *inverse trigonometric functions.*

EXAMPLE 3. Without using tables, find sec (Arctan 2/3).

Solution: If $v = \text{Arctan } 2/3$, then $\text{Tan } v = 2/3$. Since $0 < v < \pi/2$, we may write $\sec v = \sqrt{1 + \text{Tan}^2 v} = \sqrt{1 + (2/3)^2} = \sqrt{13}/3$. Hence $\sec(\text{Arctan } 2/3) = \sqrt{13}/3$.

The following examples illustrate some of the manipulations that can be carried out with the inverse trigonometric functions.

EXAMPLE 4. Evaluate sin (Arctan 1/2 − Arccos 4/5).

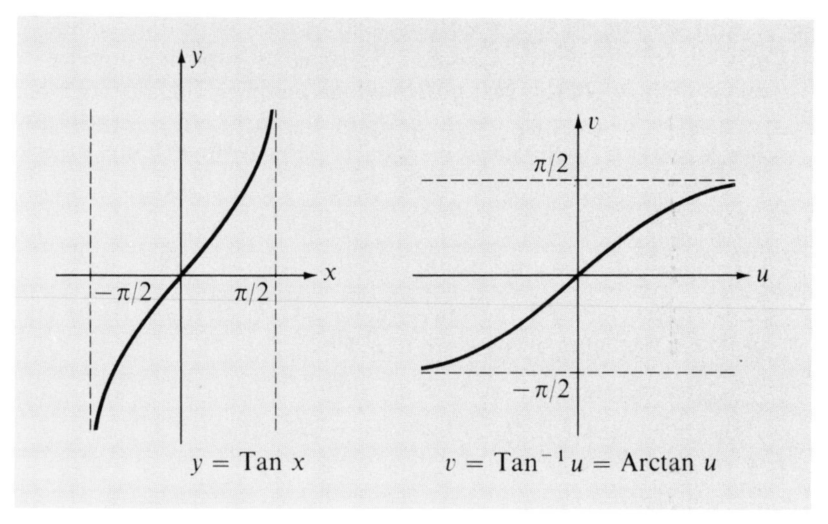

Figure 6.16

Solution: If we let $v = \text{Arctan } 1/2$ and $w = \text{Arccos } 4/5$, then $\text{Tan } v = 1/2$ and $\text{Cos } w = 4/5$. We wish to find $\sin (v - w)$. Since v and w lie in the interval $[0, \pi/2]$, they may be considered as the radian measure of positive acute angles, and other functional values of v and w may be found by referring to suitable right triangles (see Fig. 6.17). Using (5.24), we

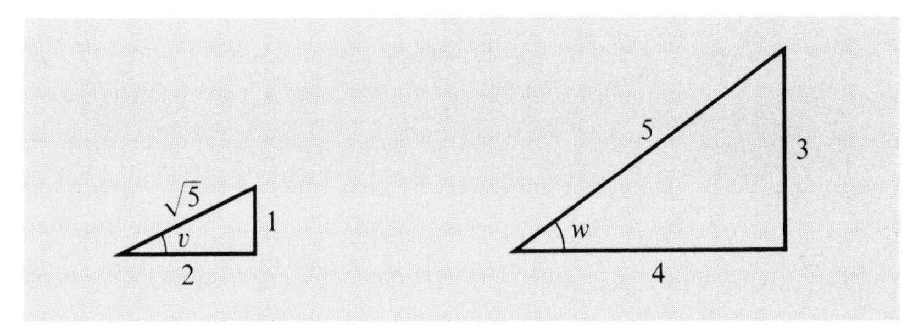

Figure 6.17

have $\sin v = 1/\sqrt{5}$, $\cos v = 2/\sqrt{5}$, and $\sin w = 3/5$. Consequently

$$\begin{aligned}\sin (v - w) &= \sin v \cos w - \cos v \sin w\\ &= (1/\sqrt{5})(4/5) - (2/\sqrt{5})(3/5)\\ &= -2\sqrt{5}/25.\end{aligned}$$

EXAMPLE 5. Express $\cos (\text{Sin}^{-1} u)$ in terms of u.

Solution: Let $v = \text{Sin}^{-1} u$, so that $\text{Sin } v = u$. We wish to find $\cos v$. Since $-\pi/2 \leq v \leq \pi/2$, $\cos v \geq 0$, and hence $\cos v = \sqrt{1 - \text{Sin}^2 v} = \sqrt{1 - u^2}$. Therefore $\cos (\text{Sin}^{-1} u) = \sqrt{1 - u^2}$.

In certain applications it is convenient to use the notation $v = \arcsin u$ (small "a") or $v = \sin^{-1} u$ (small "s") for *any* number v such that $\sin v = u$. In this case one could use arcsin (1/2) to denote $\pi/6$, $5\pi/6$, $-7\pi/6$, or, in general, any number of the form $\pi/6 + 2n\pi$ or $5\pi/6 + 2n\pi$, where $n \in \mathbf{Z}$. Similar remarks may be made for the other functions.

EXERCISES

In Exercises 1–6 find the exact values and in Exercises 7–10 find the approximate values of the given numbers.

1. $\mathrm{Cos}^{-1}(-\sqrt{3}/2)$.
2. $\mathrm{Sin}^{-1}\, 1/2$.
3. $\mathrm{Arctan}\,(-\sqrt{3})$.
4. $\mathrm{Tan}^{-1}(-1)$.
5. $\mathrm{Arcsin}\,(-1)$.
6. $\mathrm{Tan}^{-1}\sqrt{3}$.
7. $\mathrm{Sin}^{-1}\, 0.8241$.
8. $\mathrm{Arctan}\, 8.7777$.
9. $\mathrm{Cos}^{-1}\, 0.4253$.
10. $\mathrm{Arcsin}\, 2/5$.

Determine the numbers in Exercises 11–24 without using tables.

11. $\sin\,[\mathrm{Cos}^{-1}(-1/2)]$.
12. $\cos\,(\mathrm{Tan}^{-1}\, 0)$.
13. $\cos\,(\mathrm{Sin}^{-1}\, 4/5)$.
14. $\cot\,(\mathrm{Arccos}\, 2/7)$.
15. $\mathrm{Tan}^{-1}(\sin \pi/2)$.
16. $\mathrm{Arccos}\,[\tan\,(-\pi/4)]$.
17. $\cos\,[\mathrm{Sin}^{-1}\, 1/3 + \mathrm{Tan}^{-1}\, 1/3]$.
18. $\sin\,[\mathrm{Sin}^{-1}\, 1 + \mathrm{Cos}^{-1}\, 1/2]$.
19. $\tan\,[\mathrm{Arctan}\, 4/3 - \mathrm{Arcsin}\, 3/5]$.
20. $\cos\,[\mathrm{Sin}^{-1}\, 15/17 - \mathrm{Sin}^{-1}\, 8/17]$.
21. $\sin\,[2\,\mathrm{Cos}^{-1}(-3/5)]$.
22. $\cos\,[2\,\mathrm{Sin}^{-1}\, 4/5]$.
23. $\cos\,[\frac{1}{2}\,\mathrm{Arctan}\, 8/15]$.
24. $\tan\,[\frac{1}{2}\,\mathrm{Tan}^{-1}\, 3/4]$.

In Exercises 25–28 rewrite the given expressions as algebraic expressions in u.

25. $\sin\,(\mathrm{Tan}^{-1}\, u)$.
26. $\tan\,(\mathrm{Arccos}\, u)$.
27. $\cos\,(\frac{1}{2}\,\mathrm{Cos}^{-1}\, u)$.
28. $\cos\,(2\,\mathrm{Tan}^{-1}\, u)$.

Verify the identities in Exercises 29–31.

29. $\mathrm{Arcsin}\, u + \mathrm{Arccos}\, u = \pi/2$.
30. $\mathrm{Arcsin}\,(-u) = -\mathrm{Arcsin}\, u$.
31. $\mathrm{Arctan}\, 1/u = \pi/2 - \mathrm{Arctan}\, u,\ u > 0$.

10 THE LAW OF SINES

A triangle that does not contain a right angle is referred to as an *oblique triangle*. Because it is always possible to divide such triangles into two right triangles, methods developed in Chapter Five may be used for their solutions. However, sometimes it is cumbersome to proceed in this manner. In this and the next section formulas are obtained which aid in simplifying solutions of oblique triangles.

If two angles and a side of a triangle are known, or if two sides and an angle opposite one of them is known, then the remaining parts of the triangle may be found by means of the formula given below in (6.38). We shall use the letters A, B, C, a, b, c, α, β, and γ for parts of triangles as they were used in Section 7 of Chapter Five. Given triangle ABC, place angle α in standard position on a rectangular coordinate system such that B is on the positive x-axis. The case where α is obtuse is illustrated in Fig. 6.18. The type of argument we shall give may also be used when α is acute.

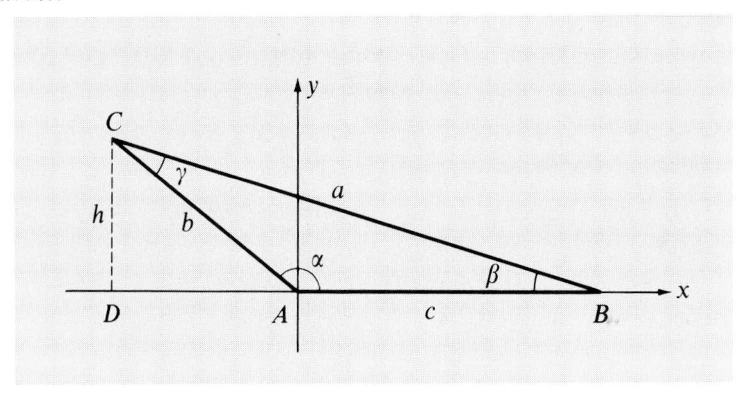

Figure 6.18

Construct a line through C parallel to the y-axis and intersecting the x-axis at point D. Suppose $d(C, D) = h$, so that the ordinate of C is h. Then by (5.23)

$$\sin \alpha = h/b \quad \text{or} \quad h = b \sin \alpha.$$

On the other hand, from triangle BDC we have, by (5.24),

$$\sin \beta = h/a \quad \text{or} \quad h = a \sin \beta.$$

Consequently

$$a \sin \beta = b \sin \alpha,$$

which may be written

$$\frac{a}{\sin \alpha} = \frac{b}{\sin \beta}.$$

If α is taken in standard position such that C is on the positive x-axis, then by the same reasoning we obtain

$$\frac{a}{\sin \alpha} = \frac{c}{\sin \gamma}.$$

The last two equalities give us the following result.

(6.38) The Law of Sines

If ABC is any oblique triangle labeled in the usual manner, then

$$\frac{a}{\sin \alpha} = \frac{b}{\sin \beta} = \frac{c}{\sin \gamma}.$$

The next example illustrates a method of applying (6.38) to the case in which two angles and a side of a triangle are known. We shall use the rules for rounding off answers discussed in Section 7 of Chapter Five.

EXAMPLE 1. Given triangle ABC with $\alpha = 48° 20'$, $\gamma = 57° 30'$ and $b = 47.3$, approximate the remaining parts.

Solution: The triangle is represented in Fig. 6.19. Since the sum of the angles is 180°, we have

$$\beta = 180° - (57° 30' + 48° 20') = 74° 10'.$$

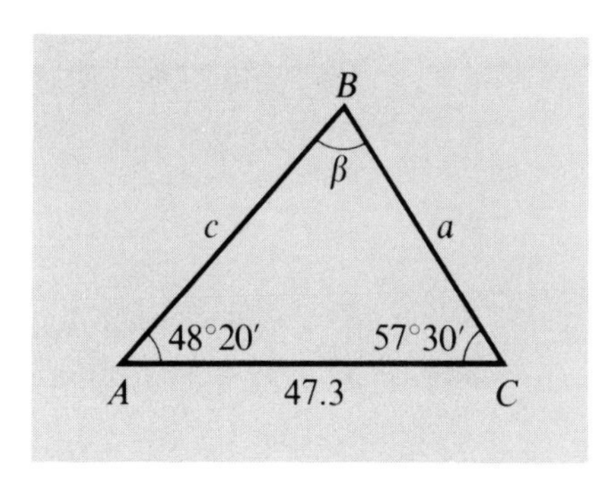

Figure 6.19

Applying (6.38), we have

$$a = \frac{b \sin \alpha}{\sin \beta} = \frac{(47.3) \sin 48° 20'}{\sin 74° 10'}.$$

Taking logarithms, we obtain

$$\log a = \log 47.3 + \log \sin 48° 20' - \log \sin 74° 10'.$$

Using Tables 1 and 4, we arrange our work as follows:

$$\begin{array}{lcr}
\log 47.3 & \doteq & 1.6749 \\
\log \sin 48° 20' & \doteq & 9.8733 - 10 \\
\hline
 & & 11.5482 - 10 \\
\log \sin 74° 10' & \doteq & 9.9832 - 10 \\
\hline
\log a & \doteq & 1.5650.
\end{array}$$

Hence, from Table 1, $a \doteq 36.7$.

Similarly,

$$c = \frac{b \sin \gamma}{\sin \beta} = \frac{(47.3) \sin 57^\circ\, 30'}{\sin 74^\circ\, 10'}.$$

Using logarithms as before, we have

$$\begin{array}{lll} \log 47.3 & \doteq & 1.6749 \\ \log \sin 57^\circ\, 30' & \doteq & 9.9260 - 10 \\ \hline & & 11.6009 - 10 \\ \log \sin 74^\circ\, 10' & \doteq & 9.9832 - 10 \\ \hline \log c & \doteq & 1.6177, \end{array}$$

which gives us $c \doteq 41.5$.

Data such as that given in Example 1 always lead to a unique triangle ABC. On the other hand, if two sides and an angle opposite one of them are given, a unique triangle is not always determined. To illustrate, suppose that two numbers a and b are to be lengths of sides of a triangle ABC. In addition, let there be given the measure of an angle α which is to be opposite the side of length a. Let us consider the case in which α is acute. Place α in standard position on a rectangular coordinate system and construct the line segment AC of length b on the terminal side of α as shown in Fig. 6.20. The third vertex B should be somewhere on the

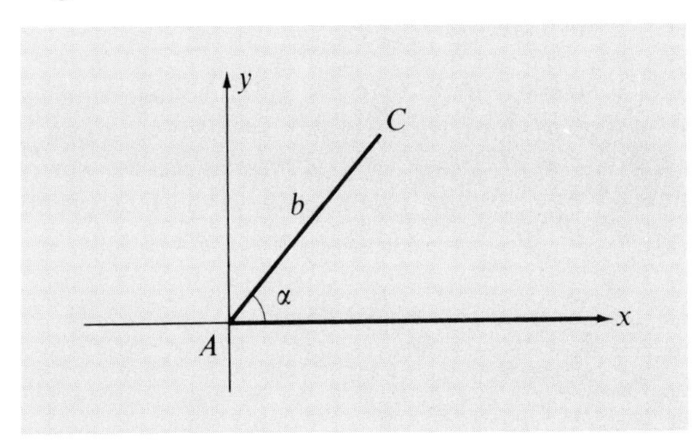

Figure 6.20

x-axis. Since the length a of the side opposite α is given, B may be found by striking off a circular arc of length a with center at C. There are four possible outcomes for this construction, as illustrated in Fig. 6.21, where the coordinate axes have been deleted. These four possibilities may be listed as follows:

(i) The arc does not intersect the x-axis and no triangle is formed.

(ii) The arc is tangent to the x-axis and a right triangle is formed.

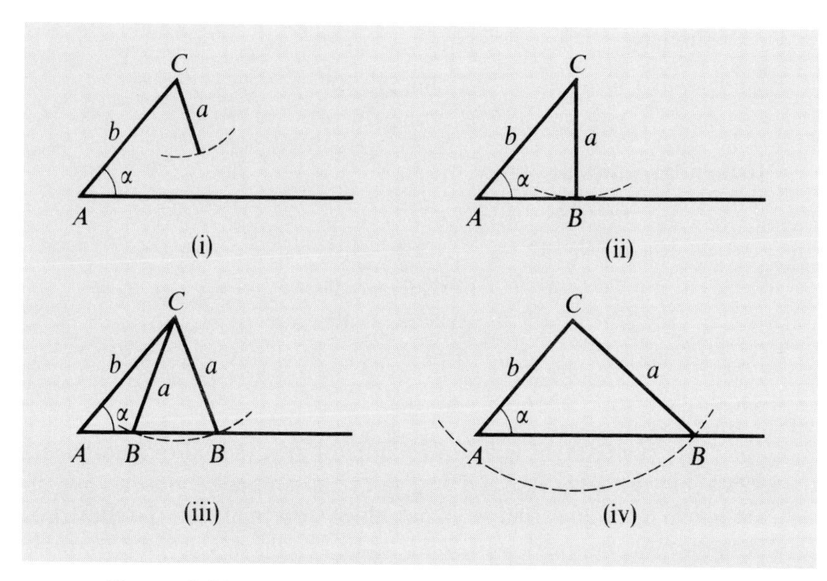

Figure 6.21

(iii) The arc intersects the positive x-axis in two distinct points and two triangles are formed.

(iv) The arc intersects both the positive and nonpositive parts of the x-axis and one triangle is formed.

Since the distance from C to the x-axis is $b \sin \alpha$ (Why?), we see that (i) occurs if $a < b \sin \alpha$, (ii) occurs if $a = b \sin \alpha$, (iii) occurs if $b \sin \alpha < a < b$, and (iv) occurs if $a \geq b$. It is unnecessary to memorize these facts since in any specific problem the case that occurs will become evident when the solution is attempted. For example, when solving the equation

$$\frac{a}{\sin \alpha} = \frac{b}{\sin \beta},$$

suppose one obtains $|\sin \beta| > 1$. This will indicate that no triangle exists. If the equation $\sin \beta = 1$ is obtained, then $\beta = 90°$ and hence case (ii) occurs. On the other hand, if $|\sin \beta| < 1$, then there are two possible choices for the angle β. By checking both possibilities, it will become apparent whether (iii) or (iv) occurs.

If the measure of α is greater than $90°$, then, as in Fig. 6.22, a triangle exists if and only if $a > b$. Needless to say, our discussion is independent of the symbols we have used — that is, there might be given b, c, β, or a, c, γ, and so on.

Because different possibilities may arise, the case in which two sides and an angle opposite one of them are given is sometimes called the *ambiguous case*.

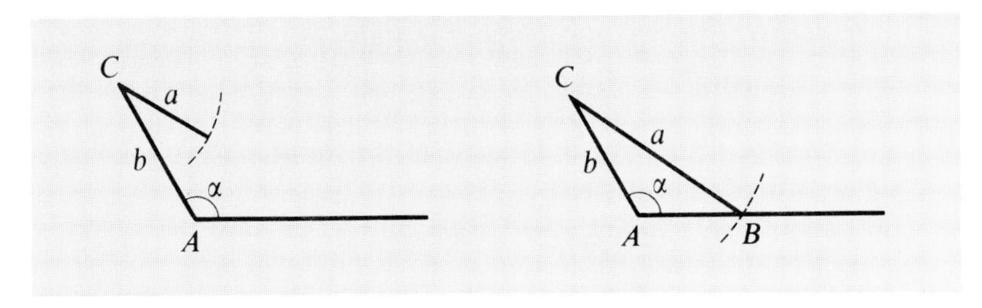

Figure 6.22

EXAMPLE 2. Solve the triangle ABC if $\alpha = 67°$, $c = 125$, and $a = 100$.

Solution: From $\sin \gamma = \dfrac{c \sin \alpha}{a}$ we have

$$\begin{aligned} \sin \gamma &= \frac{(125) \sin 67°}{100} \\ &\doteq \frac{(125)(0.9205)}{100} \\ &\doteq 1.1506. \end{aligned}$$

Since $\sin \gamma > 1$, there is no triangle with the given parts.

EXAMPLE 3. Approximate the remaining parts of triangle ABC if $a = 12.4$, $b = 8.7$, and $\beta = 36° \, 40'$.

Solution: Using $\sin \alpha = \dfrac{a \sin \beta}{b}$, we have

$$\sin \alpha = \frac{(12.4) \sin 36° \, 40'}{8.7}.$$

Hence

$$\log \sin \alpha = \log \sin 36° \, 40' + \log 12.4 - \log 8.7.$$

From Tables 1 and 4 we obtain

$$\begin{aligned} \log \sin 36° \, 40' &\doteq 9.7761 - 10 \\ \log 12.4 &\doteq 1.0934 \\ & 10.8695 - 10 \\ \log 8.7 &\doteq 0.9395 \\ \log \sin \alpha &\doteq 9.9300 - 10. \end{aligned}$$

There are two possible angles α between 0° and 180° which satisfy the last equation. If we let α' denote the reference angle for α, then from Table 4 we obtain

$$\alpha' \doteq 58° \, 20'.$$

Consequently, the two possibilities for α are

$$\alpha_1 \doteq 58^\circ\,20' \quad \text{and} \quad \alpha_2 \doteq 121^\circ\,40'.$$

Let γ_1 and γ_2 denote the third angle of the triangle corresponding to the angles α_1 and α_2, respectively. Then

$$\gamma_1 \doteq 180^\circ - (36^\circ\,40' + 58^\circ\,20') = 85^\circ$$

and

$$\gamma_2 \doteq 180^\circ - (36^\circ\,40' + 121^\circ\,40') = 21^\circ\,40'.$$

Thus there are two possible triangles which have the given parts. These are the triangles A_1BC and A_2BC shown in Fig. 6.23. Let c_1 be the side

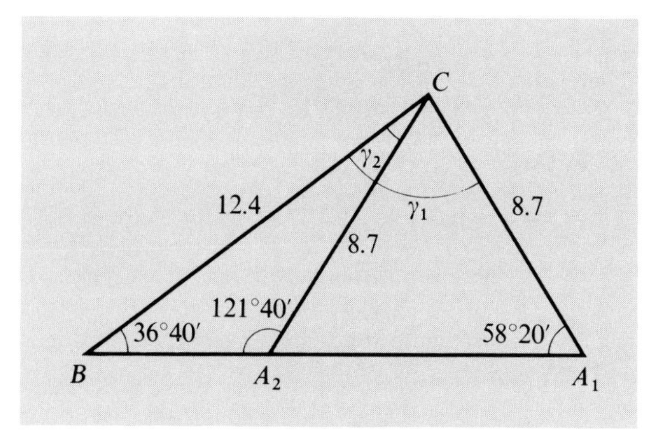

Figure 6.23

opposite γ_1 in triangle A_1BC. Then

$$c_1 \doteq \frac{a \sin \gamma_1}{\sin \alpha_1}$$

or

$$c_1 \doteq \frac{(12.4) \sin 85^\circ}{\sin 58^\circ\,20'}.$$

Taking logarithms and arranging our work in the usual way, we have

$$\begin{array}{lcl}
\log 12.4 & \doteq & 1.0934 \\
\log \sin 85^\circ & \doteq & \underline{9.9983 - 10} \\
 & & 11.0917 - 10 \\
\log \sin 58^\circ\,20' & \doteq & \underline{9.9300 - 10} \\
\log c_1 & \doteq & 1.1617.
\end{array}$$

Consulting Table 1, we obtain

$$c_1 \doteq 14.5.$$

Hence the solution for triangle A_1BC is $\alpha_1 \doteq 58° 20'$, $\gamma_1 \doteq 85°$, and $c_1 \doteq 14.5$.

Similarly, if c_2 is the side opposite angle γ_2 in triangle A_2BC, we have

$$c_2 \doteq \frac{a \sin \gamma_2}{\sin \alpha_2}$$

or

$$c_2 \doteq \frac{(12.4) \sin 21° 40'}{\sin 121° 40'}.$$

Applying logarithms, we have

$$\begin{aligned}
\log 12.4 &\doteq 1.0934 \\
\log \sin 21° 40' &\doteq 9.5673 - 10 \\
&\quad 10.6607 - 10 \\
\log \sin 121° 40' &\doteq 9.9300 - 10 \\
\log c_2 &\doteq 0.7307.
\end{aligned}$$

Using Table 1 and rounding off, we obtain $c_2 \doteq 5.4$. Consequently the solution for triangle A_2BC is $\alpha_2 \doteq 121° 40'$, $\gamma_2 \doteq 21° 40'$, and $c_2 \doteq 5.4$.

EXERCISES

1. If angle α of triangle ABC is acute, give a geometric argument which shows that $a \sin \beta = b \sin \alpha$.
2. What part of the Law of Sines is valid if triangle ABC is a right triangle?

In Exercises 3–14 approximate the remaining parts of triangle ABC.

3. $\alpha = 41°$, $\gamma = 77°$, $a = 10.5$.
4. $\beta = 20°$, $\gamma = 31°$, $b = 210$.
5. $\alpha = 27° 40'$, $\beta = 52° 10'$, $a = 32.4$.
6. $\alpha = 42° 10'$, $\gamma = 61° 20'$, $b = 19.7$.
7. $\beta = 50° 50'$, $\gamma = 70° 30'$, $c = 537$.
8. $\alpha = 7° 10'$, $\beta = 11° 40'$, $a = 2.19$.
9. $\alpha = 65° 10'$, $a = 21.3$, $b = 18.9$.
10. $\beta = 30°$, $b = 17.9$, $a = 35.8$.
11. $\gamma = 53° 20'$, $a = 140$, $c = 115$.
12. $\alpha = 27° 30'$, $c = 52.8$, $a = 28.1$.
13. $\beta = 113° 10'$, $b = 248$, $c = 195$.
14. $\gamma = 81°$, $c = 11$, $b = 12$.
15. It is desired to find the distance between two points A and B lying on opposite banks of a river. A line segment AC of length 240 yards is laid off and the measure of angles BAC and ACB are found to be $63° 20'$ and $54° 10'$, respectively. Approximate the distance from A to B.

16. In order to determine the distance between two points A and B, a surveyor chooses a point C which is 375 yards from A and 530 yards from B. If angle BAC has measure $49° 30'$, approximate the required distance.
17. When the angle of elevation of the sun is 64°, a telegraph pole that is tilted at an angle of 12° directly away from the sun casts a shadow 34 feet long on level ground. Approximate the length of the pole.
18. A straight road makes an angle of 15° with the horizontal. When the angle of elevation of the sun is 57°, a vertical pole at the side of the road casts a shadow 75 feet long directly down the road. Approximate the length of the pole.
19. The angles of elevation of a balloon from two points A and B on level ground are $24° 10'$ and $47° 40'$, respectively. If A and B are 8.4 miles apart and the balloon is between A and B in the same vertical plane, approximate the height of the balloon above the ground.
20. An airport A is 480 miles due east of airport B. A pilot flew in the direction 235° from A to C and then in the direction 320° from C to B. Approximate the total distance he flew.
21. A forest ranger at an observation point A sights a fire in the direction N27° 10′E. Another ranger at an observation point B, 6 miles due east of A, sights the same fire at N52° 40′W. Approximately how far is the fire from each of the observation points?
22. A surveyor notes that the direction from point A to point B is S63°W and the direction from A to C is S38°W. If the distance from A to B is 239 yards and the distance from B to C is 374 yards, approximate the distance and direction from A to C.

11 THE LAW OF COSINES

If two sides and the included angle or if the three sides of a triangle are given, then the Law of Sines cannot be applied directly. We may, however, use the following result.

(6.39) The Law of Cosines

If ABC is any triangle labeled in the usual manner, then

$$\begin{aligned} a^2 &= b^2 + c^2 - 2bc \cos \alpha, \\ b^2 &= a^2 + c^2 - 2ac \cos \beta, \\ c^2 &= a^2 + b^2 - 2ab \cos \gamma. \end{aligned}$$

Instead of memorizing each of the formulas given in (6.39), it is more convenient to remember the following statement, which takes all of them into account: *the square of the length of any side of a triangle equals the sum of the squares of the lengths of the other two sides minus twice the product of the lengths of the other two sides and the cosine of the angle between them.*

We shall use the distance formula to establish (6.39). We again place α in standard position on a rectangular coordinate system, as illustrated in Fig. 6.18. Although α is pictured as an obtuse angle, our development is also true if α is acute. Since segment AB has length c, the coordinates of B are $(c, 0)$. By (5.23) the coordinates of C may be written $(b \cos \alpha, b \sin \alpha)$. We also have $a = d(B, C)$. Squaring both sides of the latter equation and using the distance formula to find $d(B, C)$ produces the following equations:

$$\begin{aligned} a^2 &= (b \cos \alpha - c)^2 + (b \sin \alpha - 0)^2 \\ &= b^2 \cos^2 \alpha - 2bc \cos \alpha + c^2 + b^2 \sin^2 \alpha \\ &= b^2(\cos^2 \alpha + \sin^2 \alpha) + c^2 - 2bc \cos \alpha \\ &= b^2 + c^2 - 2bc \cos \alpha. \end{aligned}$$

This gives us the first formula of (6.39). The second and third formulas may be obtained by placing β and γ, respectively, in standard position on a rectangular coordinate system and using a similar procedure.

EXAMPLE 1. Approximate the remaining parts of triangle ABC if $a = 5.0$, $c = 8.0$, and $\beta = 77° 10'$.

Solution: Using the Law of Cosines, we have

$$\begin{aligned} b^2 &= (5.0)^2 + (8.0)^2 - 2(5.0)(8.0) \cos 77° 10' \\ &\doteq 25 + 64 - (80)(0.2221) \\ &\doteq 71.2. \end{aligned}$$

Consequently $b \doteq 8.44$. We now use the Law of Sines to find γ. Thus

$$\sin \gamma \doteq \frac{(8.0) \sin 77° 10'}{8.44}.$$

We leave it to the reader to verify that $\gamma \doteq 67° 30'$. Hence

$$\alpha \doteq 180° - (77° 10' + 67° 30') = 35° 20'.$$

EXAMPLE 2. Given sides $a = 90$, $b = 70$, and $c = 40$ of triangle ABC, approximate the angles α, β, and γ.

Solution: According to the first equation of (6.39), we may write

$$\begin{aligned} \cos \alpha &= \frac{b^2 + c^2 - a^2}{2bc} \\ &= \frac{4900 + 1600 - 8100}{5600} \\ &= -0.2857. \end{aligned}$$

From Table 3 the reference angle for α is approximately $73° 20'$. Hence $\alpha \doteq 106° 40'$.

Similarly, from the second equation of (6.39)

$$\cos \beta = \frac{a^2 + c^2 - b^2}{2ac}$$

$$= \frac{8100 + 1600 - 4900}{7200}$$

$$\doteq 0.6667.$$

Hence $\beta \doteq 48° 10'$ and $\gamma \doteq 180° - (106° 40' + 48° 10')$, or $\gamma \doteq 25° 10'$.

EXERCISES

In Exercises 1–10 approximate the remaining parts of triangle ABC.

1. $\alpha = 60°$, $b = 20$, $c = 30$.
2. $\gamma = 45°$, $b = 10$, $a = 15$.
3. $\beta = 150°$, $a = 150$, $c = 30$.
4. $\beta = 73° 50'$, $c = 14$, $a = 87$.
5. $\gamma = 115° 10'$, $a = 1.1$, $b = 2.1$
6. $\alpha = 23° 40'$, $c = 4.3$, $b = 70$.
7. $a = 2$, $b = 3$, $c = 4$.
8. $a = 10$, $b = 15$, $c = 12$.
9. $a = 25$, $b = 80$, $c = 60$.
10. $a = 20$, $b = 20$, $c = 10$.
11. A parallelogram has sides of length 30 inches and 70 inches. If one of the angles has measure 65°, approximate the lengths of each diagonal.
12. The angle at one corner of a triangular plot of ground has measure 73° 40′. If the sides that meet at this corner are 175 feet and 150 feet long, approximate the length of the third side.
13. A vertical pole 40 feet tall stands on a hillside that makes an angle of 17° with the horizontal. What is the minimal length of rope that will reach from the top of the pole to a point directly down the hill 72 feet from the base of the pole?
14. To find the distance between two points A and B, a surveyor chooses a point C which is 420 yards from A and 540 yards from B. If angle ACB has measure 63° 10′, approximate the distance.
15. Two automobiles leave from the same point and travel along straight highways which differ in direction by 84°. If their speeds are 60 miles per hour and 45 miles per hour, respectively, approximately how far apart will they be at the end of 20 minutes?
16. A triangular plot of land has sides of length 420 feet, 350 feet, and 180 feet. Find the smallest angle between the sides.
17. A ship leaves point P at 1:00 P.M. and travels S35°E at the rate of 24 miles per hour. Another ship leaves P at 1:30 P.M. and travels S20°W at 18 miles per hour. Approximately how far apart are the ships at 3:00 P.M.?

18. An airplane flies 165 miles from point A in the direction 130° and then travels in the direction 245° for 80 miles. Approximately how far is the airplane from A?

12 REVIEW EXERCISES

Oral

Define or discuss each of the following:

1. Trigonometric equation.
2. The solution set of a trigonometric equation.
3. Trigonometric identity.
4. The addition formulas.
5. The double angle formulas.
6. The half-angle formulas.
7. The product formulas.
8. The factoring formulas.
9. Reduction formulas.
10. Amplitude of a function.
11. Simple harmonic motion.
12. The inverse trigonometric functions.
13. The Law of Sines.
14. The Law of Cosines.

Written

1. Verify the following identities:
   (a) $(1 - \sin^2 t)(1 + \tan^2 t) = 1$.
   (b) $\sin x + \cos x \cot x = \csc x$.
   (c) $\dfrac{\tan^3 \phi - \cot^3 \phi}{\tan^2 \phi + \csc^2 \phi} = \tan \phi - \cot \phi$.
   (d) $\dfrac{\sin u + \sin v}{\csc u + \csc v} = \dfrac{1 - \sin u \sin v}{-1 + \csc u \csc v}$.
   (e) $\cos(x - 5\pi/2) = \sin x$.
   (f) $\sin 8\theta = 8 \sin \theta \cos \theta\,(1 - 2 \sin^2 \theta)(1 - 8 \sin^2 \theta \cos^2 \theta)$.
   (g) $\text{Arcsin} \dfrac{2u}{1 + u^2} = 2 \text{ Arctan } u, \qquad |u| \le 1$.
2. Find, in degree measure, the solutions of the following equations which are in the interval $[0°, 360°)$:
   (a) $\tan^2 \theta \cos \theta = \cos \theta$.
   (b) $2 \sin^2 x + \sin x - 6 = 0$.
   (c) $\cos 2x + 3 \cos x + 2 = 0$.
   (d) $2 \csc \alpha \cos \alpha + 2 = 4 \cos \alpha + \csc \alpha$.
   (e) $\sin \theta - \sin 2\theta = 0$.
   (f) $3 \cos x - 2 \cos^2 \frac{1}{2}x = 0$.

3. Find, without the use of tables, the exact value of:
   (a) $\sin 105°$. (b) $\tan 345°$. (c) $\cos 165°$. (d) $\sin 3\pi/8$.

4. If θ and ϕ are acute angles such that $\csc \theta = 5/3$ and $\cos \phi = 8/17$, find:
   (a) $\sin (\theta - \phi)$. (b) $\cos (\theta - \phi)$.
   (c) $\tan (\theta - \phi)$. (d) $\sin 2\phi$.
   (e) $\cos 2\phi$. (f) $\sin \theta/2$.
   (g) $\tan \theta/2$.

5. (a) Express each of the following products as a sum or difference: $\sin 5t \sin 3t$; $\cos (u/3) \cos (-u/4)$; $4 \cos x \sin 2x$.
   (b) Express each of the following as a product: $\sin 8\theta + \sin 4\theta$; $\cos 7u - \cos 5u$; $\sin (t/2) - \sin (t/3)$.

6. Find the amplitude, period, and sketch the graph of f if:
   (a) $f(x) = 4 \sin 2x$. (b) $f(x) = 2 \sin 4x$.
   (c) $f(x) = -3 \sin \frac{1}{2}x$. (d) $f(x) = -\frac{1}{2} \cos 3x$.

7. A point moves harmonically on a real axis according to the equation $y = 10 \sin 6\pi t$, where t is in seconds. Determine the amplitude, period, frequency, and describe the motion of the point.

8. Sketch the graph of the equation:
   (a) $y = 4 \cos (3x - \pi/2)$. (b) $y = -2 \sin (2x + \pi/4)$.
   (c) $y = 2 \cos x + \sin 2x$. (d) $y = 2^{-x} \sin 2x$.

9. Find each of the following without the use of tables:
   (a) $\text{Sin}^{-1} (-\sqrt{3}/2)$. (b) $\text{Cos}^{-1} (\sqrt{2}/2)$.
   (c) $\text{Arccos} (-1)$. (d) $\text{Arctan} (-\sqrt{3}/3)$.
   (e) $\sin \text{Arccos} (-\sqrt{3}/2)$. (f) $\cos [\text{Sin}^{-1} 3/5 + \text{Tan}^{-1} 4/3]$.
   (g) $\cos [2 \text{Sin}^{-1} 8/17]$. (h) $\tan [\text{Tan}^{-1} 100]$.
   (i) $\text{Tan}^{-1} (\cos 0)$.

10. Without using tables, find the remaining parts of triangle ABC if:
   (a) $\alpha = 60°$, $\beta = 45°$, $b = 100$.
   (b) $\gamma = 30°$, $a = 2\sqrt{3}$, $c = 2$.
   (c) $\alpha = 60°$, $b = 6$, $c = 7$.
   (d) $a = 2$, $b = 3$, $c = 4$.

Supplementary Questions

1. Prove that if n is any even integer, then

$$\left|\sin \left(x + n\frac{\pi}{2}\right)\right| = |\sin x| \quad \text{and} \quad \left|\cos \left(x + n\frac{\pi}{2}\right)\right| = |\cos x|.$$

Prove that if n is any odd integer, then

$$\left|\sin\left(x + n\frac{\pi}{2}\right)\right| = |\cos x| \quad \text{and} \quad \left|\cos\left(x + n\frac{\pi}{2}\right)\right| = |\sin x|.$$

2. Sketch the graph of the function f defined by $f(x) = (1/x)\sin x$ where $x > 0$ (use Table 3 to approximate ordinates of points which have abscissas close to 0). Formulate a conjecture about the behavior of $f(x)$ as x gets closer and closer to 0.
3. If f is defined by $f(x) = \sin(1/x)$ where $x > 0$, show that f takes on an infinite number of maximum and minimum values in the open interval (0, 1). Sketch the graph of f. How does the graph compare with the graph of the function g defined by $g(x) = x\sin(1/x)$ where $x > 0$?
4. Define $\text{Sec}^{-1} x$ after suitably restricting the domain of the secant function. Is it possible to obtain an inverse function by using a different domain?
5. Show that for every triangle ABC,

$$\frac{\cos\alpha}{a} + \frac{\cos\beta}{b} + \frac{\cos\gamma}{c} = \frac{a^2 + b^2 + c^2}{2abc}.$$

6. If ABC is any triangle prove the following *Law of Tangents*:

$$\frac{a - b}{a + b} = \frac{\tan\frac{1}{2}(\alpha - \beta)}{\tan\frac{1}{2}(\alpha + \beta)}.$$

State a similar formula which involves b, c, β, and γ.

chapter seven

Coordinate Geometry in Two Dimensions

In plane geometry a study is made of figures such as lines, circles, and triangles which lie in a plane. Theorems are proved by reasoning deductively from certain postulates. In *coordinate* geometry, plane geometric figures are investigated by introducing a coordinate system and using equations and formulas of various types. If the study of coordinate geometry were to be summarized by means of one statement, perhaps the following would be appropriate: "Given an equation, find its graph and, conversely, given a graph, find its equation." In this chapter we shall apply coordinate methods to several basic plane figures.

1 CIRCLES

In Chapter One (see (1.18)) we proved that an equation for the circle in a coordinate plane with center $C(h, k)$ and radius r is

(7.1) $$(x - h)^2 + (y - k)^2 = r^2.$$

If, in (7.1), we square the binomials and simplify, we obtain an equation of the form

(7.2) $$x^2 + y^2 + ax + by + c = 0,$$

where $a, b, c \in \mathbf{R}$. Conversely, if we begin with an equation in x and y such as (7.2), it is always possible, by completing the squares in x and y, to obtain an equation of the form

(7.3) $$(x - h)^2 + (y - k)^2 = l,$$

where $h, k, l \in \mathbf{R}$. The method is illustrated below in Example 1. If $l > 0$, then (7.3) is an equation of a circle with center (h, k) and radius $r = \sqrt{l}$. If $l = 0$, then since $(x - h)^2 \geq 0$ and $(y - k)^2 \geq 0$, the *only* solution of (7.3) is (h, k) and hence the graph consists of only this one point. Finally, if $l < 0$, then since the left side of (7.3) is never negative, the solution set of the equation is empty and there is no graph. This proves that the graph of (7.2) is either a circle, a point, or the empty set.

EXAMPLE 1. Find the center and radius of the circle with equation

$$x^2 + y^2 - 4x + 6y - 3 = 0.$$

Solution: We begin by arranging the given equation in the form

$$(x^2 - 4x) + (y^2 + 6y) = 3.$$

Next we complete the squares by adding appropriate numbers within the parentheses. Of course, to obtain equivalent equations we must add the numbers to *both* sides of the equation. In order to complete the square for an expression of the form $x^2 + ax$, we add the square of half the coefficient of x — that is, $(a/2)^2$ — to both sides of the equation. Similarly, for $y^2 + by$ we add $(b/2)^2$ to both sides. This leads to the equation

$$(x^2 - 4x + 4) + (y^2 + 6y + 9) = 3 + 4 + 9$$

or

$$(x - 2)^2 + (y + 3)^2 = 16.$$

By (7.1) the center is $(2, -3)$ and the radius is 4.

In the next example we use one of the fundamental principles of coordinate geometry, namely that *a point $P(x_1, y_1)$ is on the graph of an equation in x and y if and only if the pair (x_1, y_1) is in the solution set of the equation* (cf. p. 23).

EXAMPLE 2. Find the equation of the circle which contains the points $A(-1, 7)$, $B(3, 9)$, and the origin $O(0, 0)$. What is the center and radius of the circle?

Solution: The circle is sketched in Fig. 7.1. If we choose a, b, and c appropriately, then (7.2) will be an equation for the circle. Moreover, since the circle passes through the origin, $(0, 0)$ is in the solution set of (7.2) — that is,

$$0^2 + 0^2 + a \cdot 0 + b \cdot 0 + c = 0,$$

and consequently $c = 0$. Therefore, an equation for the circle is of the form

$$x^2 + y^2 + ax + by = 0. \tag{7.4}$$

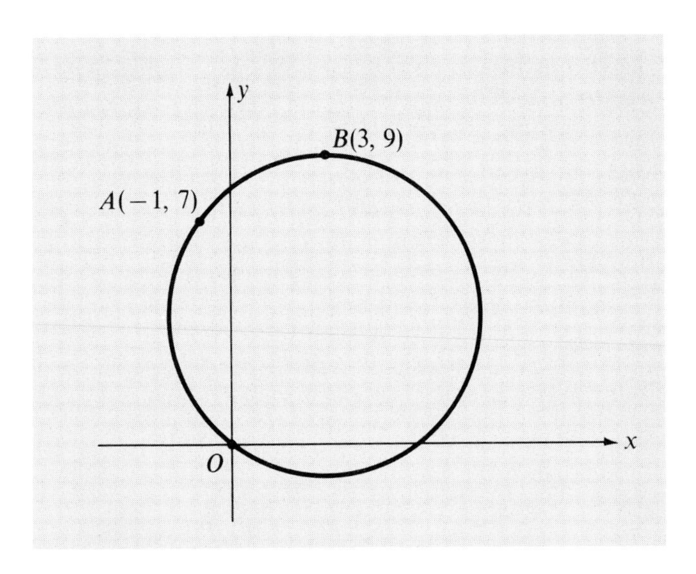

Figure 7.1

Since A and B are on the circle, $(-1, 7)$ and $(3, 9)$ are in the solution set of (7.4) — that is, $1 + 49 - a + 7b = 0$ and $9 + 81 + 3a + 9b = 0$. Simplifying, we see that the pair (a, b) must be a solution of *both* of the equations $a - 7b = 50$ and $a + 3b = -30$. Solving the first equation for a and substituting in the second equation, we obtain $(7b + 50) + 3b = -30$, or $10b = -80$. Therefore $b = -8$. Since $a = 7b + 50$, we have $a = 7(-8) + 50 = -6$. Substitution of these numbers for a and b in (7.4) leads to the equation

$$x^2 + y^2 - 6x - 8y = 0.$$

Completing the squares we obtain

$$(x^2 - 6x + 9) + (y^2 - 8y + 16) = 9 + 16,$$

or

$$(x - 3)^2 + (y - 4)^2 = 25.$$

Hence the center is $C(3, 4)$ and the radius is 5.

If P_1 and P_2 are distinct points in a coordinate plane, it will sometimes be convenient to find the coordinates of the midpoint M of the line segment P_1P_2. If the points are on the x-axis we may label them $P_1(x_1, 0)$, $P_2(x_2, 0)$, and $M(m, 0)$. This gives us one of the situations shown in Fig. 7.2, where we have not shown the y-axis since its position is im-

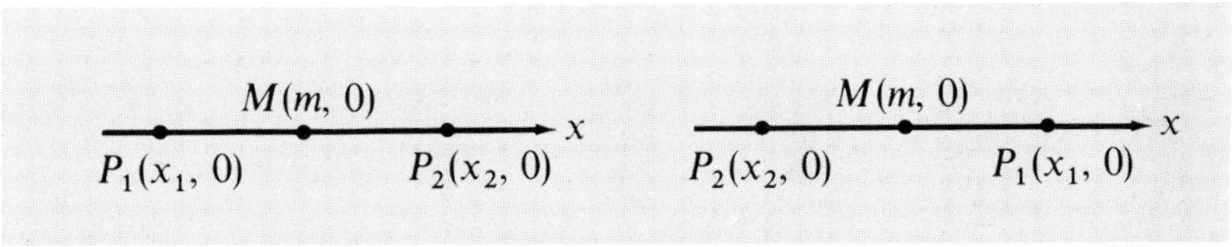

Figure 7.2

material in our discussion. Since $d(P_1, M) = d(M, P_2)$ and since the *directed* distances $\overline{P_1M}$ and $\overline{MP_2}$ are either both positive or both negative, it follows that $\overline{P_1M} = \overline{MP_2}$. Consequently, by (1.12) we have $m - x_1 = x_2 - m$. Adding $m + x_1$ to both sides of the latter equation and simplifying, we get $2m = x_1 + x_2$, and therefore $m = (x_1 + x_2)/2$. Likewise, if P_1 and P_2 are on the y-axis we may denote them by $P_1(0, y_1)$, $P_2(0, y_2)$, and a similar argument gives us the midpoint $M(0, (y_1 + y_2)/2)$ of P_1P_2.

Let us now consider arbitrary points $P_1(x_1, y_1)$ and $P_2(x_2, y_2)$ in a coordinate plane and let M be the midpoint of the segment from P_1 to P_2. Construct lines through P_1 and P_2 parallel to the y-axis and intersecting the x-axis at $A_1(x_1, 0)$ and $A_2(x_2, 0)$, respectively (see Fig. 7.3).

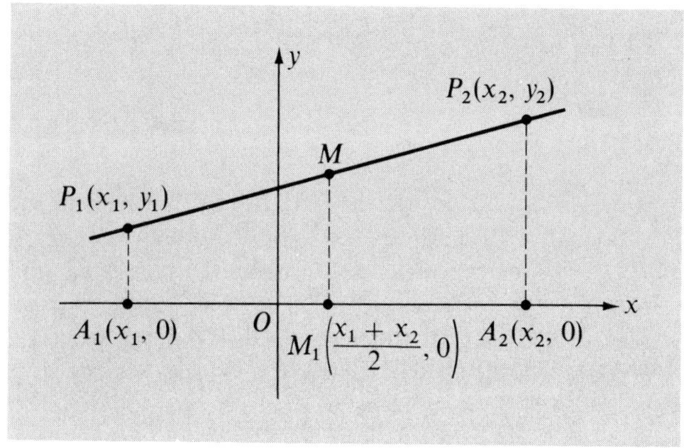

Figure 7.3

By the discussion in the preceding paragraph the midpoint M_1 of the segment from A_1 to A_2 is $((x_1 + x_2)/2, 0)$. It follows from plane geometry that the abscissa of M is $(x_1 + x_2)/2$. In similar fashion, it can be shown that the ordinate of the midpoint M is $(y_1 + y_2)/2$. This proves the following result.

(7.5) Midpoint Formula

The midpoint of the line segment from $P_1(x_1, y_1)$ to $P_2(x_2, y_2)$ is the point with coordinates

$$\left(\frac{x_1 + x_2}{2}, \frac{y_1 + y_2}{2}\right).$$

EXAMPLE 3. Find the midpoint of the segment from $(-2, 3)$ to $(4, -2)$.

Solution: By (7.5) the coordinates of the midpoint are

$$\left(\frac{-2 + 4}{2}, \frac{3 + (-2)}{2}\right), \quad \text{or} \quad \left(1, \frac{1}{2}\right).$$

In determination of the midpoint of a segment AB, a common error is to *subtract* the respective coordinates of P_1 and P_2 and divide by 2. Note that (7.5) states that the coordinates are to be *added* and the *sum* divided by 2.

EXAMPLE 4. Find an equation for the circle which has $A(8, -1)$ and $B(2, -3)$ as endpoints of a diameter.

Solution: Since AB is a diameter, the midpoint M of AB is the center of the circle. Employing (7.5) the coordinates of M are $(\frac{1}{2}(8 + 2), \frac{1}{2}(-1 - 3))$, or $(5, -2)$. It follows that the radius is

$$d(A, M) = \sqrt{(8 - 5)^2 + (-1 + 2)^2} = \sqrt{10}.$$

Letting $h = 5$, $k = -2$, and $r = \sqrt{10}$ in (7.1), we obtain

$$(x - 5)^2 + (y + 2)^2 = 10,$$

which may also be written

$$x^2 + y^2 - 10x + 4y + 19 = 0.$$

EXERCISES

In each of Exercises 1–8 find an equation of the circle which satisfies the stated conditions.

1. Center $C(-2, 5)$, radius 3.
2. Center $C(0, -4)$, radius 4.
3. Center $C(-4, -3)$, tangent to the line $x = 5$.
4. Endpoints of diameter $A(2, 5)$, $B(6, -4)$.
5. Circumscribed about the right triangle of Exercise 13, p. 20. [*Hint:* The center of the circle is the midpoint of the hypotenuse.]
6. Passing through the three points $A(-2, 3)$, $B(4, 3)$, and $C(-2, -1)$.
7. Find an equation for the family of circles with center $C(-4, 1)$. (A *family of circles* is a collection of circles, all of which have some common property.)
8. Find an equation for the family of circles with center on the line $y = x$ and tangent to both axes.

In the following exercises, find the center and radius of the circle with the given equations.

9. $x^2 + y^2 + 4x - 6y + 4 = 0.$
10. $x^2 + y^2 - 10x + 2y + 22 = 0.$

11. $x^2 + y^2 + 6x = 0$.
12. $x^2 + y^2 + x + y - 1 = 0$.
13. $2x^2 + 2y^2 - x + y - 3 = 0$.
14. $x^2 + y^2 + 8x - 12y + 52 = 0$.
15. $9x^2 + 9y^2 - 6x + 12y - 31 = 0$.
16. $x^2 + y^2 + 3y = 0$.

In Exercises 17 and 18 use the method of Example 2 to find an equation for the circle which circumscribes the triangle with the given vertices.

17. $A(-2, 1)$, $B(-4, -1)$, $C(1, 2)$.
18. $A(-3, 1)$, $B(5, 5)$, $C(-3, 9)$.
19. Prove that the midpoint of the hypotenuse of any right triangle is equidistant from the vertices. [*Hint:* Label the vertices of the triangle $O(0, 0)$, $A(a, 0)$, and $B(0, b)$.]
20. Prove that the diagonals of any parallelogram bisect each other. [*Hint:* Label three of the vertices of the parallelogram $O(0, 0)$, $A(a, b)$, and $C(0, c)$.]

2 SLOPE OF A LINE

We shall now discuss several concepts which are fundamental for the study of straight lines. All lines referred to in this section are considered to be in some fixed coordinate plane.

(7.6) Definition of Slope

If l is a line which is not parallel to the y-axis and if $P_1(x_1, y_1)$ and $P_2(x_2, y_2)$ are distinct points on l, then the *slope* m of l is given by

$$m = \frac{y_2 - y_1}{x_2 - x_1}.$$

If l is parallel to the y-axis, then it has no slope.

The numerator $y_2 - y_1$ in the formula for m, sometimes called the *rise from* P_1 *to* P_2, measures the vertical change in direction in proceeding from P_1 to P_2 and may be positive, negative, or zero. The denominator $x_2 - x_1$, sometimes called the *run from* P_1 *to* P_2, measures the amount of horizontal change in going from P_1 to P_2. The run may be positive or negative but is never zero, because l is not parallel to the y-axis. Using this terminology we could write (7.6) as

$$\text{slope of } l = \frac{\text{rise from } P_1 \text{ to } P_2}{\text{run from } P_1 \text{ to } P_2}.$$

In finding the slope of a line, it is immaterial which point is labeled P_1 and which is labeled P_2, since

$$\frac{y_2 - y_1}{x_2 - x_1} = \frac{y_1 - y_2}{x_1 - x_2}.$$

Thus we may as well assume that the points are labeled so that $x_1 < x_2$, as in Fig. 7.4. Then $x_2 - x_1 > 0$ and hence the slope is positive, negative,

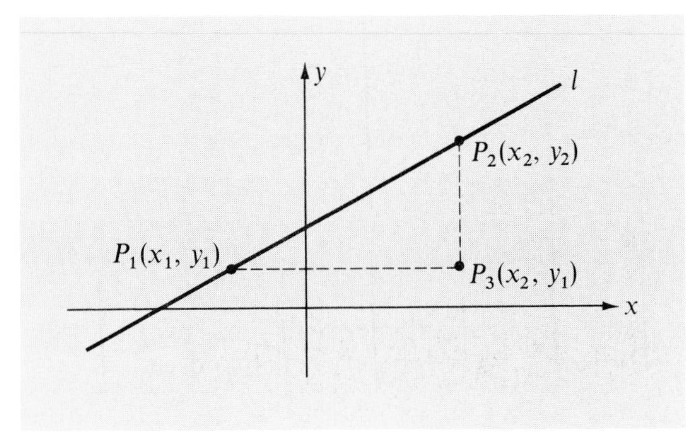

Figure 7.4

or zero according to whether $y_2 > y_1$, $y_2 < y_1$, or $y_2 = y_1$. The slope of the line shown in Fig. 7.4 is positive, whereas the slope of the line shown in Fig. 7.5 is negative. The slope is zero if and only if the line is horizontal.

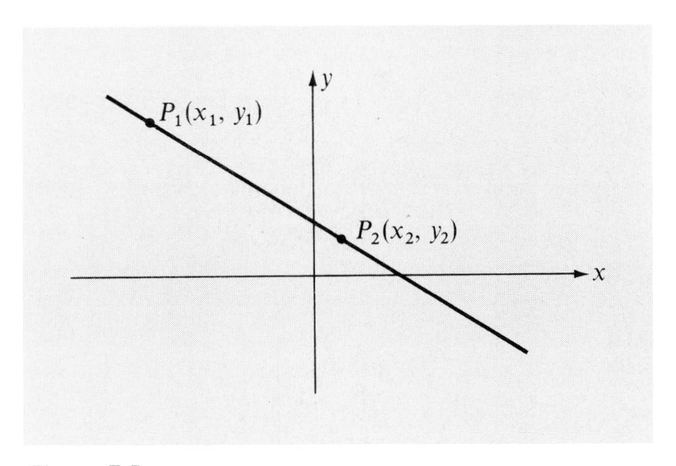

Figure 7.5

We shall occasionally use the following terminology. If the slope m of a line l is positive, then as the abscissas of points on l increase, so do ordinates, and we shall say that the line *rises*. On the other hand, if m is negative, then as abscissas of points on l increase the corresponding ordinates decrease, and we shall say that the line *falls*.

It is important to observe that the definition of slope is independent of the two points that are chosen on l, for if other points $P_1'(x_1', y_1')$ and $P_2'(x_2', y_2')$ are used, then, as in Fig. 7.6, the triangle formed by P_1', P_2',

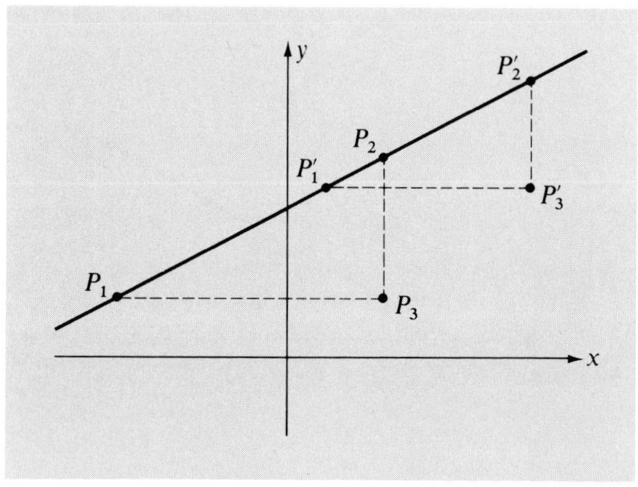

Figure 7.6

and $P_3'(x_2', y_1')$ is similar to the triangle formed by P_1, P_2, P_3, and hence the ratios of corresponding sides are equal. In particular,

$$\frac{y_2 - y_1}{x_2 - x_1} = \frac{y_2' - y_1'}{x_2' - x_1'}.$$

EXAMPLE 1. Find the slopes of the lines through the following pairs of points:

(a) $A(-1, 4)$ and $B(3, 2)$. (b) $A(6, 2)$ and $B(-8, 2)$.
(c) $A(4, -2)$ and $B(4, -4)$.

Solutions: Using (7.6), we have

(a) $$m = \frac{2 - 4}{3 - (-1)} = \frac{-2}{4} = -\frac{1}{2},$$

(b) $$m = \frac{2 - 2}{-8 - 6} = \frac{0}{-14} = 0.$$

(c) The slope does not exist, since the line is vertical. This is also seen by noting that if (7.6) is used, then the denominator is zero.

Example 2. Construct a line through $P(2, 1)$ that has (a) slope $5/3$; (b) slope $-5/3$.

Solution: From our discussion we see that if the slope of a line is a/b, where b is positive, then for every b units change in horizontal direction, the line rises or falls a units, depending on whether the quotient a/b is positive or negative, respectively. Thus if $P(2, 1)$ is on the line and $m = 5/3$, we can obtain another point on the line by starting at P and moving 3 units to the right and 5 units upward. This gives us the point $Q(5, 6)$, and the line is determined (see (i) of Fig. 7.7). Similarly, if

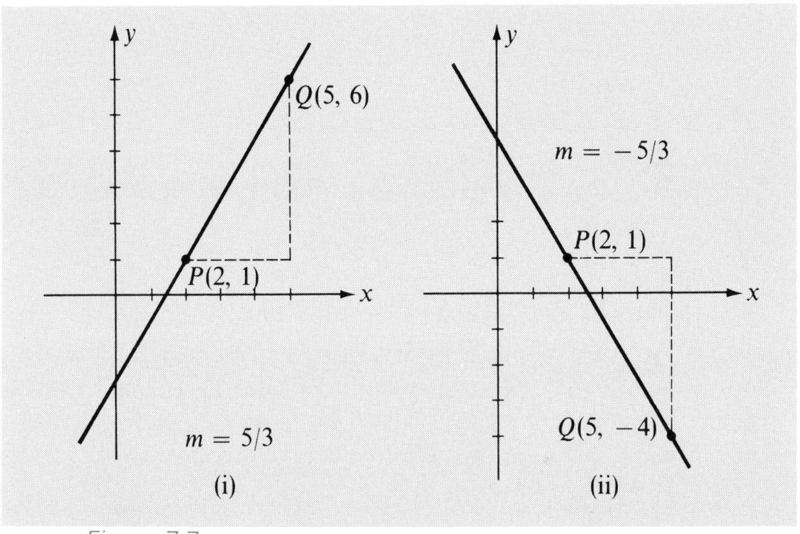

Figure 7.7

$m = -5/3$ we move 3 units to the right and 5 units downward, obtaining $Q(5, -4)$, as in (ii) of Fig. 7.7.

(7.7) Definition of Inclination

If l is a line which is not parallel to the x-axis and if P is the point of intersection of l and the x-axis, then the *inclination* of l is the smallest angle α through which the x-axis must be rotated in a counterclockwise direction about P in order to coincide with l. If l is horizontal, then $\alpha = 0°$.

We see, from (7.7), that if l is not horizontal, then $0° < \alpha < 180°$. The line shown in (i) of Fig. 7.8 illustrates the case in which $0° < \alpha < 90°$, and that in (ii) of Fig. 7.8 illustrates the case $90° < \alpha < 180°$.

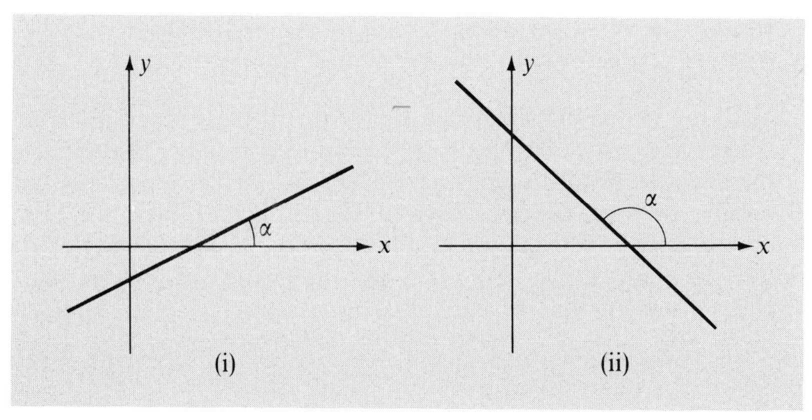

Figure 7.8

It follows from plane geometry that two lines are parallel if and only if they have the same inclination.

The next theorem states the connection between slope and inclination.

(7.8) Theorem

If l is a line with slope m and inclination α then $m = \tan \alpha$.

Proof: If l is horizontal, then $m = 0$, $\alpha = 0°$, and since $0 = \tan 0°$, the theorem is true. Next, suppose that l is not horizontal. Choose two points $P_1(x_1, y_1)$ and $P_2(x_2, y_2)$ on l, where we may assume, without loss of generality, that $x_1 < x_2$. We consider the two cases which may occur, namely (i) $y_1 < y_2$ or (ii) $y_2 < y_1$.

CASE (i): If $y_1 < y_2$, then m is positive and the line rises, as in Fig. 7.9.

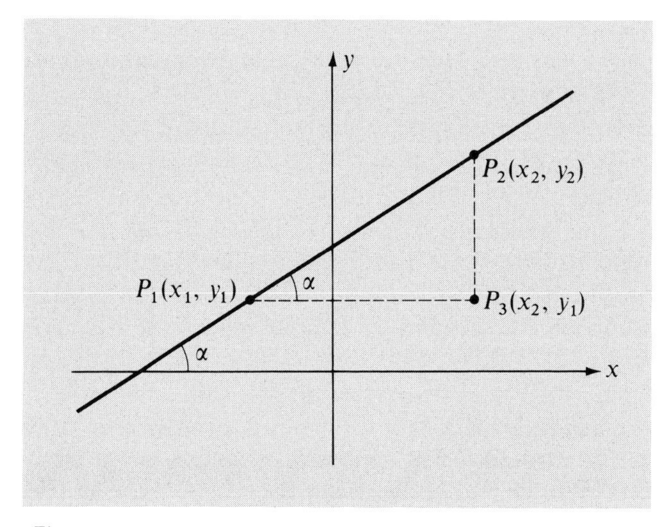

Figure 7.9

If we consider point $P_3(x_2, y_1)$, then it follows that α equals the angle at vertex P_1 in triangle $P_1P_2P_3$. Hence by (5.24)

$$\tan \alpha = \frac{d(P_2, P_3)}{d(P_1, P_3)} = \frac{|y_2 - y_1|}{|x_2 - x_1|} = \frac{y_2 - y_1}{x_2 - x_1}.$$

It now follows from (7.6) that $\tan \alpha = m$.

CASE (ii): If $y_2 < y_1$, then m is negative and the line falls, as in Fig. 7.10.

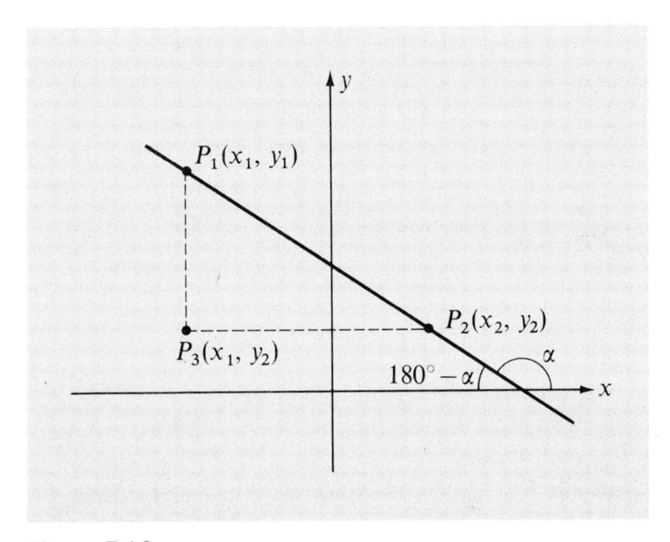

Figure 7.10

Considering $P_3(x_1, y_2)$ we see that the acute angle at vertex P_2 of triangle $P_1P_2P_3$ is $180° - \alpha$. Hence

$$\tan (180° - \alpha) = \frac{d(P_1, P_3)}{d(P_2, P_3)} = \frac{|y_2 - y_1|}{|x_2 - x_1|}.$$

Since $y_2 < y_1$, we have $y_2 - y_1 < 0$, and therefore by (1.8), $|y_2 - y_1| = -(y_2 - y_1)$. Consequently

$$\tan (180° - \alpha) = \frac{-(y_2 - y_1)}{x_2 - x_1}.$$

However, $\tan (180° - \alpha) = -\tan \alpha$ (see Exercise 25, p. 208) and hence

$$\tan \alpha = \frac{y_2 - y_1}{x_2 - x_1}$$

— that is, $\tan \alpha = m$.

(7.9) Corollary

If l_1 and l_2 are lines with slopes m_1 and m_2, respectively, then l_1 is parallel to l_2 if and only if $m_1 = m_2$.

Proof: Let α_1 and α_2 be the inclinations of l_1 and l_2, respectively. Then l_1 and l_2 are parallel if and only if $\alpha_1 = \alpha_2$ — that is, if and only if $\tan \alpha_1 = \tan \alpha_2$. The corollary now follows from (7.8).

EXAMPLE 3. (a) Find the slope m of a line whose inclination is 120°. (b) Find the inclination α of a line whose slope is 7/10.

Solutions: (a) By (7.8), $m = \tan 120° = -\sqrt{3}$.

(b) Since $\tan \alpha = 0.7$, $\alpha = \text{Tan}^{-1}(0.7)$. If an approximation is desired, then using Table 3, $\alpha \doteq 35°$.

The next theorem provides conditions for testing perpendicularity of lines.

(7.10) Theorem

If l_1 and l_2 are nonvertical lines with slopes m_1 and m_2, respectively, then l_1 and l_2 are perpendicular if and only if $m_1 m_2 = -1$.

Proof: Let α_1 and α_2 be the inclinations of l_1 and l_2, respectively. If l_1 and l_2 are perpendicular, then $\alpha_1 \neq \alpha_2$, and we may assume, without loss of generality, that $\alpha_2 > \alpha_1$. This gives us a situation similar to that illustrated in Fig. 7.11 and hence $\alpha_2 = \alpha_1 + 90°$ (Why?). Therefore, $\tan \alpha_2 = \tan(\alpha_1 + 90°) = -\cot \alpha_1 = -1/\tan \alpha_1$ (see Exercise 26, p. 208). Now using (7.8) we obtain $m_2 = -1/m_1$ — that is, $m_1 m_2 = -1$.

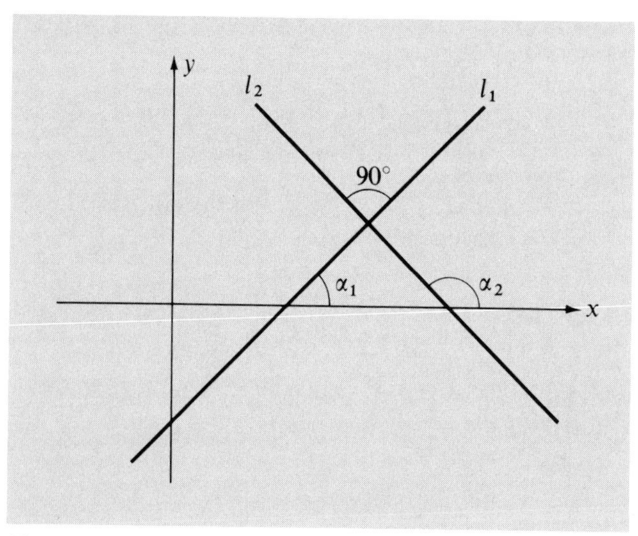

Figure 7.11

Conversely, if $m_1m_2 = -1$, then one slope is positive and the other is negative. If we assume that m_1 is positive and m_2 negative, then $0° < \alpha_1 < 90°$ and $90° < \alpha_2 < 180°$. From $m_2 = -1/m_1$ we obtain

$$\tan \alpha_2 = -1/\tan \alpha_1 = -\cot \alpha_1 = \tan (\alpha_1 + 90°).$$

Since $0° < \alpha_1 < 90°$, we have $90° < \alpha_1 + 90° < 180°$, and consequently $\alpha_2 = \alpha_1 + 90°$. Therefore the lines are perpendicular.

Another way of stating (7.10) is to say that l_1 and l_2 are perpendicular if and only if $m_1 = -1/m_2$ — that is, m_1 and m_2 are negative reciprocals of one another.

EXAMPLE 4. Prove that the triangle with vertices $A(-1, -3)$, $B(6, 1)$, and $C(2, -5)$ is a right triangle (cf. Example 3, p. 19).

Solution: Let m_1 denote the slope of the line through B and C, and let m_2 denote the slope of the line through A and C. By (7.6) we have

$$m_1 = \frac{1 - (-5)}{6 - 2} = \frac{3}{2}, \qquad m_2 = \frac{-3 - (-5)}{-1 - 2} = -\frac{2}{3}.$$

Thus $m_1m_2 = -1$, and hence the angle at C is a right angle.

EXERCISES

In each of Exercises 1–6 plot the points A and B and find the slope of the line through A and B.

1. $A(-4, 6)$, $B(-1, 18)$.
2. $A(6, -2)$, $B(-3, 5)$.
3. $A(2/3, 1/2)$, $B(2, 1)$.
4. $A(\sqrt{8}, \sqrt{12})$, $B(-\sqrt{2}, -\sqrt{27})$.
5. $A(-1, -3)$, $B(-1, 2)$.
6. $A(-3, 4)$, $B(2, 4)$.
7. Use (7.6) to show that $A(-3, 1)$, $B(5, 3)$, $C(3, 0)$, and $D(-5, -2)$ are vertices of a parallelogram.
8. Show that $A(2, 3)$, $B(5, -1)$, $C(0, -6)$, and $D(-6, 2)$ are vertices of a trapezoid.
9. Prove that the following points are vertices of a rectangle: $A(6, 15)$, $B(11, 12)$, $C(-1, -8)$, $D(-6, -5)$.
10. Prove that the points $A(1, 4)$, $B(6, -4)$, and $C(-15, -6)$ are vertices of a right triangle.

In Exercises 11 and 12 find the inclination of the line which has the given slope.

11. (a) -1. (b) $\sqrt{3}/3$. (c) 2. (d) $\tan 20°$.
12. (a) 1. (b) $\sqrt{3}$. (c) -4. (d) $\cot 55°$.

In Exercises 13 and 14 find the slope of the line with the given inclination.

13. (a) $0°$. (b) $120°$. (c) Arctan 1. (d) Arctan 4/3.
14. (a) $135°$. (b) $5\pi/6$. (c) Arcsin 1/2. (d) Arctan 1/2.

In Exercises 15 and 16 use slopes to show that the points A, B, and C lie on a straight line.

15. $A(1, -3)$, $B(-3, -11)$, $C(3, 1)$.
16. $A(0, 4)$, $B(6, 1)$, $C(-2, 5)$.

17–20. Find the slope of the perpendicular bisector of the line segment AB if A and B are the points given in Exercises 1–4.

21. If three consecutive vertices of a parallelogram are $A(-1, -3)$, $B(4, 2)$, and $C(-7, 5)$, find the fourth vertex.

22. Let $A(x_1, y_1)$, $B(x_2, y_2)$, $C(x_3, y_3)$, and $D(x_4, y_4)$ denote the vertices of an arbitrary quadrilateral. Prove that the line segments joining midpoints of adjacent sides form a parallelogram.

3 EQUATIONS OF LINES

The equation $y = b$, where b is a real number, may be considered as an equation in two variables x and y by writing it in the form $0 \cdot x + y = b$. Evidently, the solution set of the latter equation consists of all pairs of the form (x, b), where x may have *any* value and b is fixed. It follows that the graph of $y = b$ is a straight line parallel to the x-axis with y-intercept b. Conversely, every horizontal line is the graph of an equation of the form $y = b$. A similar argument gives us the result that the graph of the equation $x = a$ is a line parallel to the y-axis with x-intercept a. The graphs of these lines are illustrated in Fig. 7.12.

Let us now find an equation of a line l through a point $P_1(x_1, y_1)$ with slope m (only one such line exists). If $P(x, y)$ is any point with

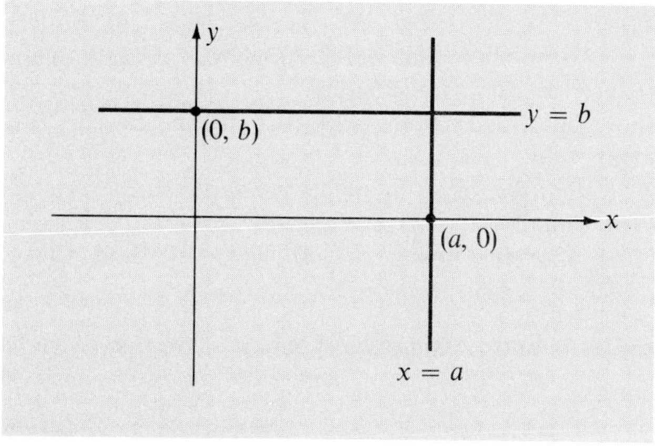

Figure 7.12

$x \neq x_1$ (see Fig. 7.13), then P is on l if and only if the slope of the line through P_1 and P is m — that is, if and only if

$$\frac{y - y_1}{x - x_1} = m.$$

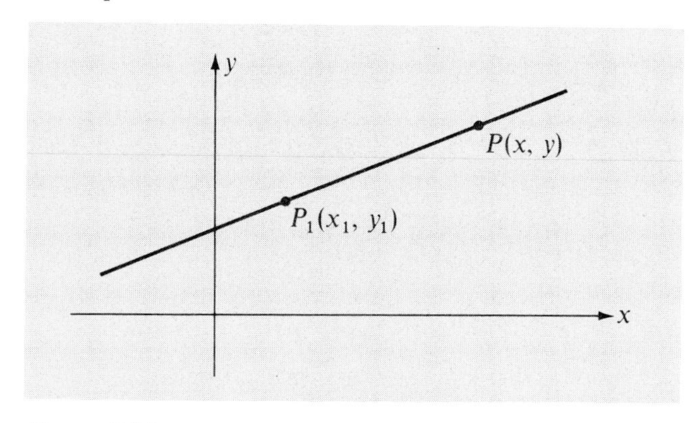

Figure 7.13

This equation may be written in the form

(7.11) $$y - y_1 = m(x - x_1).$$

Note that (x_1, y_1) is also a solution of the latter equation, and hence the points on l are precisely the points which correspond to the solution set of (7.11). Consequently, (7.11) is an equation for l which we shall refer to as the *point-slope form* for the equation of a line.

EXAMPLE 1. Find an equation of the line through the points $A(1, 7)$ and $B(-3, 2)$.

Solution: The slope m is given by

$$m = \frac{7 - 2}{1 - (-3)} = \frac{5}{4}.$$

Using the coordinates of A in the point-slope form gives us

$$y - 7 = \tfrac{5}{4}(x - 1),$$

which is equivalent to

$$4y - 28 = 5x - 5$$

or

$$5x - 4y + 23 = 0.$$

The same equation would have been obtained if the coordinates of point B had been substituted in (7.11).

EXAMPLE 2. Find an equation for the perpendicular bisector of the line segment AB, where A and B are the points given in Example 1.

Solution: By (7.5), the midpoint M of AB has coordinates

$$(\tfrac{1}{2}(-3 + 1), \tfrac{1}{2}(7 + 2)), \quad \text{or} \quad (-1, 9/2).$$

Since the slope of AB is 5/4, then by (7.10) the slope of the perpendicular bisector is $-4/5$. Applying the point-slope formula (7.11) we have

$$y - \frac{9}{2} = \left(-\frac{4}{5}\right)(x + 1).$$

Multiplying both sides by 10 and simplifying gives us $8x + 10y - 37 = 0$.

Equation (7.11) may be rewritten as

$$y = mx - mx_1 + y_1,$$

which is of the form

$$y = mx + b, \tag{7.12}$$

where $b \in \mathbf{R}$. The real number b is the y-intercept of the graph, as may be seen by setting $x = 0$. Since (7.12) displays the slope and y-intercept of l, it is called the *slope-intercept form* for the equation of a line. Conversely, given an equation of the form (7.12) with $m \neq 0$, we may write

$$y - 0 = m(x + b/m),$$

which we see, upon comparison with (7.11), has for its graph a straight line with slope m.

EXAMPLE 3. Find an equation of a line parallel to the line with equation $6x + 3y - 4 = 0$ and passing through the point $(5, -7)$.

Solution: The equation of the given line may be rewritten

$$3y = -6x + 4.$$

Dividing both sides by 3, we have

$$y = -2x + 4/3.$$

This is in slope-intercept form (7.12), with $m = -2$ and $b = 4/3$. Hence the slope is -2. By (7.9) the required line also has slope -2. Applying the point-slope formula, we obtain the equation

$$y + 7 = -2(x - 5),$$

or equivalently,

$$2x + y - 3 = 0.$$

The work we have done in this section shows that every straight line is the graph of an equation of the form

$$ax + by + c = 0, \tag{7.13}$$

where $a, b, c \in \mathbf{R}$, and a and b are not both zero. Such an equation is called a *linear equation* in x and y, or an *equation of the first degree* in x and y. We may show, conversely, that a linear equation in x and y has, for its graph, a straight line. Thus, given (7.13), with $b \neq 0$, we may write

$$y = (-a/b)x + (-c/b),$$

which, by the slope-intercept form, is the equation of a line with slope $-a/b$ and y-intercept $-c/b$. On the other hand, if $b = 0$ but $a \neq 0$, then we may write (7.13) as $x = -c/a$, which is the equation of a line parallel to the y-axis with x-intercept $-c/a$. Summarizing, we have proved the following theorem:

(7.14) Theorem

The graph of a linear equation $ax + by + c = 0$ is a straight line, and conversely every straight line is the graph of a linear equation.

As a corollary to the previous theorem it follows that the graph of a linear function is always a straight line. This proves the remark made on p. 34.

Example 4. Sketch the graph of the equation $2x - 5y = 8$.

Solution: We know from (7.14) that the graph is a straight line, and hence it is sufficient to find two points on the graph. Let us find the x- and y-intercepts. Substituting $y = 0$ in the given equation we obtain the x-intercept 4, whereas substituting $x = 0$ we see that the y-intercept is $-8/5$. This leads to the graph in Fig. 7.14. Another method of solution is to express the given equation in the slope-intercept form $y = (2/5)x - (8/5)$, and sketch a line through the point $(0, -8/5)$ with slope $2/5$.

If a line has slope m, we have seen that its equation may be written in the form $y = mx + b$, for some b. By letting b take on all real values, we obtain *all* lines with slope m. A collection of lines all of which have some common property is sometimes referred to as a *family of lines.*

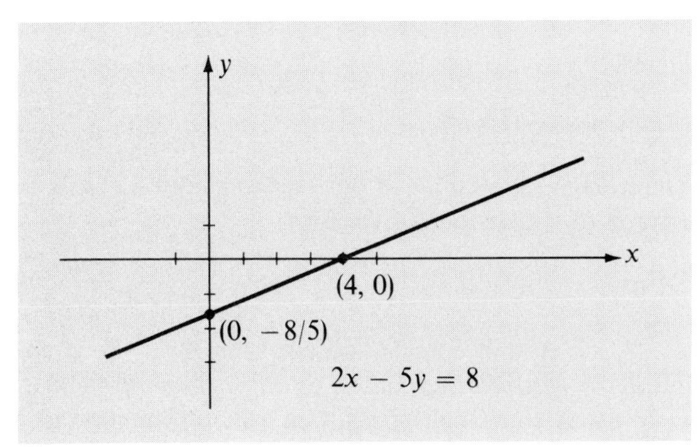

Figure 7.14

Figure 7.15 shows a few of the lines in the family $y = \frac{1}{2}x + b$. As another illustration, if b is held fixed and m is allowed to vary, then the equation

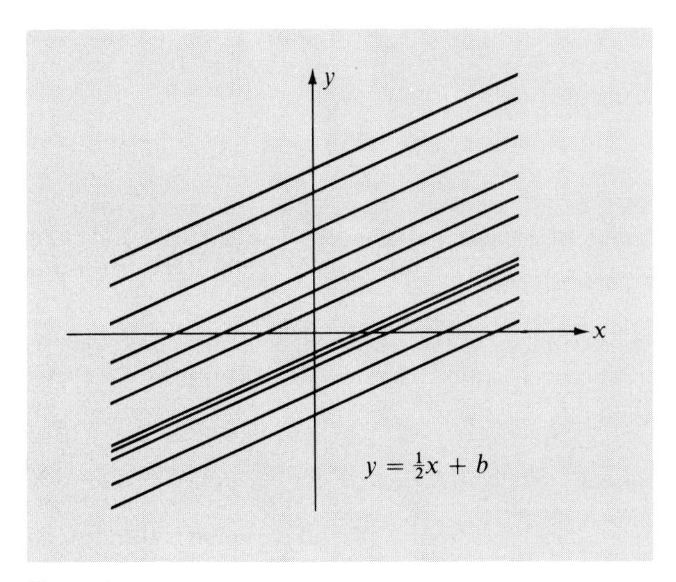

Figure 7.15

$y = mx + b$ defines the family of lines with y-intercept b. Several lines of the family $y = mx + 2$ are shown in Fig. 7.16.

Example 5. Find an equation of the family of lines which pass through the point $(-2, 1)$. Find the member of this family with x-intercept 5.

Solution: By the point-slope form (7.11), an equation of the family is

$$y - 1 = m(x + 2).$$

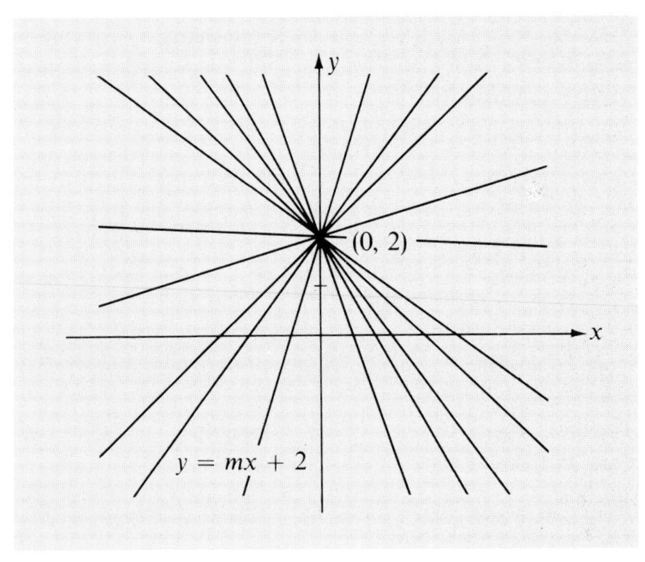

Figure 7.16

To find the member with x-intercept 5, we use the fact that if the point $P(5, 0)$ lies on the graph, then the pair $(5, 0)$ is in the solution set of the equation. Substituting, we obtain $0 - 1 = m(5 + 2)$, or $m = -1/7$. Hence the member of the family with x-intercept 5 is given by $y - 1 = (-1/7)(x + 2)$.

EXERCISES

In each of Exercises 1–10 find an equation for the line satisfying the given conditions.

1. Through $A(2, -6)$, slope $1/2$.
2. Slope -3, y-intercept 5.
3. Through $A(-5, -7)$ and $B(3, -4)$.
4. x-intercept -4, y-intercept 8.
5. Through $A(8, -2)$, y-intercept -3.
6. Slope 6, x-intercept -2.
7. Through $A(10, -6)$, parallel to (a) the y-axis, (b) the x-axis.
8. Through $A(-5, 1)$, perpendicular to (a) the y-axis, (b) the x-axis.
9. Through $A(7, -3)$, perpendicular to the line with equation $2x - 5y = 8$.
10. Through $(-\frac{3}{4}, -\frac{1}{2})$, parallel to the line with equation $x + 3y = 1$.
11. Given $A(3, -1)$ and $B(-2, 6)$, find an equation for the perpendicular bisector of the line segment AB.
12. Find an equation for the line which bisects the second and fourth quadrants.

13. Find equations for the altitudes of the triangle with vertices $A(-3, 2)$, $B(5, 4)$, $C(3, -8)$, and find the point at which they intersect.
14. Find equations for the medians of the triangle in Exercise 13, and find their point of intersection.

In each of Exercises 15–22 use the slope-intercept form to find the slope and y-intercept of the line with the given equation and sketch the graph.

15. $3x - 4y + 8 = 0$.
16. $2y - 5x = 1$.
17. $x + 2y = 0$.
18. $8x = 1 - 4y$.
19. $y = 4$.
20. $x + 2 = \frac{1}{2}y$.
21. $5x + 4y = 20$.
22. $y = 0$.

In each of Exercises 23–26 find an equation for the family of lines satisfying the given condition.

23. Through $A(-5, 7)$.
24. Slope 4.
25. y-intercept 3/5.
26. Parallel to the line with equation $3x - 2y + 4 = 0$.

In each of Exercises 27–30 find the common property of the lines in the family with the given equation.

27. $4x + 5y - k = 0$.
28. $y - k = 3(x - 1)$.
29. $y - 3 = -2(x - b)$.
30. $y = ax - 1$.

4 CONIC SECTIONS

Each of the geometric figures to be discussed in the remainder of this chapter can be obtained by intersecting a double-napped right circular cone with a plane. For this reason they are called *conic sections*, or sometimes simply *conics*. If, as in Fig. 7.17, the plane cuts entirely across one

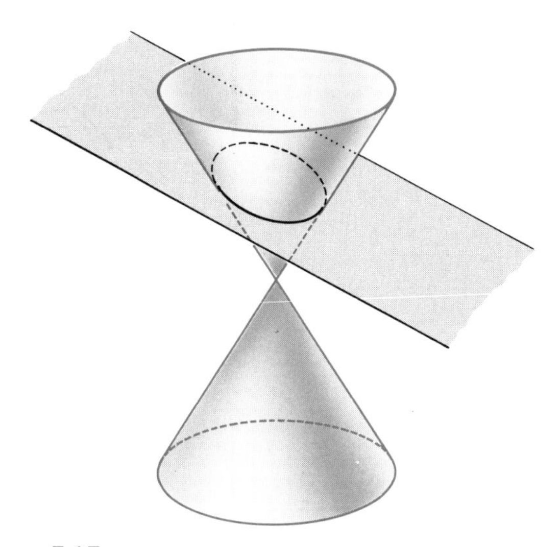

Figure 7.17

nappe of the cone, then the curve of intersection is called an *ellipse*, except for the case in which the plane is perpendicular to the axis of the cone, in which case it is a circle. If the plane does not cut entirely across one nappe and does not intersect both nappes, as illustrated in Fig. 7.18,

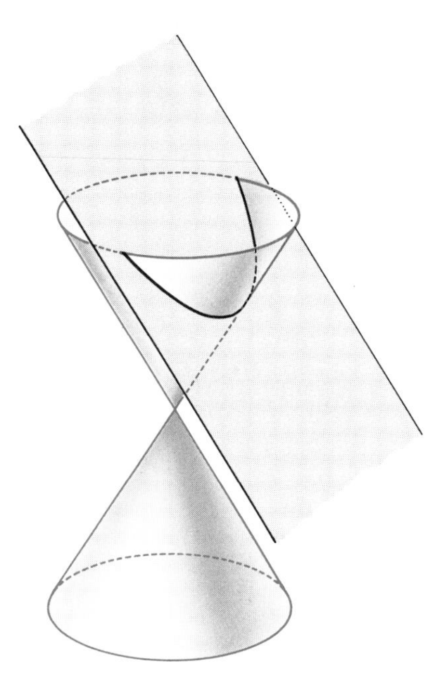

Figure 7.18

then the curve of intersection is a *parabola*. If the plane cuts through both nappes of the cone, as in Fig. 7.19, then the resulting figure is called a *hyperbola*.

By changing the position of the plane and the shape of the cone, conics can be made to vary considerably. For certain positions of the plane there result what are called *degenerate conics*. For example, if the plane intersects the cone only at the vertex, then the conic consists of one point. If the axis of the cone lies on the plane, then the points of intersection lie on a pair of intersecting lines. Finally, if we begin with the parabolic case, as in Fig. 7.18, and move the plane parallel to its initial position until it coincides with one of the generators of the cone, a straight line results.

The conic sections were studied extensively by the early Greek mathematicians, who used the methods of Euclidean geometry. They discovered the properties which enable us to define conics in terms of points (foci) and lines (directrices) in the plane of the conic. These definitions will be introduced in subsequent sections. Reconciliation of the latter definitions with the point of view of intersections of a plane with a cone requires proofs which we shall not go into here.

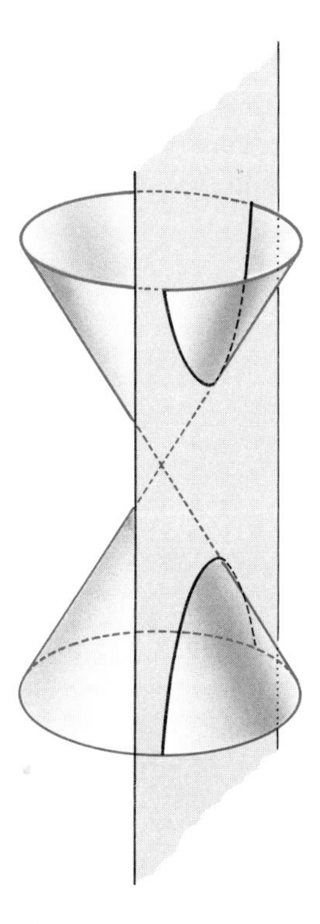

Figure 7.19

A remarkable fact about conic sections is that, although they were studied thousands of years ago, they are far from obsolete. Indeed, they are important tools for present-day investigations in outer space and for the study of the behavior of atomic particles. It is shown in physics that if a mass moves under the influence of what is called an *inverse square force field*, then the path may be described by means of a conic section. Examples of inverse square fields, which are very common in physics, are gravitational and electromagnetic fields. Planetary orbits are elliptical. If the ellipse is very "flat," the curve resembles the path of a comet. The hyperbola is useful for describing the path of an alpha particle in the electric field of the nucleus of an atom. The parabola is also useful in atomic physics for describing the paths of certain particles. The interested person can find hundreds of other applications of conic sections.

5 PARABOLAS

The notion of parabola was mentioned briefly in Chapter Two (see p. 61). Parabolas are very useful in applications of mathematics to the physical world. For example, it can be shown that if a projectile is fired and it is assumed that it is acted upon only by the force of gravity (that is, air resistance and other outside factors are ignored), then the path of the projectile is parabolic. Properties of parabolas are used in the design of mirrors for telescopes and searchlights. These are only a few of many physical applications.

(7.15) Definition of Parabola

A *parabola* is the set of all points in a plane equidistant from a fixed point F (the *focus*) and a fixed line l (the *directrix*) in the plane.

We shall assume that F is not on l, for otherwise the parabola degenerates into a straight line. If P is any point in the plane, let P' denote the point on l determined by a line through P which is perpendicular to l. A typical situation is illustrated in Fig. 7.20, where dashes indicate possible positions for P. According to (7.15), P is on the parabola if and only if $d(P, F) = d(P, P')$.

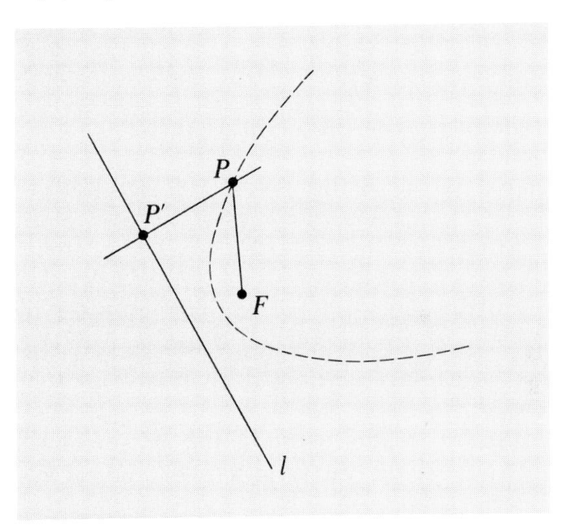

Figure 7.20

The line l' through F, perpendicular to l, is called the *axis* of the parabola. If Q is the point of intersection of l' and l, then the midpoint V of the segment from Q to F is on the parabola, since it is equidistant from l and F (see Fig. 7.21). The point V is called the *vertex* of the parabola.

In order to use the methods of coordinate geometry, it is necessary to introduce a coordinate system. For simplicity, we choose the y-axis along the axis of the parabola and with the origin at the vertex V, as in Fig. 7.22. The focus F then has coordinates $(0, p)$ for some real number $p \neq 0$, and the equation of the directrix is $y = -p$. By the distance formula (1.15) a point $P(x, y)$ is on the parabola if and only if

$$\sqrt{(x - 0)^2 + (y - p)^2} = \sqrt{(x - x)^2 + (y + p)^2}.$$

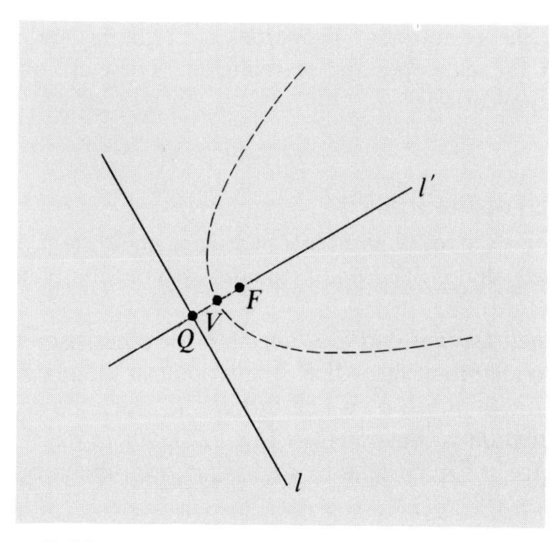

Figure 7.21

This equation is equivalent to

$$(x - 0)^2 + (y - p)^2 = (y + p)^2,$$

which simplifies to

$$x^2 = 4py. \tag{7.16}$$

The latter equation is called the *standard form* for the equation of a parabola with focus at $F(0, p)$ and directrix $y = -p$. If $p > 0$, the parabola *opens upward*, as in Fig. 7.22, whereas if $p < 0$, the parabola *opens downward.*

An analogous situation exists if the axis of the parabola is taken along the x-axis. If the vertex is $V(0, 0)$, the focus $F(p, 0)$, and the

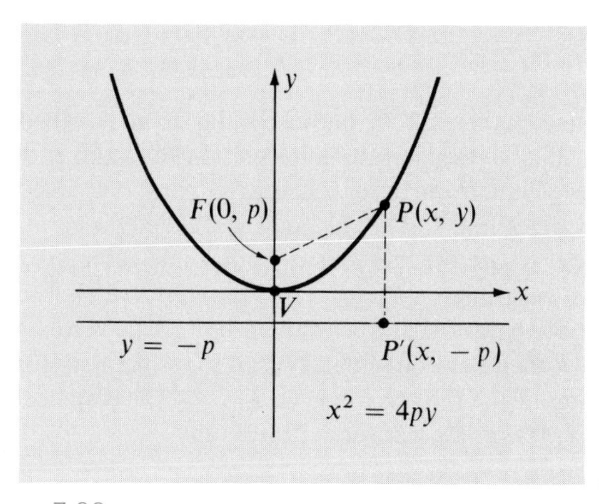

Figure 7.22

directrix $x = -p$ (see Fig. 7.23), then, using the same type of argument, we obtain the *standard form*

$$y^2 = 4px. \tag{7.17}$$

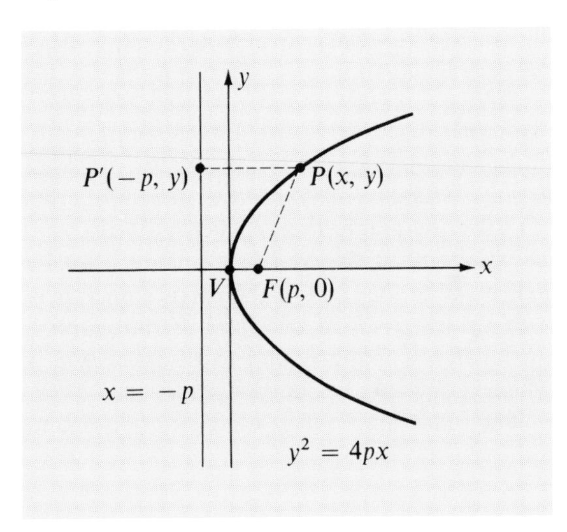

Figure 7.23

EXAMPLE 1. Find the focus and directrix of the parabola that has the equation $y^2 = -6x$, and sketch the graph of the equation.

Solution: The equation has the form (7.17) with $4p = -6$, and hence $p = -3/2$. Consequently, the focus is $F(-3/2, 0)$ and the equation of the directrix is $x = 3/2$. For the sketch see Fig. 7.24.

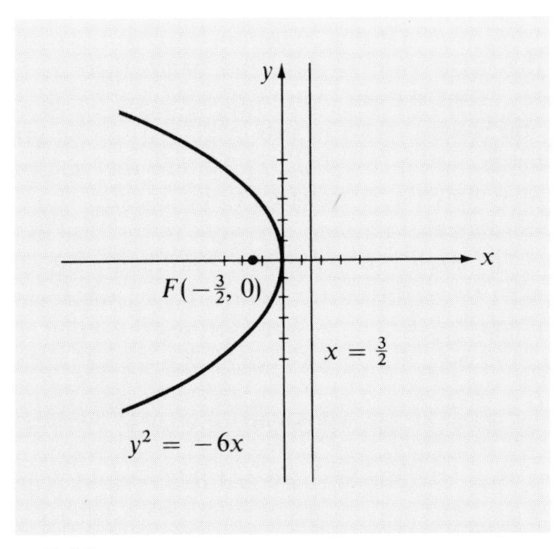

Figure 7.24

EXAMPLE 2. Find an equation of the parabola that has vertex at the origin, opens upward, and passes through the point $P(-3, 7)$.

Solution: The general form of the equation is given by (7.16). If P is on the parabola, then $(-3, 7)$ is in the solution set of the equation. Hence we must have $(-3)^2 = 4p(7)$, or $p = 9/28$. Substituting for p in (7.16) leads to the desired equation $x^2 = (9/7)y$, or $7x^2 = 9y$.

The parabolas discussed above are symmetric with respect to one of the coordinate axes, in the sense of the following discussion. A graph in a coordinate plane is said to be *symmetric with respect to the y-axis* if whenever $P(x, y)$ is on the graph, then $P'(-x, y)$ is also on the graph. From an intuitive viewpoint, if the coordinate plane is folded along the y-axis until the half-plane to the right of the y-axis coincides with the half-plane to the left of the y-axis, then the parts of the graph in each half-plane coincide. Similarly, a graph is said to be *symmetric with respect to the x-axis* if whenever $P(x, y)$ lies on the graph, then $P''(x, -y)$ also lies on the graph. The following result is useful for determining symmetries.

(7.18) Tests for Symmetry

(i) The graph of an equation is symmetric with respect to the y-axis if and only if substitution of $-x$ for x does not change the solution set of the equation.

(ii) The graph of an equation is symmetric with respect to the x-axis if and only if substitution of $-y$ for y does not change the solution set of the equation.

The proof of (7.18) is fairly straightforward and will be omitted. As an illustration, replacing x by $-x$ in (7.16) gives us $(-x)^2 = 4py$, which is the same as $x^2 = 4py$, and hence there is no change in the solution set. Thus the parabola with equation (7.16) is symmetric with respect to the y-axis. This is also evident from the graph in Fig. 7.22. In like manner, using (ii) of (7.18), we see that the parabola given by $y^2 = 4px$ is symmetric with respect to the x-axis.

It is not difficult to extend our work to the case in which the axis of the parabola is parallel to one of the coordinate axes. Thus in Fig. 7.25 we have taken the vertex at the point $V(h, k)$, the focus at $F(h, k + p)$, and the directrix $y = k - p$. As before, the point $P(x, y)$ is on the parabola if and only if $d(P, F) = d(P, P')$— that is, if and only if

$$\sqrt{(x - h)^2 + (y - k - p)^2} = \sqrt{(x - x)^2 + (y - k + p)^2}.$$

We leave it as an exercise to show that this equation can be simplified to

$$(x - h)^2 = 4p(y - k), \tag{7.19}$$

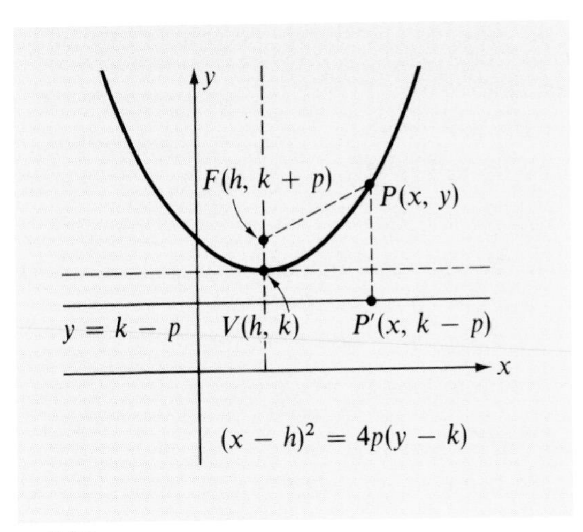

Figure 7.25

which is called the *standard form* for the equation of a parabola with vertex at (h, k) and axis parallel to the y-axis. As a special case, when $(h, k) = (0, 0)$, then (7.19) reduces to (7.16).

If we square the left side of (7.19) and simplify, we obtain an equation of the form

$$y = ax^2 + bx + c,$$

where a, b, and c are real numbers. Conversely, given such an equation, then by completing the square in x, it is possible to arrive at the standard form (7.19). This proves the remark made earlier (see p. 61) that the graph of a quadratic function is a parabola with vertical axis. An illustration of this appears in Example 4 below.

In similar fashion, the graph of the equation

$$(y - k)^2 = 4p(x - h) \qquad (7.20)$$

is a parabola with horizontal axis and vertex $V(h, k)$. The parabola opens to the right or left according to whether p is positive or negative.

Example 3. Find an equation of the parabola with vertex $(4, -1)$, with axis parallel to the y-axis, and which passes through the origin.

Solution: By (7.19) the equation is of the form

$$(x - 4)^2 = 4p(y + 1).$$

If the origin is on the parabola, then $(0, 0)$ is in the solution set of this equation, and hence $(0 - 4)^2 = 4p(0 + 1)$. Thus $16 = 4p$ and $p = 4$. Therefore the desired equation is $(x - 4)^2 = 16(y + 1)$.

EXAMPLE 4. Discuss and sketch the graph of the equation

$$2y = x^2 + 8x + 22.$$

Solution: By our previous remarks, the graph of the equation is a parabola with a vertical axis. Writing

$$x^2 + 8x = 2y - 22,$$

we complete the square on the left by adding 16 to both sides. This gives us

$$x^2 + 8x + 16 = 2y - 6.$$

The latter equation may be written

$$(x + 4)^2 = 2(y - 3),$$

which is in the form (7.19) with $h = -4$, $k = 3$, and $p = 1/2$. Hence the vertex is $(-4, 3)$. Since $p = 1/2 > 0$, the parabola opens upward with focus at $(-4, 3 + 1/2)$, or $(-4, 7/2)$. The equation of the directrix is $y = k - p$, or $y = 5/2$. The parabola is sketched in Fig. 7.26.

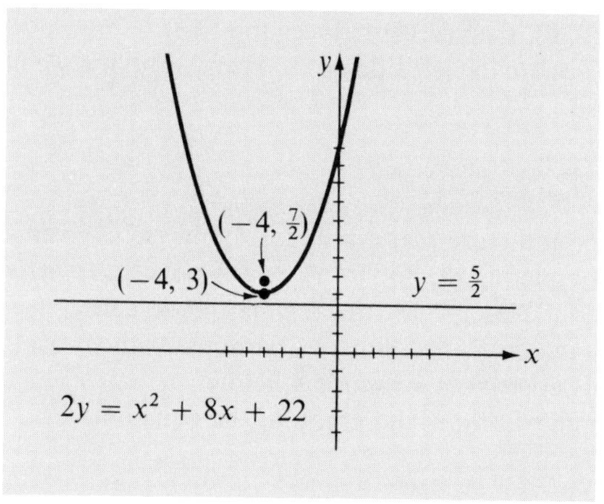

Figure 7.26

In a manner similar to our work with lines, we can discuss *families of parabolas*. As an illustration, the equation $(x - 4)^2 = 4p(y + 1)$ employed in the solution of Example 3 defines a family of parabolas with vertex at $(4, -1)$.

EXERCISES

In each of Exercises 1–12 find the focus and directrix of the parabola with the given equation and sketch the graph.

1. $y^2 = 8x.$
2. $y^2 = (1/2)x.$
3. $x^2 = -12y.$
4. $x^2 = -3y.$

5. $2y^2 = -3x$.
6. $y^2 = -100x$.
7. $8x^2 = y$.
8. $x^2 = y$.
9. $(y - 4)^2 = 4(x + 1)$.
10. $4(x - 5)^2 + y = 3$.
11. $y = x^2 - 4x + 2$.
12. $y = 8x^2 + 16x + 10$.

In each of Exercises 13–18 find an equation for the parabola that satisfies the given conditions.

13. Focus (2, 0), directrix $x = -2$.
14. Focus $(0, -4)$, directrix $y = 4$.
15. Focus (6, 4), directrix $y = -2$.
16. Focus $(-3, -2)$, directrix $y = 1$.
17. Vertex at the origin, symmetric to the y-axis, and passing through the point $A(2, -3)$.
18. Vertex $V(-3, 5)$, axis parallel to the x-axis, and passing through $A(5, 9)$.
19. Find an equation for the family of parabolas which have vertex at $V(6, -11)$ and axis parallel to the x-axis.
20. Find an equation for the family of parabolas having focus on the y-axis, opening upward, and such that the distance between the focus and directrix is 2.
21. A searchlight reflector is designed so that a cross section through its axis is a parabola and the light source is at the focus. Find the focus if the reflector is 3 feet across at the opening and 1 foot deep.
22. Prove that the point on a parabola that is closest to the focus is the vertex.
23. Show that the equation on line 36 of p. 264 may be written in the form (7.19).
24. Derive (7.17).
25. Derive (7.20).
26. If f is an even function (cf. Exercise 17, p. 33), prove that the graph of f is symmetric with respect to the y-axis.

6 ELLIPSES

Another useful geometric figure, the ellipse, may be defined as follows.

(7.21) Definition of Ellipse

An *ellipse* is the set of all points in a plane, the sum of whose distances from two fixed points in the plane (the *foci*) is constant.

It is known that the orbits of planets in the solar system are elliptical, with the sun at one of the foci. This is only one of many important applications of ellipses.

There is an easy way to construct an ellipse on paper. Insert two thumbtacks in the paper at points labeled F and F' and fasten the ends of a piece of string to the thumbtacks. If the string is now looped around a pencil and drawn taut at point P, as in Fig. 7.27, then moving the pencil

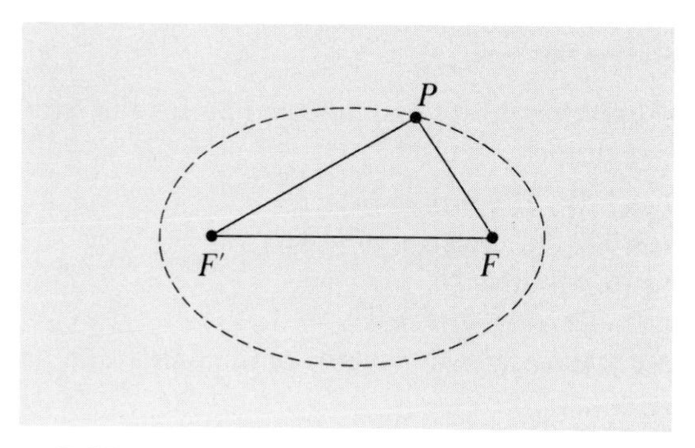

Figure 7.27

and at the same time keeping the string taut, the sum of the distances $d(F, P)$ and $d(F', P)$ is the length of the string, and hence is constant. The pencil therefore traces out a figure which resembles an ellipse with foci at F and F'. By varying the positions of F and F' but keeping the length of string fixed, the shape of the ellipse can be made to change considerably. Thus if F and F' are far apart, in the sense that $d(F, F')$ is almost the same as the length of the string, then the ellipse is quite "flat." On the other hand, if $d(F, F')$ is close to zero, the ellipse is almost circular. Indeed, if one takes $F = F'$, a circle is obtained.

By introducing suitable coordinate systems we may derive simple equations for ellipses. Let us choose the x-axis as the line through the two foci F and F', with the origin at the midpoint of the segment $F'F$. This point is called the *center* of the ellipse. If F has coordinates $(c, 0)$, where $c > 0$, then, as shown in Fig. 7.28, F' has coordinates $(-c, 0)$ and hence the distance between F and F' is $2c$. Let the constant sum of the distances of P from F and F' be denoted by $2a$, where in order to get points not on

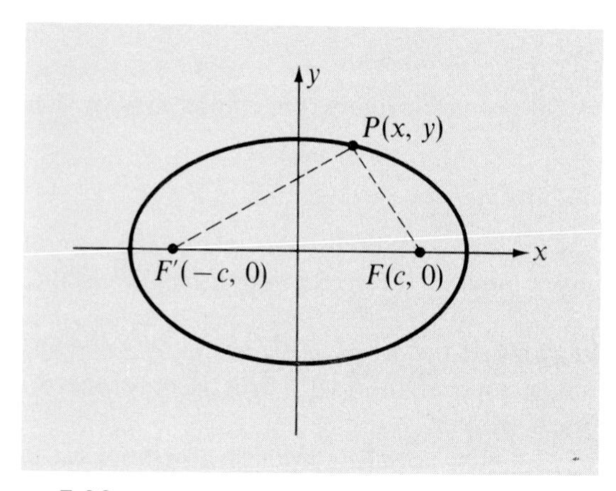

Figure 7.28

the x-axis we must have $2a > 2c$, that is, $a > c$ (Why?). By (7.21), $P(x, y)$ is on the ellipse if and only if

$$d(P, F) + d(P, F') = 2a,$$

or, by the distance formula,

$$\sqrt{(x - c)^2 + (y - 0)^2} + \sqrt{(x + c)^2 + (y - 0)^2} = 2a.$$

Writing this equation as

$$\sqrt{(x - c)^2 + y^2} = 2a - \sqrt{(x + c)^2 + y^2}$$

and squaring both sides, we obtain

$$x^2 - 2cx + c^2 + y^2 = 4a^2 - 4a\sqrt{(x + c)^2 + y^2} + x^2 + 2cx + c^2 + y^2,$$

which simplifies to

$$a\sqrt{(x + c)^2 + y^2} = a^2 + cx.$$

Squaring both sides of the last equation gives us

$$a^2(x^2 + 2cx + c^2 + y^2) = a^4 + 2a^2cx + c^2x^2,$$

which may be written in the form

$$x^2(a^2 - c^2) + a^2y^2 = a^2(a^2 - c^2).$$

We now divide both sides by $a^2(a^2 - c^2)$, obtaining

$$\frac{x^2}{a^2} + \frac{y^2}{a^2 - c^2} = 1.$$

For convenience, we let

(7.22) $$b^2 = a^2 - c^2, \quad b > 0,$$

in the preceding equation, obtaining

(7.23) $$\frac{x^2}{a^2} + \frac{y^2}{b^2} = 1.$$

Since $c > 0$, we have from (7.22) that $a^2 > b^2$ and hence $a > b$. Equation (7.23) is called the *standard form* for the equation of an ellipse with foci on the x-axis and center at the origin.

The x-intercepts may be found by setting $y = 0$. Doing so gives us $x^2/a^2 = 1$, or $x^2 = a^2$, and consequently the x-intercepts are a and $-a$. The corresponding points $V(a, 0)$ and $V'(-a, 0)$ on the graph are called the *vertices* of the ellipse, and the line segment $V'V$ is referred to as the *major axis*. Setting $x = 0$ in (7.23), we obtain the y-intercepts b and $-b$. The segment from $M'(0, -b)$ to $M(0, b)$ is called the *minor axis* of the ellipse. Note that the major axis is longer than the minor axis, since $a > b$.

By the test for symmetry (7.18) we see that the ellipse is symmetric to both the x-axis and the y-axis. It is also *symmetric to the origin*, in the sense that when the point $P(x, y)$ is on the graph, then the point $P'(-x, -y)$ is also on the graph. Note that P' can be found by extending the line segment from P to the origin O *through* O a distance equal to $d(O, P)$. The following result is useful for investigating symmetry with respect to the origin.

(7.24) Test for Symmetry

The graph of an equation is symmetric with respect to the origin if and only if the simultaneous substitution of $-x$ for x and $-y$ for y does not change the solution set of the equation.

EXAMPLE 1. Discuss and sketch the graph of the equation

$$4x^2 + 18y^2 = 36.$$

Solution: To obtain the standard form, we divide both sides of the given equation by 36 and simplify. This leads to

$$\frac{x^2}{9} + \frac{y^2}{2} = 1,$$

which is in the form (7.23) with $a^2 = 9$ and $b^2 = 2$. Thus $a = 3$ and $b = \sqrt{2}$, and so the endpoints of the major axis are $(\pm 3, 0)$ and the endpoints of the minor axis are $(0, \pm\sqrt{2})$. From (7.22) we obtain $c^2 = a^2 - b^2 = 9 - 2 = 7$, or $c = \sqrt{7}$, and hence the foci are $(\pm\sqrt{7}, 0)$. The graph is shown in Fig. 7.29.

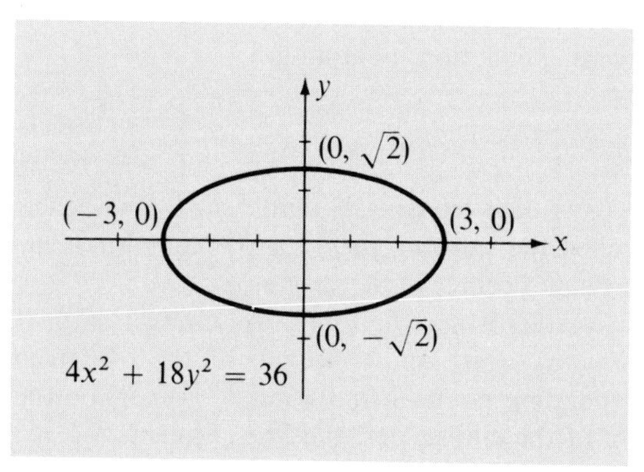

Figure 7.29

EXAMPLE 2. Find an equation of the ellipse with vertices $(\pm 4, 0)$ and foci $(\pm 2, 0)$.

Solution: Substituting $a = 4$ and $c = 2$ in (7.22), we have $b^2 = 16 - 4 = 12$. Then (7.23) gives us the equation

$$\frac{x^2}{16} + \frac{y^2}{12} = 1.$$

Multiplying by the number 48 produces the equivalent equation

$$3x^2 + 4y^2 = 48.$$

It is sometimes convenient to choose the major axis of the ellipse along the y-axis. If the foci are $(0, \pm c)$, then, by the same type of argument used to derive (7.23), we obtain the equation

$$\frac{x^2}{b^2} + \frac{y^2}{a^2} = 1, \tag{7.25}$$

where $a > b$. As before, the connection between a, b, and c is given by $b^2 = a^2 - c^2$. Thus we see that an equation of an ellipse with center at the origin and foci on a coordinate axis can always be written in the form

$$\frac{x^2}{p} + \frac{y^2}{q} = 1$$

or

$$qx^2 + py^2 = pq,$$

where p and q are positive. If $p > q$, then the x-axis is the major axis, whereas if $q > p$, then the y-axis is the major axis. It is unnecessary to memorize these facts, since in working problems the major axis can be determined by examining the x- and y-intercepts.

EXAMPLE 3. Sketch the graph of the equation $9x^2 + 4y^2 = 25$.

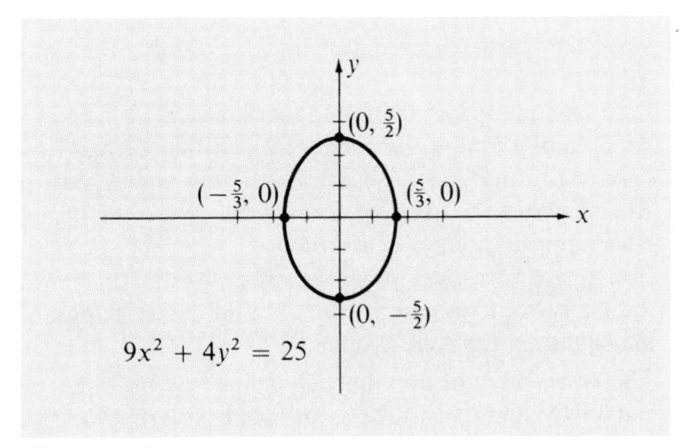

Figure 7.30

Solution: The graph is an ellipse with center at the origin and foci on one of the coordinate axes. To find the x-intercepts, we let $y = 0$, obtaining $9x^2 = 25$, or $x = \pm 5/3$. Similarly, to find the y-intercepts, we let $x = 0$, obtaining $4y^2 = 25$, or $y = \pm 5/2$. This enables us to sketch the ellipse (see Fig. 7.30). Since $5/3 < 5/2$, the major axis is on the y-axis.

EXERCISES

In each of Exercises 1–8 sketch the graph of the equation and give coordinates of the vertices and foci.

1. $\dfrac{x^2}{9} + \dfrac{y^2}{4} = 1.$
2. $\dfrac{x^2}{25} + \dfrac{y^2}{16} = 1.$
3. $4x^2 + y^2 = 16.$
4. $y^2 + 9x^2 = 9.$
5. $5x^2 + 2y^2 = 10.$
6. $(1/2)x^2 + 2y^2 = 8.$
7. $4x^2 + 25y^2 = 1.$
8. $10y^2 + x^2 = 5.$

In each of Exercises 9–14 find an equation for the ellipse satisfying the given conditions.

9. Vertices $V(\pm 8, 0)$, foci $F(\pm 5, 0)$.
10. Vertices $V(0, \pm 7)$, foci $F(0, \pm 2)$.
11. Vertices $V(0, \pm 5)$, length of minor axis 3.
12. Foci $F(\pm 3, 0)$, length of minor axis 2.
13. Vertices $V(0, \pm 6)$, passing through (3, 2).
14. Center at the origin, symmetric with respect to both axes, passing through $A(2, 3)$ and $B(6, 1)$.

In each of Exercises 15 and 16 find the points of intersection of the graphs of the given equations. Sketch both graphs on the same coordinate axes, showing points of intersection.

15. $\begin{cases} x^2 + 4y^2 = 20 \\ x + 2y = 6. \end{cases}$
16. $\begin{cases} x^2 + 4y^2 = 36 \\ x^2 + y^2 = 12. \end{cases}$
17. Find an equation of the family of ellipses with vertices $V(\pm 10, 0)$. Find the particular member of this family that passes through $A(5, 2)$.
18. Find an equation of the family of ellipses with foci $F(\pm 1, 0)$.
19. An arch of a bridge is semielliptical with major axis horizontal. The base of the arch is 30 feet across and the highest part of the arch is 10 feet above the horizontal roadway. Find the height of the arch 6 feet from the center of the base.
20. If a square with sides parallel to the coordinate axes is inscribed in the ellipse with equation $x^2/a^2 + y^2/b^2 = 1$, express the area A of the square in terms of a and b.
21. The *eccentricity* of an ellipse is defined as the ratio $(\sqrt{a^2 - b^2})/a$. If a is fixed and b varies, describe the general shape of the ellipse when the eccentricity is close to 1 and when it is close to zero.

7 HYPERBOLAS

The definition of hyperbola is similar to that of an ellipse. The only change is that instead of using the *sum* of distances from two fixed points we use the *difference*.

(7.26) Definition of Hyperbola

A *hyperbola* is the set of all points in a plane, the difference of whose distances from two fixed points in the plane (the *foci*) is constant.

To find a simple equation for a hyperbola, we choose a coordinate system with foci at $F(c, 0)$ and $F'(-c, 0)$ and denote the (constant) distance by $2a$. Referring to Fig. 7.31, we see that a point $P(x, y)$ is on the hyperbola if and only if either one of the following is true:

(7.27)
$$d(P, F) - d(P, F') = 2a,$$
$$d(P, F') - d(P, F) = 2a.$$

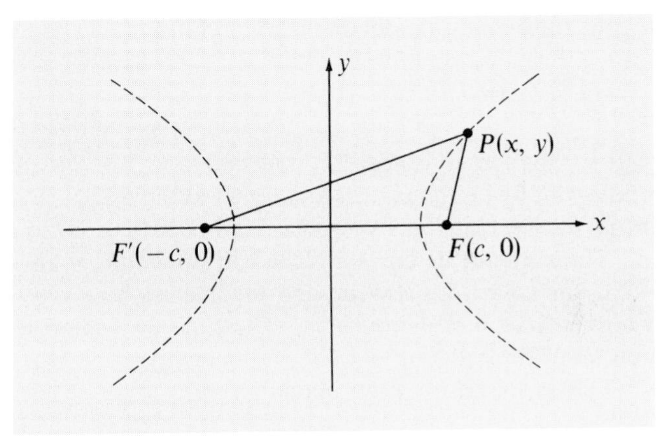

Figure 7.31

For hyperbolas (unlike ellipses) we need $a < c$ in order to obtain points on the hyperbola which are not on the x-axis, for if P is such a point, then from Fig. 7.31 we see that

$$d(P, F) < d(F', F) + d(P, F'),$$

since the length of one side of a triangle is always less than the sum of the lengths of the other two sides. Similarly,

$$d(P, F') < d(F', F) + d(P, F).$$

Equivalent forms for the previous two inequalities are

$$d(P, F) - d(P, F') < d(F', F),$$
$$d(P, F') - d(P, F) < d(F', F).$$

From (7.27) and the fact that $d(F', F) = 2c$, the latter inequalities imply that $2a < 2c$, or $a < c$.

Equations (7.27) may be replaced by the single equation

$$|d(P, F) - d(P, F')| = 2a.$$

It then follows from the distance formula that the equation of the hyperbola is given by

$$|\sqrt{(x - c)^2 + (y - 0)^2} - \sqrt{(x + c)^2 + (y - 0)^2}| = 2a.$$

Employing the type of simplification procedure used to derive the equation of an ellipse, we arrive at the equivalent equation

$$\frac{x^2}{a^2} - \frac{y^2}{c^2 - a^2} = 1.$$

For convenience, let

(7.28) $$b^2 = c^2 - a^2, \quad b > 0,$$

in the preceding equation, obtaining

(7.29) $$\frac{x^2}{a^2} - \frac{y^2}{b^2} = 1.$$

The latter equation is called the *standard form* for the equation of a hyperbola with foci on the x-axis and center at the origin. By the tests for symmetry we see that this hyperbola is symmetric to both axes and the origin. The x-intercepts are $\pm a$. The corresponding points $V(a, 0)$ and $V'(-a, 0)$ are called the *vertices*, and the line segment $V'V$ is known as the *transverse axis* of the hyperbola. There are no y-intercepts, since the solution set of the equation $-y^2/b^2 = 1$ is empty.

If (7.29) is solved for y, we obtain

(7.30) $$y = \pm \frac{b}{a}\sqrt{x^2 - a^2}.$$

Hence an excluded region of the plane consists of those points (x, y) such that $x^2 - a^2 < 0$, that is, $-a < x < a$. On the other hand, there *are* points on the graph where $x \geq a$ and $x \leq -a$.

In order to arrive at a precise description of the graph, it is necessary to investigate the position of the point $P(x, y)$ on the hyperbola when x is numerically very large. If $x \geq a$, we may write (7.30) in the form

$$y = \pm \frac{b}{a} x \sqrt{1 - \frac{a^2}{x^2}}.$$

It follows that when x is large (in comparison to a), the radicand is close to 1, and hence the ordinate y of the point $P(x, y)$ on the hyperbola is close to either $(b/a)x$ or $-(b/a)x$. This means that the point $P(x, y)$ is close to the line with equation $y = (b/a)x$ when y is positive, or the line with equation $y = -(b/a)x$ when y is negative. As x increases (or decreases), we say that the point $P(x, y)$ *approaches* one of these lines. A corresponding situation exists when $x \leq -a$. The lines

(7.31) $$y = \pm \frac{b}{a} x$$

are called the *asymptotes* of the hyperbola with equation (7.29). The asymptotes serve as an excellent guide for sketching the graph. This is illustrated in Fig. 7.32, where we have represented the asymptotes by

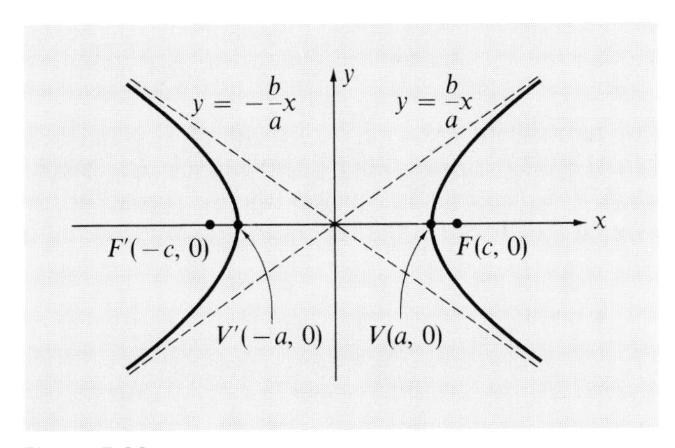

Figure 7.32

dashed lines and indicated the manner in which the points on the hyperbola approach the asymptotes as x increases or decreases. The two curves which make up the hyperbola are called the *branches* of the hyperbola.

There is a convenient way to sketch asymptotes. We plot the vertices $V(a, 0)$ and $V'(-a, 0)$ and the points $W(0, b)$ and $W'(0, -b)$. The line segment $W'W$ of length $2b$ is called the *conjugate axis* of the hyperbola. If horizontal and vertical lines are drawn through the endpoints of the conjugate and transverse axes respectively, then the diagonals of the resulting rectangles have slopes b/a and $-b/a$ (see Fig. 7.33), and hence, by extending these diagonals, we obtain the asymptotes. The hyperbola is then sketched by using the asymptotes as a guide.

EXAMPLE 1. Discuss and sketch the graph of the equation

$$9x^2 - 4y^2 = 36.$$

Solution: Dividing both sides by 36, we have

$$\frac{x^2}{4} - \frac{y^2}{9} = 1,$$

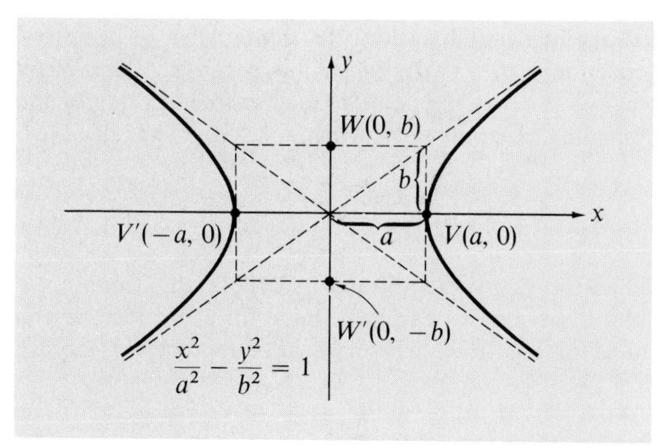

Figure 7.33

which is in the standard form (7.29) with $a^2 = 4$ and $b^2 = 9$. Hence $a = 2$ and $b = 3$. The vertices $(\pm 2, 0)$ and the endpoints $(0, \pm 3)$ of the conjugate axis determine a rectangle whose diagonals (extended) give us the asymptotes. The graph of the equation is sketched in Fig. 7.34. The

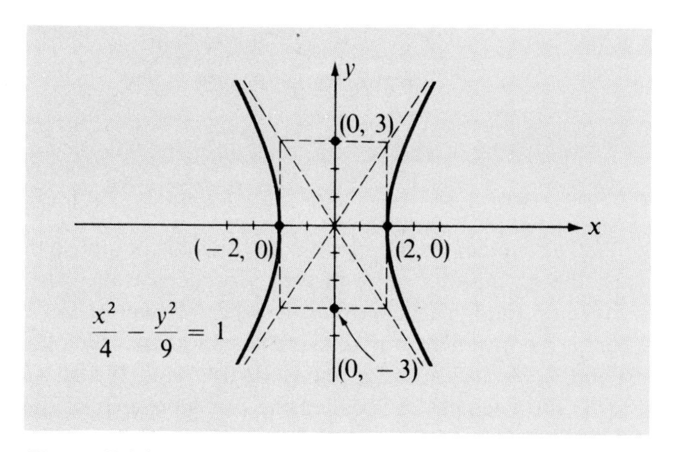

Figure 7.34

equations of the asymptotes, $y = \pm\frac{3}{2}x$, can be found by referring to the graph or to (7.31). From (7.28) we have $c^2 = a^2 + b^2 = 4 + 9 = 13$, and consequently the foci are $(\pm\sqrt{13}, 0)$.

The preceding example indicates that for hyperbolas it is not always true that $a > b$, as was the case for ellipses. Indeed, we may have $a < b$, $a > b$, or $a = b$.

EXAMPLE 2. Find an equation, the foci, and the asymptotes of a hyperbola which has vertices $(\pm 3, 0)$ and passes through the point $P(5, 2)$.

Solution: Substituting $a = 3$ in (7.29), we obtain the equation

$$\frac{x^2}{9} - \frac{y^2}{b^2} = 1.$$

If (5, 2) is in the solution set of this equation, then

$$\frac{25}{9} - \frac{4}{b^2} = 1.$$

Solving, we obtain $b^2 = 9/4$ and hence the desired equation is

$$\frac{x^2}{9} - \frac{4y^2}{9} = 1,$$

or equivalently, $x^2 - 4y^2 = 9$.

From (7.28), $c^2 = a^2 + b^2 = 9 + 9/4 = 45/4$, and therefore the foci are $(\pm\frac{3}{2}\sqrt{5}, 0)$. Substituting for b and a in (7.31) and simplifying, we obtain equations of the asymptotes, $y = \pm\frac{1}{2}x$.

If the foci of a hyperbola are at $(0, \pm c)$ on the y-axis, then we may show that an equation for the hyperbola is

$$\frac{y^2}{a^2} - \frac{x^2}{b^2} = 1, \tag{7.32}$$

where again $b^2 = c^2 - a^2$. This equation is called the *standard form* for the equation of a hyperbola with foci on the y-axis and center at the origin. The points $V(0, a)$ and $V'(0, -a)$ are the vertices of the hyperbola, and the endpoints of the conjugate axis are now $W(b, 0)$ and $W'(-b, 0)$. The asymptotes are found, as before, by using the diagonals of the rectangle

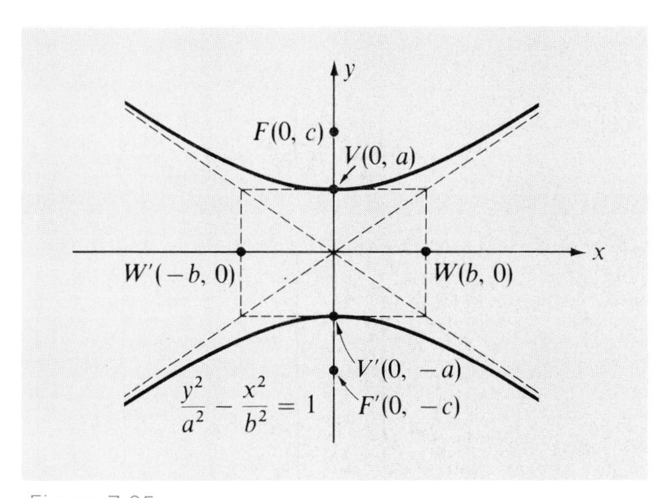

Figure 7.35

determined by these points and lines parallel to the coordinate axes. The graph is sketched in Fig. 7.35. The equations of the asymptotes are

$$y = \pm \frac{a}{b} x. \tag{7.33}$$

Note the difference between this formula and (7.31).

EXAMPLE 3. Discuss and sketch the graph of the equation

$$4y^2 - 2x^2 = 1.$$

Solution: The standard form is $y^2/(1/4) - x^2/(1/2) = 1$. Thus $a^2 = 1/4$, $b^2 = 1/2$, and $c^2 = 1/4 + 1/2 = 3/4$. Consequently $a = 1/2$, $b = \sqrt{2}/2$, and $c = \sqrt{3}/2$. The vertices are $(0, \pm 1/2)$ and the foci are $(0, \pm\sqrt{3}/2)$. The graph is sketched in Fig. 7.36. Equations of the asymptotes are $y = \pm\sqrt{2}/2x$.

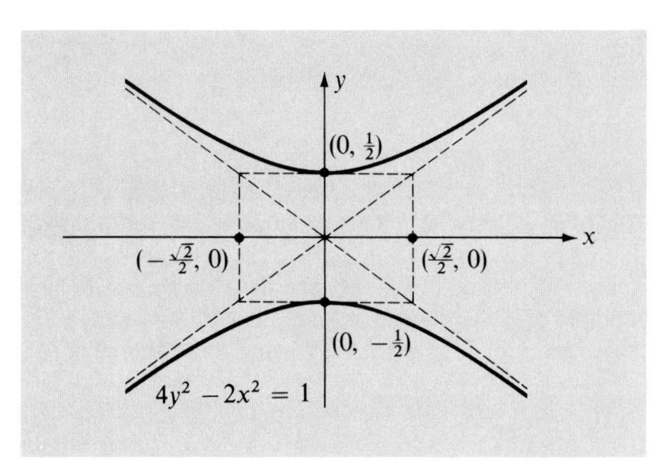

Figure 7.36

EXERCISES

In Exercises 1–8 sketch the graph of the equation, find the coordinates of the vertices and foci, and write equations for the asymptotes.

1. $\frac{x^2}{9} - \frac{y^2}{4} = 1.$
2. $\frac{y^2}{49} - \frac{x^2}{16} = 1.$
3. $\frac{y^2}{9} - \frac{x^2}{4} = 1.$
4. $\frac{x^2}{49} - \frac{y^2}{16} = 1.$
5. $y^2 - 4x^2 = 16.$
6. $x^2 - 2y^2 = 8.$
7. $x^2 - y^2 = 1.$
8. $y^2 - 16x^2 = 1.$

In Exercises 9–14 find an equation for the hyperbola satisfying the given conditions.

9. Foci $F(\pm 5, 0)$, vertices $V(\pm 3, 0)$.
10. Foci $F(0, \pm 3)$, vertices $V(0, \pm 2)$.
11. Foci $F(0, \pm 5)$, length of conjugate axis 4.
12. Vertices $V(\pm 4, 0)$, passing through $P(8, 2)$.
13. Vertices $V(\pm 3, 0)$, equations of asymptotes $y = \pm 2x$.
14. Foci $F(0, \pm 10)$, equations of asymptotes $y = \pm(1/3)x$.

In each of Exercises 15 and 16 find the points of intersection of the graphs of the given equations and sketch both graphs on the same coordinate axes, showing points of intersection.

15. $\begin{cases} y^2 - 4x^2 = 16 \\ y - x = 4. \end{cases}$

16. $\begin{cases} x^2 - y^2 = 4 \\ y^2 - 3x = 0. \end{cases}$

17. Find an equation for the family of hyperbolas having asymptotes $y = \pm 5x$ and foci on the x-axis.
18. Show how to obtain the equation

$$\frac{x^2}{a^2} - \frac{y^2}{c^2 - a^2} = 1$$

from the equation

$$\left|\sqrt{(x - c)^2 + (y - 0)^2} - \sqrt{(x + c)^2 + (y - 0)^2}\right| = 2a.$$

19. The graphs of the equations

$$\frac{x^2}{a^2} - \frac{y^2}{b^2} = 1 \quad \text{and} \quad \frac{x^2}{a^2} - \frac{y^2}{b^2} = -1$$

are called *conjugate hyperbolas*. Sketch the graphs of both equations on the same coordinate system with $a = 2$ and $b = 5$. Describe the relationship between the two graphs.

20. Find an equation for the family of hyperbolas with vertices $V(0, \pm 4)$. Find the particular member of this family that passes through $A(2, -8)$.

8 TRANSLATION OF AXES

Suppose $C(h, k)$ is an arbitrary point in an xy-coordinate plane, and let us introduce a new $x'y'$-coordinate system with origin O' at C such that the x'- and y'-axes are parallel to and have the same unit lengths and positive directions as the x- and y-axes, respectively. A typical situation of this type is illustrated in Fig. 7.37, where, for simplicity, we have placed C in the first quadrant. We shall use primes on letters which denote coordinates of points in the $x'y'$-coordinate system in order to distinguish

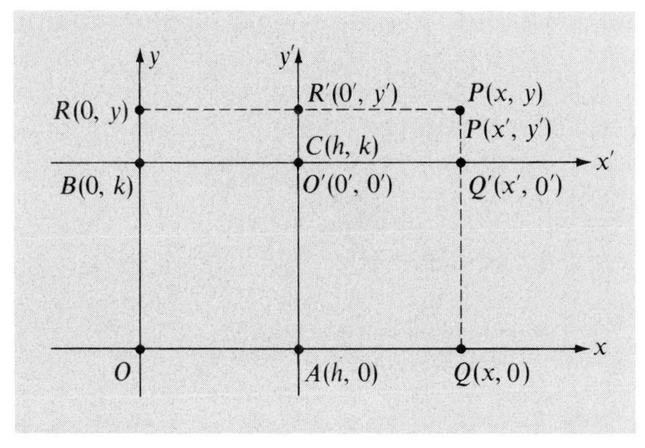

Figure 7.37

them from coordinates with respect to the xy-coordinate system. Thus the point $P(x, y)$ in the xy-system will be denoted by $P(x', y')$ in the $x'y'$-system. If we label projections of P on the various axes as indicated in Fig. 7.37, and let A and B denote projections of C on the x- and y-axes, respectively, then using directed distances we have

$$x = \overline{OQ} = \overline{OA} + \overline{AQ} = \overline{OA} + \overline{O'Q'} = h + x'$$

$$y = \overline{OR} = \overline{OB} + \overline{BR} = \overline{OB} + \overline{O'R'} = k + y'.$$

To summarize, if (x, y) are the coordinates of a point P relative to the xy-coordinate system, and if (x', y') are the coordinates of P relative to an $x'y'$-coordinate system with origin at the point $C(h, k)$ of the xy-system, then

(7.34) $$x = x' + h, \qquad y = y' + k$$

or equivalently,

(7.35) $$x' = x - h, \qquad y' = y - k.$$

The formulas given in (7.34) and (7.35) enable us to go from either coordinate system to the other. Their major use is to change the form of equations of graphs. To be specific: if, in the xy-plane, a certain curve is the graph of an equation in x and y, then to find the equation in x' and y' which has the same graph in the $x'y'$-plane we may substitute $x' + h$ for x and $y' + k$ for y in the given equation. Conversely, if a curve in the $x'y'$-plane is the graph of an equation in x' and y', then to find the corresponding equation in x and y we substitute $x - h$ for x' and $y - k$ for y'.

As a simple illustration of the preceding remarks, the equation $(x')^2 + (y')^2 = r^2$ has, for its graph in the $x'y'$-plane, a circle of radius r with center at the origin O'. Using (7.35), an equation for this circle in the xy-plane is $(x - h)^2 + (y - k)^2 = r^2$, which is in agreement with formula (7.1) for a circle of radius r with center at $C(h, k)$ in the xy-plane. As another illustration, we know by (7.16) that $(x')^2 = 4py'$ is an equation of a parabola with vertex at the origin O' of the $x'y'$-plane. Using (7.35) we see that $(x - h)^2 = 4p(y - k)$ is an equation of the same parabola in the xy-plane. This checks with formula (7.19), which was derived for a parabola with vertex at the point $V(h, k)$ in the xy-plane. It should now be evident how this technique can be applied to all the conics. For example, by (7.23), the graph of $(x')^2/a^2 + (y')^2/b^2 = 1$ is an ellipse with center at O' in the $x'y'$-plane, as illustrated in Fig. 7.38. According to (7.35) its equation relative to the xy-coordinate system is

$$\frac{(x - h)^2}{a^2} + \frac{(y - k)^2}{b^2} = 1. \tag{7.36}$$

A similar situation exists for hyperbolas.

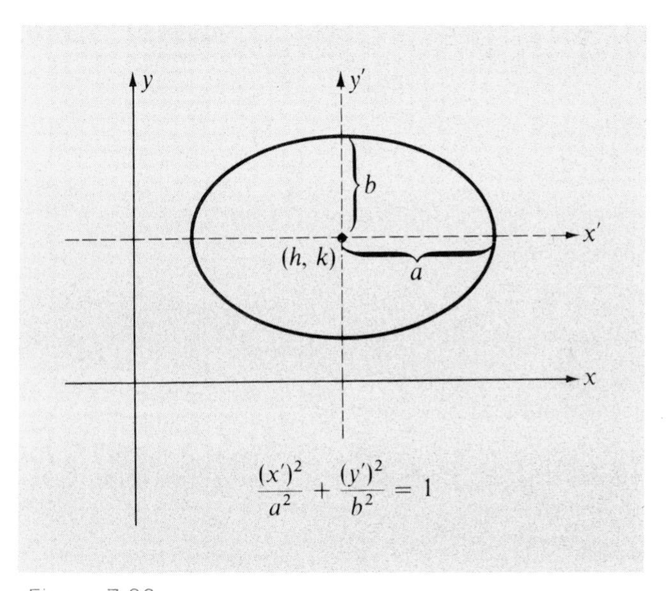

Figure 7.38

In certain cases, given an equation in x and y, we may, by a proper translation of axes, obtain a simpler equation in x' and y' which has the same graph. In particular this is true for an equation of degree 2 in x and y of the form

$$Ax^2 + Cy^2 + Dx + Ey + F = 0, \tag{7.37}$$

where the coefficients are real numbers. The graph of (7.37) is a conic — except for the degenerate cases in which points, lines, or no graphs are obtained. We shall not give a general proof of this fact, but will instead illustrate by means of examples the procedure which is used.

EXAMPLE 1. Discuss and sketch the graph of the equation

$$16x^2 + 9y^2 + 64x - 18y - 71 = 0.$$

Solution: In order to determine the origin of a new $x'y'$-coordinate system which will enable us to simplify the given equation, we begin by writing the equation in the form

$$16(x^2 + 4x) + 9(y^2 - 2y) = 71.$$

Next, we complete the squares for the expressions within parentheses, obtaining

$$16(x^2 + 4x + 4) + 9(y^2 - 2y + 1) = 71 + 64 + 9.$$

Note that by adding 4 to the expression within the first parentheses we have added 64 to the left side of the equation and hence must compensate by adding 64 to the right side. Similarly, by adding 1 to the expression within the second parentheses, 9 is added to the left side and consequently 9 must also be added to the right side. This gives us the equation

$$16(x + 2)^2 + 9(y - 1)^2 = 144.$$

Dividing by 144 we have

$$\frac{(x + 2)^2}{9} + \frac{(y - 1)^2}{16} = 1,$$

which is of the form

$$\frac{(x')^2}{9} + \frac{(y')^2}{16} = 1, \tag{7.38}$$

where $x' = x + 2$ and $y' = y - 1$. This shows that if we let $h = -2$ and $k = 1$ in (7.35), then (7.38) reduces to the given equation. Since the graph of (7.38) is an ellipse with center at the origin O' in the $x'y'$-plane, it follows that the given equation is an ellipse with center $C(-2, 1)$ in the xy-plane and axes parallel to the coordinate axes. The graph is sketched in Fig. 7.39.

EXAMPLE 2. Discuss and sketch the graph of the equation

$$9x^2 - 4y^2 - 54x - 16y + 29 = 0.$$

Solution: As in Example 1 we arrange our work as follows:

$$9(x^2 - 6x) - 4(y^2 + 4y) = -29$$

$$9(x^2 - 6x + 9) - 4(y^2 + 4y + 4) = -29 + 81 - 16$$

$$9(x - 3)^2 - 4(y + 2)^2 = 36$$

$$\frac{(x - 3)^2}{4} - \frac{(y + 2)^2}{9} = 1.$$

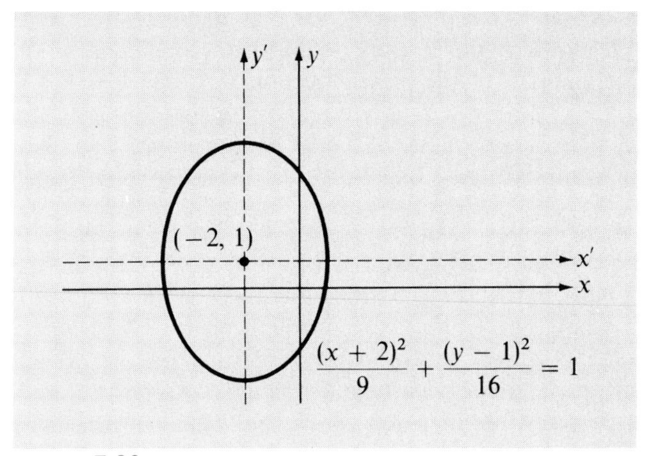

Figure 7.39

If we substitute $h = 3$ and $k = -2$ in (7.35), then the given equation reduces to the standard form (7.29) for the equation of a hyperbola, namely

$$\frac{(x')^2}{4} - \frac{(y')^2}{9} = 1.$$

By translating the x- and y-axes to the new origin $C(3, -2)$ we obtain the sketch shown in Fig. 7.40.

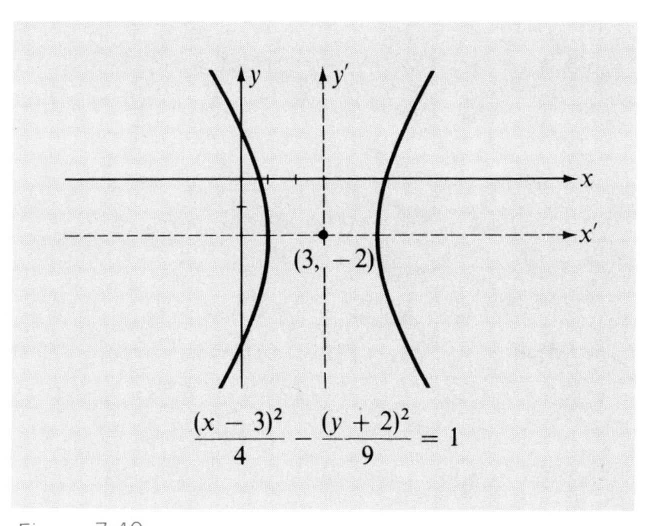

Figure 7.40

EXAMPLE 3. Discuss and sketch the graph of the equation

$$2y = x^2 + 8x + 22.$$

Solution: This example is the same as Example 4 on p. 266, where we completed the square in x and obtained the equation $(x + 4)^2 = 2(y - 3)$. By the methods of the present section, if we let $h = -4$ and $k = 3$ in (7.35), then the given equation reduces to the standard form $(x')^2 = 2y'$ of a parabola with vertex at O'. Consequently, a translation of axes to the new origin $C(-4, 3)$ leads to the sketch shown in Fig. 7.26.

Although we have only considered special examples of (7.37), our methods are perfectly general. If A and C are equal and not zero, then the graph of (7.37), when it exists, is a circle, or in exceptional cases, a point (see p. 240). If A and C are unequal but have the same sign, then by completing squares and properly translating axes we obtain an equation whose graph, when it exists, is an ellipse (or a point). If A and C have opposite signs, the equation of a hyperbola is obtained, or possibly, in the degenerate case, two intersecting straight lines. Finally, if either A or C is zero (but not both), the graph is a parabola, or in certain cases, a pair of parallel straight lines.

EXERCISES

Discuss and sketch the graph of each of the following equations after making a suitable translation of axes.

1. $(y - 5)^2 = 8(x + 1)$.
2. $(x + 3)^2 = -5(y - 4)$.
3. $\dfrac{(x - 4)^2}{9} + \dfrac{(y - 2)^2}{4} = 1$.
4. $4(x + 1)^2 + 8(y - 5)^2 = 32$.
5. $\dfrac{(x + 5)^2}{16} - \dfrac{(y - 1)^2}{25} = 1$.
6. $\dfrac{(y - 6)^2}{4} - (x + 2)^2 = 1$.
7. $4(x + 5)^2 + (y - 3)^2 = 1$.
8. $(x - 3)^2 + 4y^2 = 16$.
9. $100y^2 - 16(x - 5)^2 = 1600$.
10. $(x - 5)^2 - (y + 1)^2 = 1$.
11. $9x^2 + 16y^2 + 54x - 32y - 47 = 0$.
12. $4x^2 + 9y^2 + 24x + 18y + 9 = 0$.
13. $25x^2 + 4y^2 - 250x - 16y + 541 = 0$.
14. $4x^2 + y^2 = 2y$.
15. $4x^2 - 32y + 4x - 49 = 0$.
16. $4y^2 - x - 16y + 13 = 0$.
17. $4y^2 - x^2 + 40y - 4x + 60 = 0$.
18. $25x^2 - 9y^2 - 100x - 54y + 10 = 0$.
19. $9y^2 - x^2 - 36y + 12x - 36 = 0$.
20. $4x^2 - y^2 + 32x - 8y + 49 = 0$.

9 ROTATION OF AXES

The $x'y'$-coordinate system discussed in Section 8 may be thought of as having been obtained by moving the origin O of the xy-system to a new position $C(h, k)$ while, at the same time, not changing the positive directions of the axes or the units of length. We shall now introduce a new coordinate system by keeping the origin O fixed and rotating the x- and y-axes about O to another position denoted by x' and y'. A transformation of this type will be referred to as a *rotation of axes.*

Let us consider a rotation of axes and, as shown in Fig. 7.41, let ϕ denote the angle through which the positive x-axis must be rotated in

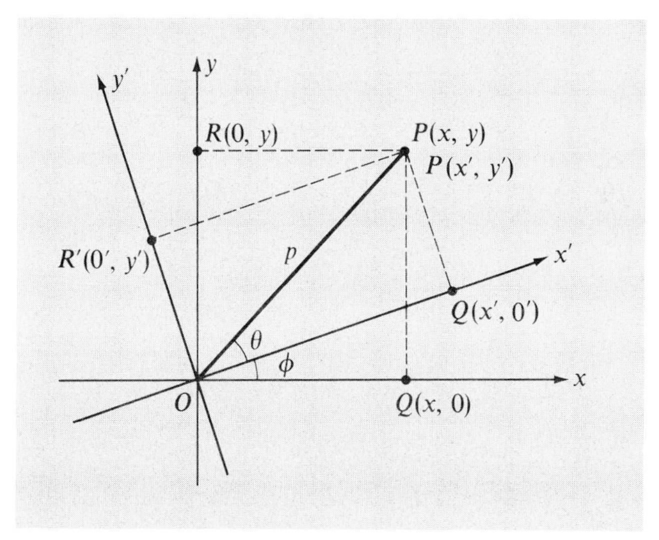

Figure 7.41

order to coincide with the positive x'-axis. If (x, y) are the coordinates of a point P relative to the xy-plane, then as before (x', y') will denote its coordinates relative to the new $x'y'$-coordinate system. Let the projections of P on the various axes be denoted as in Fig. 7.41 and let θ denote angle POQ'. If $p = d(O, P)$, then by (5.23) we obtain

$$x' = p \cos \theta, \qquad y' = p \sin \theta, \tag{7.39}$$

$$x = p \cos (\theta + \phi), \qquad y = p \sin (\theta + \phi). \tag{7.40}$$

Applying the addition formulas to (7.40) we have

$$x = p \cos \theta \cos \phi - p \sin \theta \sin \phi$$

$$y = p \sin \theta \cos \phi + p \cos \theta \sin \phi.$$

By (7.39) we may replace $p \cos \theta$ and $p \sin \theta$ by x' and y', respectively.

This gives us

$$\text{(7.41)} \qquad \begin{aligned} x &= x' \cos \phi - y' \sin \phi \\ y &= x' \sin \phi + y' \cos \phi. \end{aligned}$$

If the equations in (7.41) are solved for x' and y' we obtain

$$\text{(7.42)} \qquad \begin{aligned} x' &= x \cos \phi + y \sin \phi \\ y' &= -x \sin \phi + y \cos \phi. \end{aligned}$$

The formulas in (7.41) and (7.42) can be used to go from either coordinate system to the other. We shall call them the *formulas for rotation of axes through the angle* ϕ.

EXAMPLE 1. The graph of the equation $xy = 1$, or equivalently $y = 1/x$, was determined in Chapter One (see Fig. 1.21). If the coordinate axes are rotated through an angle of 45°, find the equation of the graph relative to the new $x'y'$-coordinate system.

Solution: Letting $\phi = 45°$ in (7.41), we obtain

$$\begin{aligned} x &= x'(\sqrt{2}/2) - y'(\sqrt{2}/2) = (\sqrt{2}/2)(x' - y') \\ y &= x'(\sqrt{2}/2) + y'(\sqrt{2}/2) = (\sqrt{2}/2)(x' + y'). \end{aligned}$$

Substitution for x and y in the equation $xy = 1$ gives us

$$(\sqrt{2}/2)(x' - y') \cdot (\sqrt{2}/2)(x' + y') = 1.$$

This reduces to

$$\frac{(x')^2}{2} - \frac{(y')^2}{2} = 1,$$

which we recognize as the standard equation of a hyperbola with vertices $(\pm\sqrt{2}, 0)$ on the $x'y'$-axes. Fig. 7.42 shows the graph, together with the new coordinate axes. Note that the asymptotes for the hyperbola have equations $y' = \pm x'$ in the new system. These correspond to the original x- and y-axes.

Example 1 illustrates a method for eliminating a term of an equation which contains the product xy. This method can be used to transform any equation of degree 2 in x and y of the form

$$\text{(7.43)} \qquad Ax^2 + Bxy + Cy^2 + Dx + Ey + F = 0,$$

where $B \neq 0$, into an equation in x' and y' which contains no $x'y'$ term. Let us prove that this may always be done. If we rotate the axes through

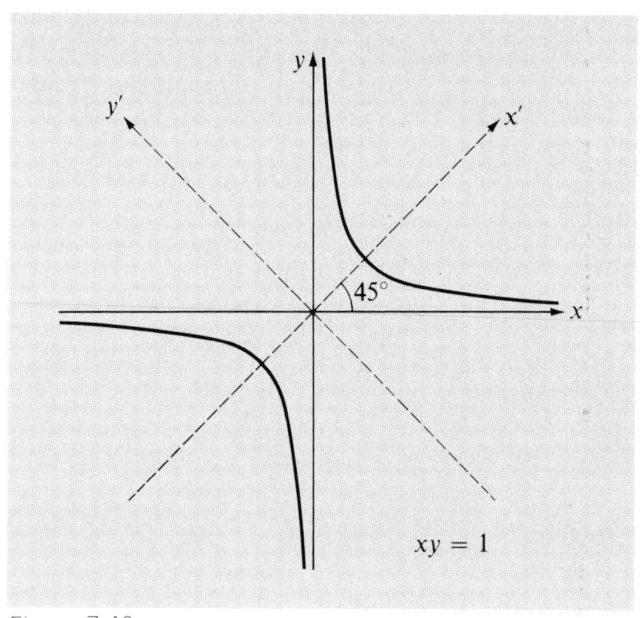

Figure 7.42

an angle ϕ, then substituting the expressions of (7.41) for x and y in (7.43) gives us

$$\begin{aligned} &A(x' \cos \phi - y' \sin \phi)^2 \\ &\qquad + B(x' \cos \phi - y' \sin \phi)(x' \sin \phi + y' \cos \phi) \\ &\qquad + C(x' \sin \phi + y' \cos \phi)^2 + D(x' \cos \phi - y' \sin \phi) \\ &\qquad + E(x' \sin \phi + y' \cos \phi) + F = 0. \end{aligned}$$

This equation may be written in the form

$$A'(x')^2 + B'x'y' + C'(y')^2 + D'x' + E'y' + F = 0, \tag{7.44}$$

where the coefficient B' of $x'y'$ is given by

$$B' = 2(C - A) \sin \phi \cos \phi + B(\cos^2 \phi - \sin^2 \phi).$$

In order to eliminate the $x'y'$ term in (7.44) we must select ϕ so that

$$2(C - A) \sin \phi \cos \phi + B(\cos^2 \phi - \sin^2 \phi) = 0. \tag{7.45}$$

Using the double angle formulas (6.10) and (6.11), the last equation may be written in the form

$$(C - A) \sin 2\phi + B \cos 2\phi = 0,$$

which is equivalent to

$$\cot 2\phi = \frac{A - C}{B}, \quad B \neq 0. \tag{7.46}$$

Thus, to eliminate the xy term in (7.43), we may choose ϕ such that (7.46) is true and then employ the rotation of axes formulas (7.41). The resulting equation will contain no $x'y'$ term and therefore can be analyzed by previous methods. This proves that if the graph of (7.43) exists it is a conic (except for degenerate cases).

EXAMPLE 2. Discuss and sketch the graph of the equation

$$41x^2 - 24xy + 34y^2 - 25 = 0.$$

Solution: Using the notation of (7.43) we have $A = 41$, $B = -24$, and $C = 34$. Applying (7.46), $\cot 2\phi = (41 - 34)/(-24) = -7/24$. Since $\cot 2\phi$ is negative we may choose 2ϕ such that $90° < 2\phi < 180°$, and consequently $\cos 2\phi = -7/25$ (Why?). We now use the half-angle formulas (6.16) with $v = 2\phi$. Since $45° < \phi < 90°$ this gives us

$$\sin \phi = \sqrt{\frac{1 - \cos 2\phi}{2}} = \sqrt{\frac{1 - (-7/25)}{2}} = 4/5$$

$$\cos \phi = \sqrt{\frac{1 + \cos 2\phi}{2}} = \sqrt{\frac{1 + (-7/25)}{2}} = 3/5.$$

Therefore, the desired rotation formulas (7.41) are

$$x = \tfrac{3}{5}x' - \tfrac{4}{5}y', \quad y = \tfrac{4}{5}x' + \tfrac{3}{5}y'.$$

We leave it to the reader to show that after substituting for x and y in the given equation and simplifying, one obtains the equation

$$(x')^2 + 2(y')^2 = 1.$$

Thus the graph is an ellipse with vertices at $(\pm 1, 0)$ on the x'-axis. Since $\tan \phi = \sin \phi/\cos \phi = (4/5)/(3/5) = 4/3$, we obtain $\phi = \text{Tan}^{-1}(4/3)$. If an approximation is desired, then by Table 3, $\phi \doteq 53°\,8'$. The graph is sketched in Fig. 7.43.

EXERCISES

After a suitable rotation of axes, describe and sketch the graph of each of the following equations.

1. $32x^2 - 72xy + 53y^2 = 80$.
2. $7x^2 - 48xy - 7y^2 = 225$.
3. $11x^2 + 10\sqrt{3}xy + y^2 = 4$.
4. $x^2 - xy + y^2 = 3$.
5. $5x^2 - 8xy + 5y^2 = 9$.
6. $11x^2 - 10\sqrt{3}xy + y^2 = 20$.
7. $16x^2 - 24xy + 9y^2 - 60x - 80y + 100 = 0$.
8. $64x^2 - 240xy + 225y^2 + 1020x - 544y = 0$.

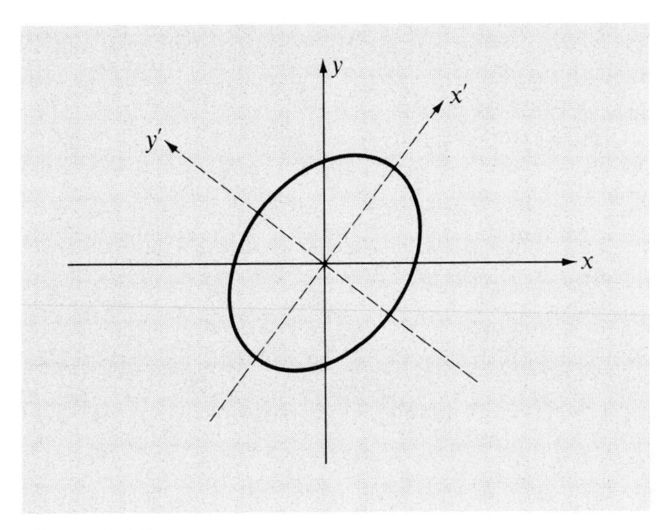

Figure 7.43

10 POLAR COORDINATES

We have previously specified points in a plane in terms of rectangular coordinates, using the ordered pair (a, b) to denote the point whose directed distance from the x- and y-axes are a and b, respectively. Another important method for representing points is by means of *polar coordinates*. In this case we again use ordered pairs; however, one of the numbers represents the measure of an angle instead of a directed distance. In order to introduce a system of polar coordinates in a plane we begin with a fixed point O (called the *origin*, or *pole*) and a directed half-line (called the *polar axis*) with endpoint O. Next we consider any point P in the plane different from O. If, as illustrated in Fig. 7.44, $r = d(O, P)$ and θ denotes

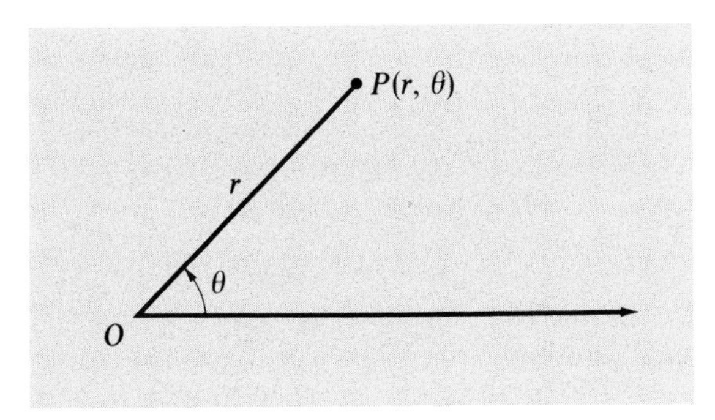

Figure 7.44

the measure of any angle determined by the polar axis and OP, then r and θ are called *polar coordinates* of P and the symbols (r, θ) or $P(r, \theta)$ are used to denote P. As in our discussion on p. 160, θ is considered positive if the angle is generated by a counterclockwise rotation of the polar axis and negative if the rotation is clockwise. Either radians or degrees may be used for the measure of θ. Since there are many angles with the same terminal side, the polar coordinates of a point are not unique. For example, $(3, \pi/4)$, $(3, 9\pi/4)$, and $(3, -7\pi/4)$ all represent the same point (see Fig. 7.45). We shall also allow r to be negative. In this

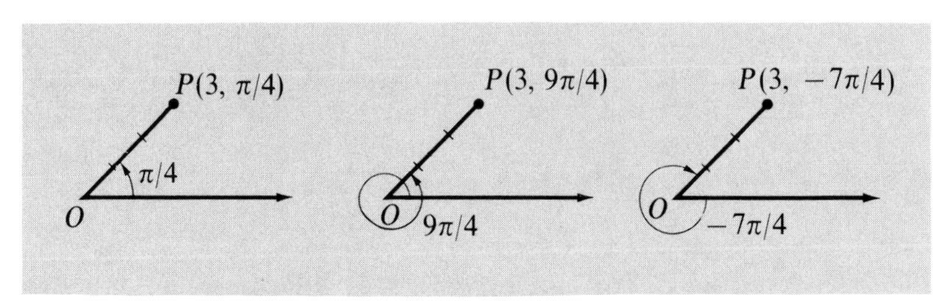

Figure 7.45

event, instead of measuring $|r|$ units along the terminal side of the angle θ, we measure along the half-line with endpoint O which has the direction opposite to that of the terminal side. Figure 7.46 contains illustrations for

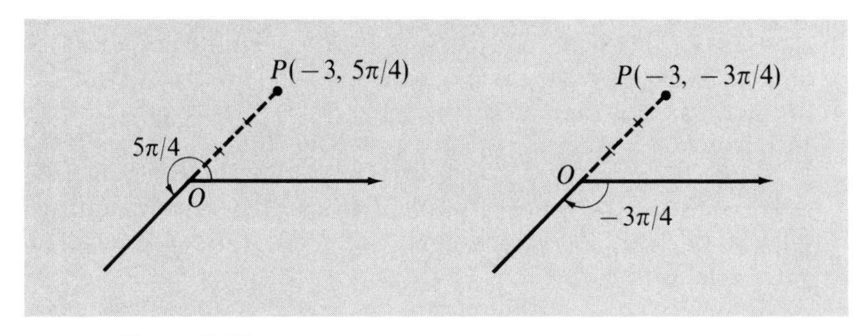

Figure 7.46

the pairs $(-3, 5\pi/4)$ and $(-3, -3\pi/4)$. Finally we agree that the pole O has polar coordinates $(0,\theta)$ for *any* θ. An assignment of ordered pairs of the form (r, θ) to points in a plane will be referred to as a *polar coordinate system* and the plane will be called an $r\theta$-plane.

In a manner analogous to our work with equations in x and y, we now consider *polar equations*; that is, equations in r and θ. A solution of such an equation is an ordered pair (a, b) which leads to equality when a is substituted for r and b for θ. As usual, the solution set is the set of all solutions and the graph is the set of all points (in an $r\theta$-plane) which correspond to the solutions. Although many polar equations contain trigonometric expressions, their graphs will differ from those discussed in Chapters Five and Six since points are plotted in a *polar* coordinate system instead of a rectangular coordinate system.

EXAMPLE 1. Sketch the graph of the equation $r = 4 \sin \theta$.

Solution: The following table contains some solutions of the equation.

θ	0	$\pi/6$	$\pi/4$	$\pi/3$	$\pi/2$	$2\pi/3$	$3\pi/4$	$5\pi/6$	π
r	0	2	$2\sqrt{2}$	$2\sqrt{3}$	4	$2\sqrt{3}$	$2\sqrt{2}$	2	0

We know that in rectangular coordinates, the graph of the given equation consists of sine waves of amplitude 4 and period 2π. However, if polar coordinates are used, then the points which correspond to the pairs in the table appear to lie on a circle of radius 2 and we draw the graph accordingly (see Fig. 7.47). The proof that the graph is a circle will be given in

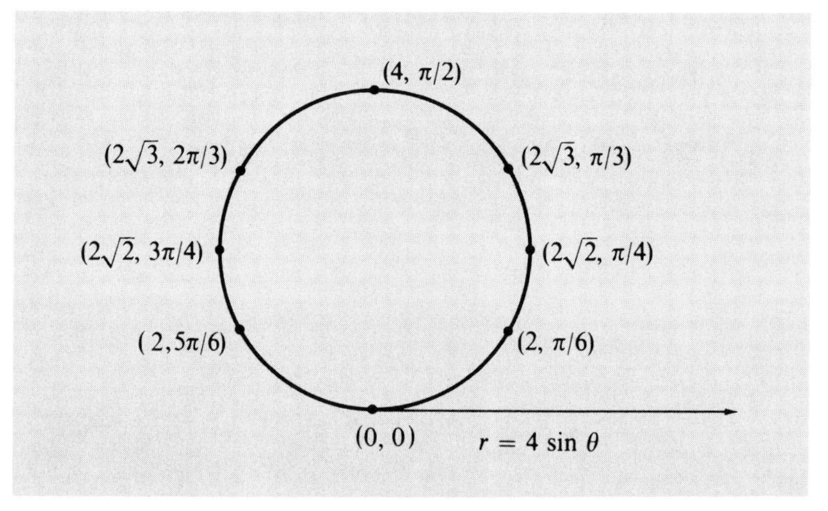

Figure 7.47

Example 4. Additional points obtained by letting θ vary from π to 2π lie on the same circle. For example, the solution $(-2, 7\pi/6)$ gives us the same point as $(2, \pi/6)$; the point corresponding to $(-2\sqrt{2}, 5\pi/4)$ is the same as that obtained from $(2\sqrt{2}, \pi/4)$; and so on. Due to the periodicity of the sine function, if we let θ increase through all real numbers we obtain the same points over and over.

EXAMPLE 2. Sketch the graph of the equation $r = 2 + 2\cos\theta$.

Solution: Since the cosine function decreases from 1 to -1 as θ varies from 0 to π, it follows that r decreases from 4 to 0 in this θ-interval. The following table exhibits some solutions of the given equation.

θ	0	$\pi/6$	$\pi/4$	$\pi/3$	$\pi/2$	$2\pi/3$	$3\pi/4$	$5\pi/6$	π
r	4	$2+\sqrt{3}$	$2+\sqrt{2}$	3	2	1	$2-\sqrt{2}$	$2-\sqrt{3}$	0

If θ increases from π to 2π, then $\cos\theta$ increases from -1 to 1 and consequently r increases from 0 to 4. Plotting points and connecting them with a smooth curve leads to the sketch shown in Fig. 7.48. The graph is

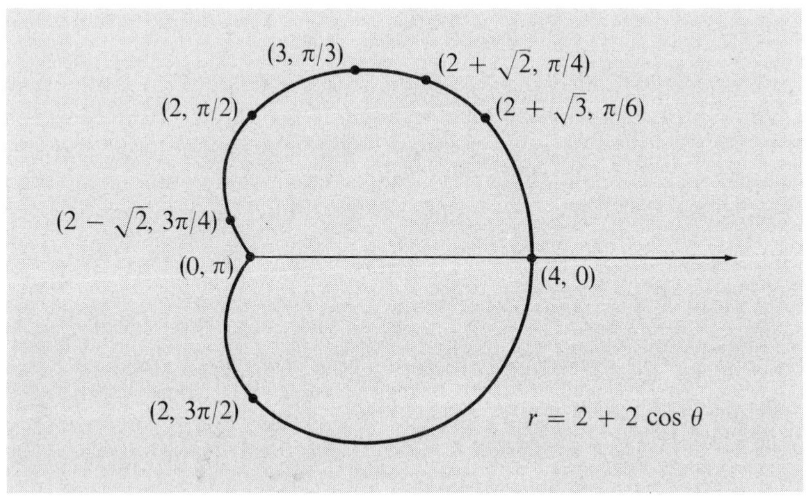

Figure 7.48

called a *cardioid.* The same graph may be obtained by taking other intervals for θ.

EXAMPLE 3. Sketch the graph of the equation $r = a \sin 2\theta$, where $a > 0$.

Solution: Instead of tabulating solutions, let us reason as follows. If θ increases from 0 to $\pi/4$, then 2θ varies from 0 to $\pi/2$ and hence $\sin 2\theta$ increases from 0 to 1. It follows that r increases from 0 to a in the θ-interval $[0, \pi/4]$. If we next let θ increase from $\pi/4$ to $\pi/2$, then 2θ changes from $\pi/2$ to π and consequently r decreases from a to 0 in the θ-interval $[\pi/4, \pi/2]$ (Why?). The corresponding points on the graph constitute a "loop" as illustrated in Fig. 7.49. We shall leave it to the reader to show that as θ increases from $\pi/2$ to π, a similar "loop" is obtained directly *below* the first "loop" (note that for this range of θ we have $\pi < 2\theta < 2\pi$ and hence $\sin 2\theta$ is negative). Similar "loops" are obtained for the θ-intervals $[\pi, 3\pi/2]$ and $[3\pi/2, 2\pi]$. We have plotted only those points on the graph which correspond to the largest numerical values of r. The graph is called a *four-leaved rose.*

Many other interesting graphs result from polar equations. Some of these are included in the exercises at the end of this section. Polar coordinates are very useful in applications involving circles with centers at the origin or lines that pass through the origin since the equations which have these graphs may be written in the simple forms $r = k$ or $\theta = k$ for some $k \in \mathbf{R}$ (Verify this fact!).

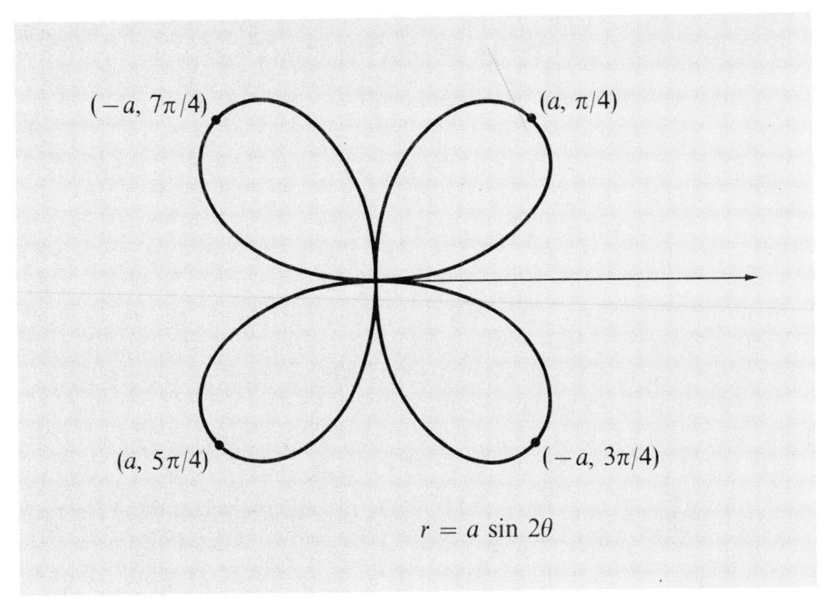

Figure 7.49

Let us now superimpose an xy-plane on an $r\theta$-plane in such a way that the positive x-axis coincides with the polar axis. Any point P in the plane may then be assigned rectangular coordinates (x, y) or polar coordinates (r, θ). It is not difficult to obtain formulas which specify the relationship between the two coordinate systems. Thus, if $r > 0$, we have a situation similar to that illustrated in (i) of Fig. 7.50, whereas if $r < 0$ we have that shown in (ii) where, for later purposes, we have also plotted the point P' which has polar coordinates $(|r|, \theta)$ and rectangular coordinates $(-x, -y)$. Although we have pictured θ as an acute angle,

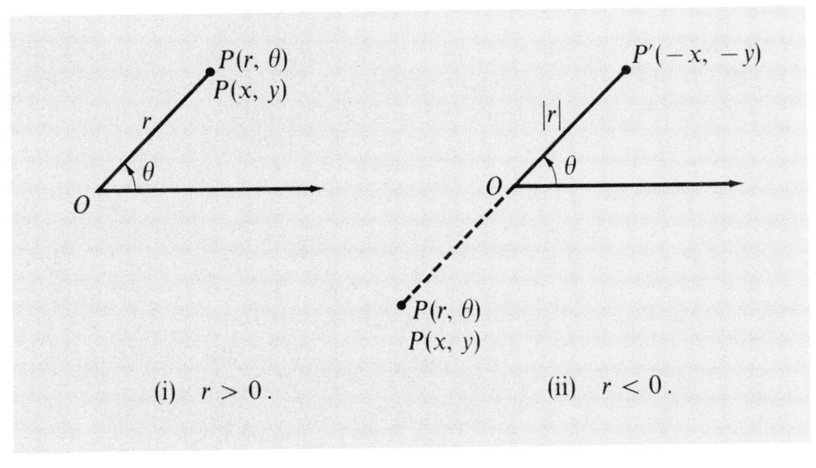

Figure 7.50

the discussion which follows is valid for all angles. On the one hand, if $r > 0$, then using (5.23) we obtain

$$x = r \cos \theta, \quad y = r \sin \theta. \tag{7.47}$$

On the other hand, if $r < 0$, then applying (5.23) to (ii) of Fig. 7.50 and using the fact that $|r| = -r$ we have

$$\cos \theta = -x/|r| = -x/(-r) = x/r,$$
$$\sin \theta = -y/|r| = -y/(-r) = y/r.$$

Multiplication by r produces (7.47) and, therefore, the latter formulas hold whether r is positive or r is negative. If $r = 0$, then the point is the pole and we again see that (7.47) is true. The following formulas are further consequences of our discussion.

$$\tan \theta = y/x, \quad r^2 = x^2 + y^2. \tag{7.48}$$

It is possible to use (7.47) and (7.48) to change the rectangular coordinates of a point to polar coordinates and vice versa. A more important use is for transforming a polar equation to an equation in x and y and vice versa. This is illustrated in the next two examples.

Example 4. Find an equation in x and y which has the same graph as $r = 4 \sin \theta$.

Solution: It is convenient to multiply both sides of the given equation by r, obtaining $r^2 = 4r \sin \theta$. Applying (7.47) and (7.48) gives us $x^2 + y^2 = 4y$. The latter equation is equivalent to $x^2 + (y - 2)^2 = 4$ whose graph is a circle of radius 2 with center at $(0, 2)$ in the xy-plane. This proves the remark made on p. 291.

Example 5. Find the general polar equation of a straight line.

Solution: According to (7.14), every straight line in an xy-coordinate system is the graph of a linear equation $ax + by + c = 0$. Substituting for x and y from (7.47) leads to the polar equation $r(a \cos \theta + b \sin \theta) + c = 0$.

EXERCISES

Sketch the graph of each of the following equations.

1. $r = 5$.
2. $\theta = \pi/4$.
3. $r = 4 \cos \theta$.
4. $r = -2 \sin \theta$.
5. $r = 4(1 - \sin \theta)$ (cardioid).
6. $r = 1 + 2 \cos \theta$ (limaçon).
7. $r = a \cos 3\theta$ (three-leaved rose).

8. $r = a \sin 4\theta$ (eight-leaved rose).
9. $r^2 = a^2 \cos 2\theta$ (lemniscate).
10. $r = a \sin^2 (\frac{1}{2}\theta)$ (cardioid).
11. $r = 4 \csc \theta$.
12. $r = -3 \sec \theta$.
13. $r^2\theta = a^2$, $r > 0$ (lituus).
14. $r = 2 + 2 \sec \theta$ (conchoid).
15. $r = 2^\theta$, $\theta \geq 0$ (spiral).
16. $r\theta = 1$, $\theta > 0$ (spiral).

In Exercises 17–22 find a polar equation which has the same graph as the given equation.

17. $x^2 + y^2 = 16$.
18. $x^2 = 8y$.
19. $y = 6$.
20. $y = 6x$.
21. $x^2 - y^2 = 16$.
22. $9x^2 + 4y^2 = 36$.

In Exercises 23–28 find an equation in x and y which has the same graph as the given polar equation.

23. $r - 6 \sin \theta = 0$.
24. $r = 2(1 + \cos \theta)$.
25. $r = a$.
26. $\theta = \pi/4$.
27. $r = \tan \theta$.
28. $r = 4 \sec \theta$.
29. If $P_1(r_1, \theta_1)$ and $P_2(r_2, \theta_2)$ are points in an $r\theta$-plane, use the Law of Cosines to prove that $[d(P_1, P_2)]^2 = r_1^2 + r_2^2 - 2r_1r_2 \cos (\theta_2 - \theta_1)$.
30. Prove that the graph of the polar equation $r = a \sin \theta + b \cos \theta$ is a circle and find its center and radius.

11 REVIEW EXERCISES

Oral

Define or discuss each of the following:

1. The midpoint formula.
2. The slope of a line.
3. The inclination of a line.
4. The point-slope form.
5. The slope-intercept form.
6. Families of lines.
7. Conic sections.
8. A parabola.
9. Focus, directrix, vertex, and axis of a parabola.
10. Symmetry of a graph with respect to a coordinate axis or with respect to the origin.
11. Tests for symmetry.
12. An ellipse.
13. Major and minor axes of an ellipse.
14. Foci and vertices of an ellipse.
15. A hyperbola.
16. Transverse and conjugate axes of a hyperbola.
17. Foci and vertices of a hyperbola.
18. Asymptotes of a hyperbola.
19. Translation of axes.
20. Rotation of axes.
21. Polar coordinates of a point.

Written

1. (a) Find the center and radius of the circle which has equation $x^2 + y^2 - 10x + 14y - 7 = 0$.
   (b) Find an equation for the circle concentric to the circle of part (a) and passing through the origin.
   (c) Find an equation for the circle of radius 4 with center in the fourth quadrant and tangent to both axes.

2. Given the points $A(-4, 2)$, $B(3, 6)$, and $C(2, -5)$, find:
   (a) an equation for the line through B which is parallel to the line through A and C.
   (b) an equation for the line through B which is perpendicular to the line through A and C.

(c) an equation for the line through C and the midpoint of the line segment AB.
(d) an equation for the line through A which is parallel to the y-axis.
(e) an equation for the line through C which is perpendicular to the line with equation $3x - 10y + 7 = 0$.

3. Find the foci, vertices and sketch the graphs of the conics which have the following equations:
(a) $y^2 = 64x$.
(b) $y - 1 = 8(x + 2)^2$.
(c) $9y^2 = 144 - 16x^2$.
(d) $9y^2 = 144 + 16x^2$.
(e) $x^2 - y^2 - 4 = 0$.

4. Find an equation for:
(a) the parabola with focus $(0, -10)$ and directrix $y = 10$.
(b) the parabola with vertex at the origin, symmetric to the x-axis and passing through the point $(5, -1)$.
(c) the ellipse with vertices $V(0, \pm 10)$ and foci $F(0, \pm 5)$.
(d) the hyperbola with foci $F(\pm 10, 0)$ and vertices $V(\pm 5, 0)$.
(e) the hyperbola with vertices $V(0, \pm 6)$ and asymptotes which have equations $y = \pm 9x$.

5. Discuss and sketch the graph of each of the following equations after making a suitable translation of axes:
(a) $4x^2 + 9y^2 + 24x - 36y + 36 = 0$.
(b) $4x^2 - y^2 - 40x - 8y + 88 = 0$.
(c) $y^2 - 8x + 8y + 32 = 0$.

6. Sketch the graphs of the following equations:
(a) $r = -4 \cos \theta$.
(b) $r = 4 \cos 5\theta$.
(c) $r = 3 - 2 \cos \theta$.
(d) $r^2 = 4 \sin 2\theta$.
(e) $r = \theta$.

7. Change the following equations to polar equations:
(a) $y^2 = 4x$.
(b) $x^2 + y^2 - 3x + 4y = 0$.
(c) $2x - 3y = 8$.

8. Change each of the following to equations in x and y:
(a) $r^2 = \tan \theta$.
(b) $r = 2 \cos \theta + 3 \sin \theta$.
(c) $r^2 = 4 \sin 2\theta$.

Supplementary Questions

1. Concepts developed in coordinate geometry may be employed to establish properties of geometric figures. Prove each of the following

statements by using an appropriate figure on a coordinate plane.

(a) The sum of the squares of the lengths of the sides of a parallelogram is equal to the sum of the squares of the lengths of the diagonals.

(b) If the diagonals of a rectangle are perpendicular, then it is a square.

(c) The lines which join the midpoints of the sides of a triangle divide the triangle into four congruent triangles.

(d) The diagonals of a rhombus are perpendicular.

2. Prove that, except for degenerate cases, the graph of equation (7.43) is:

(a) a parabola if $B^2 - 4AC = 0$.

(b) an ellipse if $B^2 - 4AC < 0$.

(c) a hyperbola if $B^2 - 4AC > 0$.

3. Let F be a fixed point in a plane and l a fixed line. A *conic section* with focus F and directrix l may be defined as the set of all points P in the plane such that the ratio $d(F, P)/s$ is a positive constant, where s denotes the distance from P to l. The positive constant e is called the *eccentricity* of the conic. Derive the polar equation of a conic if F is the pole and l is perpendicular to the polar axis at the point $(d, 0)$ where $d > 0$. Prove that the conic is an ellipse if $e < 1$, a hyperbola if $e > 1$, and a parabola if $e = 1$. What is the equation of the conic if $d < 0$? What if l is parallel to the polar axis?

chapter eight

Coordinate Geometry in Three Dimensions

In this chapter concepts developed in Chapter Seven are extended to solid, or three-dimensional, geometry.

1 RECTANGULAR COORDINATE SYSTEMS

Coordinate systems in space are introduced by means of ordered triples of real numbers. In analogy with the concept of ordered pair given on p. 15, an *ordered triple* (a, b, c) is considered as a set $\{a, b, c\}$ of three numbers in which a is considered as the "first" element of the set, b as the "second" element, and c as the "third" element. The totality of all such ordered triples is often denoted by $\mathbf{R} \times \mathbf{R} \times \mathbf{R}$. Two ordered triples (a_1, a_2, a_3) and (b_1, b_2, b_3) are said to be *equal*, and we write $(a_1, a_2, a_3) = (b_1, b_2, b_3)$, if and only if $a_1 = b_1$, $a_2 = b_2$, and $a_3 = b_3$.

In order to specify points in space we choose a fixed point O (called the *origin*) and consider three mutually perpendicular coordinate lines (called the x-, y-, and z-axes) with common origin O, as illustrated in Fig. 8.1. To visualize this configuration, one may regard the y- and z-axes as lying in the plane of the paper and the x-axis as projecting out from the paper toward the reader. The coordinate plane determined by the x- and y-axes is called the *xy-plane*. Similarly, the coordinate plane determined by the y- and z-axes is called the *yz-plane*, whereas that determined by the x- and z-axes is called the *xz-plane*.

If P is a point, then the projection of P on the x-axis has some coordinate a, which we call the *x-coordinate* of P. The x-coordinate of P may also be thought of as a directed distance from the yz-plane to P. Similarly, the coordinates b and c of the projections of P on the y- and

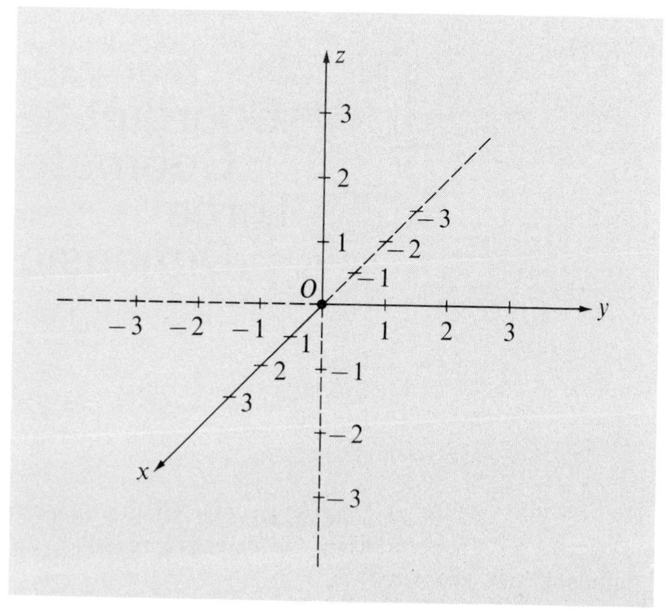

Figure 8.1

z-axes, respectively, are called the *y-coordinate* and *z-coordinate* of P. We shall denote the coordinates of P, as well as P itself, by the ordered triple (a, b, c). The notation $P(a, b, c)$ will also be used for the point P with coordinates a, b, and c. If P is not on a coordinate plane, then the three planes through P which are parallel to the coordinate planes, together with the coordinate planes, form a rectangular parallelepiped. This is illustrated in Fig. 8.2, where we have labeled the eight vertices in the manner just described. Conversely, to each ordered triple (a, b, c) of real numbers, there corresponds a point P having coordinates a, b, and c.

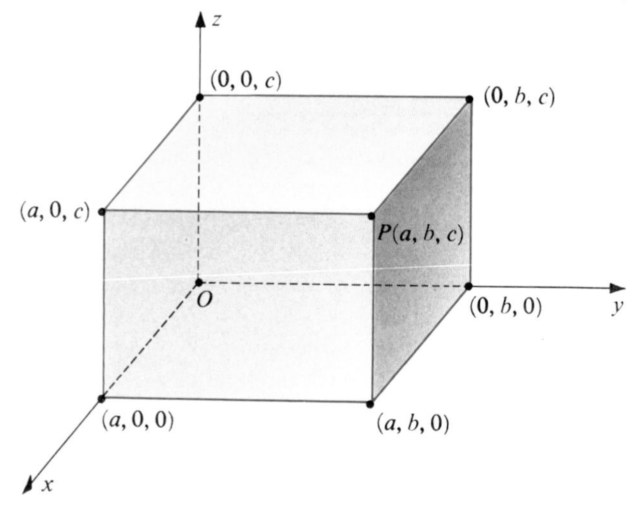

Figure 8.2

The concept of plotting points is similar to that used in two dimensions. As an aid to plotting, it is often convenient to construct a parallelepiped as shown in Fig. 8.2. The points $A(-1, -3, 1)$ and $B(3, 4, -2)$ are plotted in Fig. 8.3.

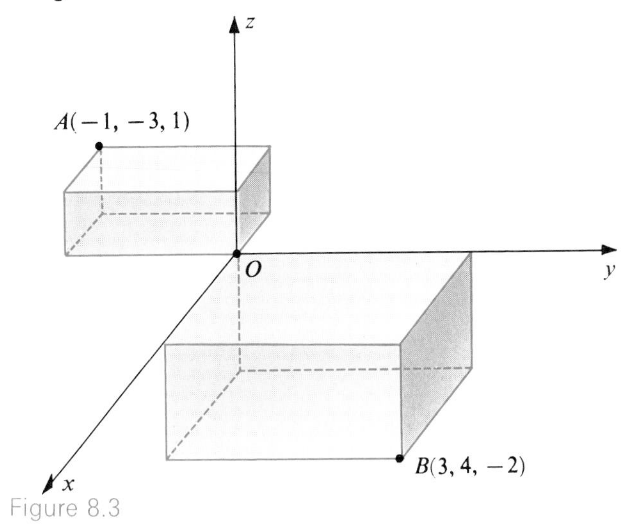

Figure 8.3

A one-to-one correspondence between the points in space and ordered triples of real numbers of the type we have described is called a *rectangular coordinate system in three dimensions.* The three coordinate planes partition space into eight parts, called *octants.* The part consisting of all points $P(a, b, c)$ whose three coordinates a, b, and c are positive is called the *first octant.*

It is not difficult to derive a formula for the distance $d(P_1, P_2)$ between two points P_1 and P_2. If P_1 and P_2 are on a line parallel to the z-axis, as illustrated in Fig. 8.4, and if their projections on the z-axis are

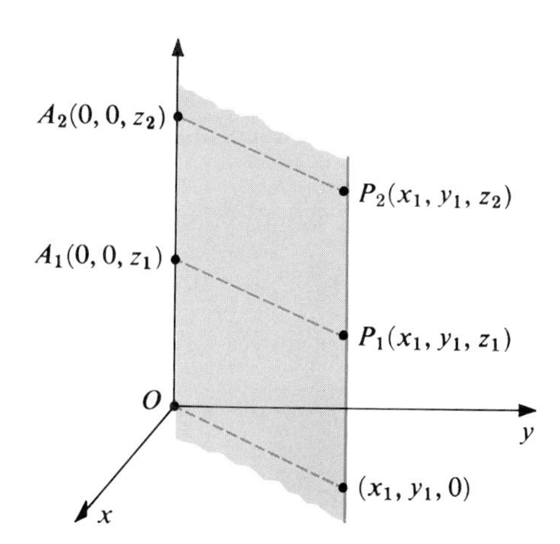

Figure 8.4

$A_1(0, 0, z_1)$ and $A_2(0, 0, z_2)$, respectively, then evidently $d(P_1, P_2) = d(A_1, A_2) = |z_2 - z_1|$. Similar formulas hold if the line through P_1 and P_2 is parallel to the x- or y-axes.

On the other hand, if we have a situation of the type illustrated in Fig. 8.5, then triangle P_1AP_2 is a right triangle, and hence by the Pythagorean Theorem,

$$d(P_1, P_2) = \sqrt{[d(P_1, A)]^2 + [d(A, P_2)]^2}. \tag{8.1}$$

Since P_1 and A are in a plane parallel to the xy-plane, it follows from the distance formula in two dimensions that $[d(P_1, A)]^2 = (x_2 - x_1)^2 + (y_2 - y_1)^2$, whereas from our previous remarks $[d(A, P_2)]^2 = (z_2 - z_1)^2$.

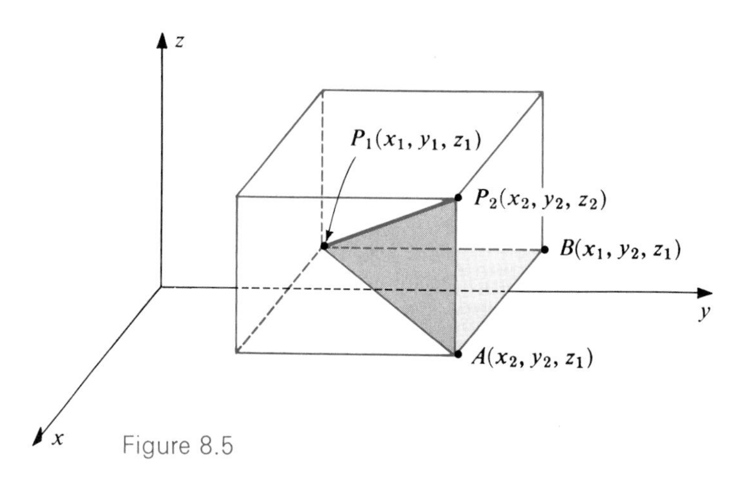

Figure 8.5

Substituting in (8.1) we obtain the following *distance formula in three dimensions*:

$$d(P_1, P_2) = \sqrt{(x_2 - x_1)^2 + (y_2 - y_1)^2 + (z_2 - z_1)^2}. \tag{8.2}$$

Note that if P_1 and P_2 are on the xy-plane, so that $z_1 = z_2 = 0$, then (8.2) reduces to the two-dimensional distance formula (1.15).

EXAMPLE 1. Find the distance between $A(-1, -3, 1)$ and $B(3, 4, -2)$.

Solution: Points A and B are plotted in Fig. 8.3. Using (8.2) we have

$$\begin{aligned} d(A, B) &= \sqrt{(3 + 1)^2 + (4 + 3)^2 + (-2 - 1)^2} \\ &= \sqrt{16 + 49 + 9} = \sqrt{74}. \end{aligned}$$

By referring to Fig. 8.5 and using similar triangles the following result can be proved:

(8.3) Midpoint Formula

The coordinates of the midpoint of the line segment from $P_1(x_1, y_1, z_1)$ to $P_2(x_2, y_2, z_2)$ are

$$\left(\frac{x_1 + x_2}{2}, \frac{y_1 + y_2}{2}, \frac{z_1 + z_2}{2}\right).$$

As an illustration of (8.3), if A and B are the points in Example 1, then the midpoint of segment AB has coordinates $(1, 1/2, -1/2)$.

The graph of an equation in three variables x, y, and z is defined as the set of all points $P(a, b, c)$ in a rectangular coordinate system such that the ordered triple (a, b, c) is a solution of the equation — that is, equality is obtained when a, b, and c are substituted for x, y, and z, respectively. Graphs of such equations are *surfaces* of some type. It is easy to derive an equation which has, for its graph, a sphere of radius r with center at the point $C(h, k, l)$. As illustrated in Fig. 8.6, a point $P(x, y, z)$ is on the

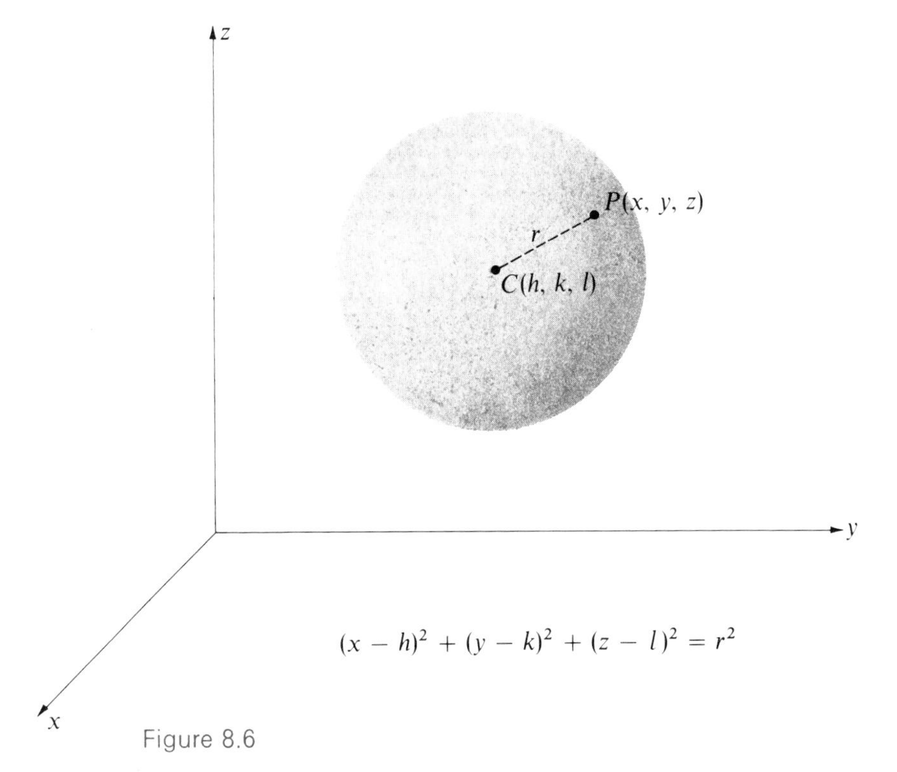

Figure 8.6

sphere if and only if $[d(C, P)]^2 = r^2$. Using the distance formula (8.2) this leads to the following *standard equation of a sphere of radius r with center C(h, k, l)*:

(8.4) $$(x - h)^2 + (y - k)^2 + (z - l)^2 = r^2.$$

If we square the indicated expressions and simplify, then (8.4) may be written in the form

$$x^2 + y^2 + z^2 + ax + by + cz + d = 0, \tag{8.5}$$

where the coefficients are real numbers. Conversely, in a manner similar to our work with circles, if one begins with an equation of the form (8.5), then by completing squares the left side may be written as in (8.4), and hence the graph, if it exists, is a sphere or a point.

EXAMPLE 2. Discuss the graph of the equation

$$x^2 + y^2 + z^2 - 6x + 8y + 4z + 4 = 0.$$

Solution: We arrange our work as follows:

$$(x^2 - 6x) + (y^2 + 8y) + (z^2 + 4z) = -4$$

$$(x^2 - 6x + 9) + (y^2 + 8y + 16) + (z^2 + 4z + 4) = -4 + 9 + 16 + 4$$

$$(x - 3)^2 + (y + 4)^2 + (z + 2)^2 = 25.$$

Comparing the last equation with (8.4) we see that the graph is a sphere of radius 5 with center $C(3, -4, -2)$.

EXERCISES

In Exercises 1–4 plot the points A and B and find:

(a) $d(A, B)$. (b) the midpoint of AB.

1. $A(2, 4, -5)$, $B(4, -2, 3)$.
2. $A(1, -2, 7)$, $B(2, 4, -1)$.
3. $A(-4, 0, 1)$, $B(3, -2, 1)$.
4. $A(0, 5, -4)$, $B(1, 1, 0)$.

In Exercises 5 and 6, if A and B are opposite vertices of a parallelepiped having its faces parallel to the coordinate planes, find the coordinates of the other vertices.

5. $A(2, 5, -3)$, $B(-4, 2, 1)$.
6. $A(-3, 2, 6)$, $B(1, 5, -1)$.

In Exercises 7 and 8 prove that A, B, and C are vertices of a right triangle and find its area.

7. $A(2, 0, 1)$, $B(3, 1, 2)$, $C(1, 2, 0)$.
8. $A(4, -3, 2)$, $B(6, -2, 1)$, $C(7, -6, 5)$.

In Exercises 9–12 find the equation of the sphere having center C and radius r.

9. $C(3, -1, 2)$, $r = 3$.
10. $C(4, -5, 1)$, $r = 5$.
11. $C(-5, 0, 1)$, $r = 1/2$.
12. $C(0, -3, -6)$, $r = \sqrt{3}$.

In Exercises 13–16 find the center and radius of the sphere having the given equation.

13. $x^2 + y^2 + z^2 + 4x - 2y + 2z + 2 = 0.$
14. $x^2 + y^2 + z^2 - 6x - 10y + 6z + 34 = 0.$
15. $x^2 + y^2 + z^2 - 8x + 8z + 16 = 0.$
16. $4x^2 + 4y^2 + 4z^2 - 4x + 8y - 3 = 0.$
17. Find an equation for the set of all points equidistant from $A(2, -1, 3)$ and $B(-1, 5, 1)$. Describe the graph of the equation.
18. Same as for Exercise 17 using $A(5, 0, -4)$ and $B(2, -1, 7)$.
19. Describe the graph of each of the following equations:
(a) $z = 5$. (b) $y = 2$. (c) $x = 0$.
20. Describe the graph of the equation $xyz = 0$.

2 PLANES

In two dimensions the graph of the equation $x = a$ is a line parallel to the y-axis (cf. p. 252). We shall now determine its graph in three dimensions. Since the equation may be written in the form $x + 0 \cdot y + 0 \cdot z = a$, we see that the solutions consist of all ordered triples (a, y, z), where a is fixed, but y and z may have any values. It follows that the graph is the set of all points which are a (directed) distance a from the yz-plane. This set of points is a plane parallel to the yz-plane. In (i) of Fig. 8.7 we have

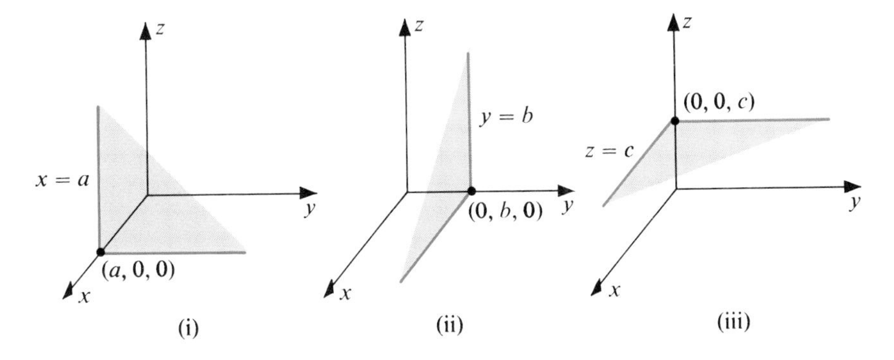

Figure 8.7

sketched a portion of this plane which lies in the first octant. Similarly, the graphs of the equations $y = b$ and $z = c$ are planes parallel to the xz-plane and xy-plane, respectively. Parts of the graphs are shown in (ii) and (iii) of Fig. 8.7 for the cases $b > 0$ and $c > 0$.

It can be proved that the graph of every equation of the form

$$ax + by + cz + d = 0, \tag{8.6}$$

where the coefficients are real numbers, is a plane, and conversely every plane is the graph of such an equation. The proof of this fact requires concepts which are not given in this text. The discussion in the first paragraph applies to the cases in which two of the coefficients a, b, or c are zero.

In order to sketch the graph of an equation of the form (8.6) we often find the *traces* of the graph in the coordinate planes — that is, the lines in which the graph intersects the coordinate planes. To find the trace in the xy-plane we substitute 0 for z in (8.6), since this will lead to all points of the graph which lie on the xy-plane. Similarly, to find the trace in the yz-plane and the xz-plane we let $x = 0$ and $y = 0$, respectively, in (8.6).

EXAMPLE 1. Sketch the graph of the equation

$$2x + 3y + 4z = 12.$$

Solution: There are three points on the plane which are easily found, namely the points of intersection of the plane with the coordinate axes. Substituting 0 for both y and z in the equation, we obtain $2x = 12$, or $x = 6$. Thus the point (6, 0, 0) is on the graph. As in two dimensions, 6 is called the *x-intercept* of the graph. Similarly, substitution of 0 for x and z gives us the y-intercept 4, and hence the point (0, 4, 0) is on the graph. The point (0, 0, 3) (or z-intercept 3) is obtained in like manner. The trace in the xy-plane is found by substituting 0 for z in the given equation. This leads to $2x + 3y = 12$, which has, for its graph in the xy-plane, a straight line with x-intercept 6 and y-intercept 4. This trace, and the traces of the graph in the xz- and yz-planes, are illustrated in Fig. 8.8.

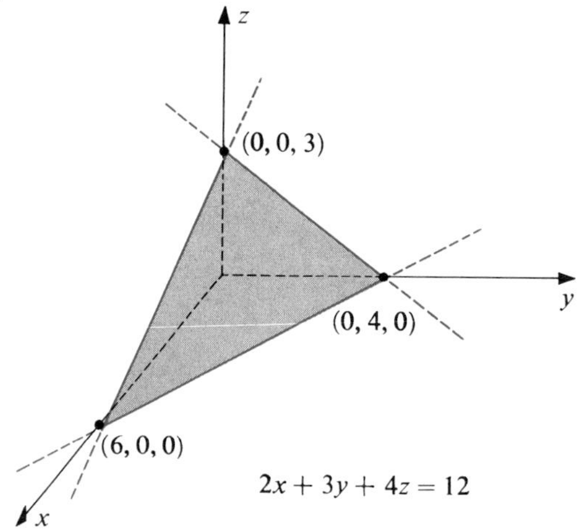

Figure 8.8

If exactly one of the coefficients a, b, or c in (8.6) is zero, then the graph is a plane perpendicular to the coordinate plane of the variables with the nonzero coefficients. This is illustrated in the next example.

EXAMPLE 2. Sketch the graph of the equation

$$3x + 5z = 10.$$

Solution: The x- and z-intercepts are $10/3$ and 2, respectively, and the trace in the xz-plane has equation $3x + 5z = 10$ (Why?). To find the trace in the xy-plane we substitute 0 for z, obtaining the equation $3x = 10$, or $x = 10/3$. Thus the trace is a straight line which is parallel to the y-axis with x-intercept $10/3$. Similarly, to find the trace in the yz-plane we substitute 0 for x, obtaining the equation $z = 2$, which has, for its graph in the yz-plane, a line parallel to the y-axis with z-intercept 2. There is no y-intercept, since substituting 0 for x and z simultaneously produces the absurdity $0 = 10$. The graph of the given equation is a plane which is perpendicular to the xz-plane. A portion of the graph, showing the traces in the three coordinate planes, is sketched in Fig. 8.9.

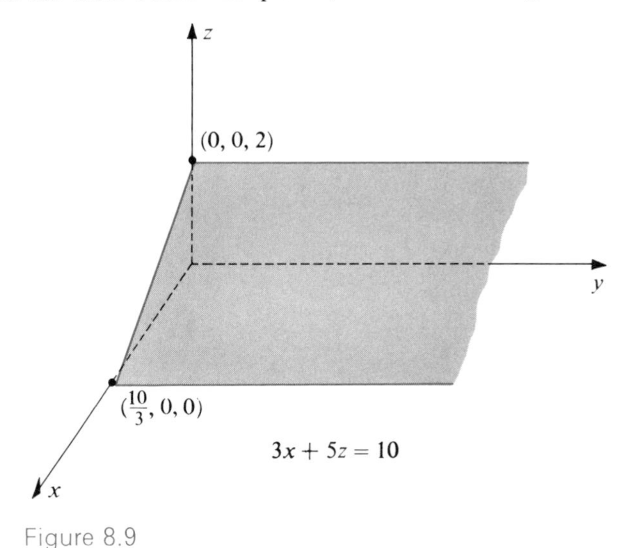

Figure 8.9

EXAMPLE 3. Find an equation for the plane through the three points $A(1, 2, -1)$, $B(3, 1, 0)$, and $C(0, 0, 0)$.

Solution: The required plane has an equation of the form (8.6) for certain choices of a, b, c, and d. Moreover, the coordinates of A, B, and C must be solutions of this equation (Why?). In particular, using the coordinates of C we have

$$a \cdot 0 + b \cdot 0 + c \cdot 0 + d = 0,$$

from which it follows that $d = 0$. Hence there is an equation for the plane of the form

$$ax + by + cz = 0. \tag{8.7}$$

Since the coordinates of A and B are solutions of this equation we must have

$$a \cdot 1 + b \cdot 2 + c \cdot (-1) = 0$$

$$a \cdot 3 + b \cdot 1 + c \cdot 0 = 0.$$

Thus a, b, and c satisfy both of the equations

$$a + 2b - c = 0$$

$$3a + b = 0.$$

From the second equation we obtain $b = -3a$, which, when substituted in the first equation, gives us $a + 2(-3a) - c = 0$, or $c = -5a$. Substituting for b and c in (8.7) we obtain

$$ax + (-3a)y + (-5a)z = 0.$$

The coordinates of A, B, and C are solutions of the latter equation for *all* choices of a (Check!). In particular, if we let $a = 1$, there results $x - 3y - 5z = 0$, which is one form of the desired equation.

EXERCISES

In Exercises 1–10 sketch the graph of the given equations.

1. (a) $x = 3$. (b) $y = -2$. (c) $z = 5$.
2. (a) $x = -4$. (b) $y = 0$. (c) $z = -2/3$.
3. $2x + y - 6 = 0$.
4. $3x - 2z - 24 = 0$.
5. $4y - 2z - 15 = 0$.
6. $3x + 7y + 21 = 0$.
7. $x + 2y + 3z - 12 = 0$.
8. $5x + y - 4z + 20 = 0$.
9. $2x - y + 5z + 10 = 0$.
10. $x + y + z = 0$.

In Exercises 11–16 find an equation of the plane which satisfies the stated conditions.

11. Through $A(6, -7, 4)$, parallel to

    (a) the xy-plane. (b) the yz-plane. (c) the xz-plane.

12. Same as Exercise 11 for $A(-4, 1/2, -2/3)$.
13. Through $A(4, 5, 0)$ and $B(-2, 3, 0)$, perpendicular to the xy-plane.
14. Through $A(-1, 0, 7)$ and $B(3, 0, 2)$, perpendicular to the xz-plane.
15. Through the origin and the points $A(0, 2, 5)$ and $B(1, 4, 0)$.
16. Through $A(3, 2, 1)$, $B(-1, 1, -2)$, and $C(3, -4, 1)$.

3 GRAPHS OF EQUATIONS

In this section and the next we shall consider equations in x, y, and z which have, for their graphs, surfaces which are fundamental in the study of coordinate geometry. We previously defined the trace of a surface in a coordinate plane. More generally, the trace of a surface in *any* plane is the intersection of the surface and the plane. To find the shape of a surface from its equation, we shall make considerable use of traces in planes which are parallel to coordinate planes.

EXAMPLE 1. Find traces, in various planes, of the surface having equation $z = x^2 + y^2$, and sketch the graph of the equation.

Solution: Substitution of 0 for x in the stated equation gives us $z = y^2$, and hence the trace of the surface in the yz-plane is a parabola with vertex at the origin and opening upward, as shown in Fig. 8.10. Similarly, the

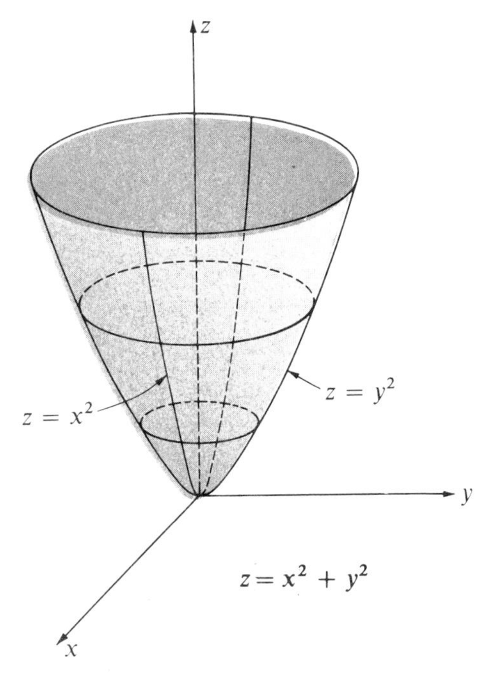

Figure 8.10

result of substituting 0 for y is the equation $z = x^2$, and hence the trace in the xz-plane is also a parabola. It is instructive to find traces in planes parallel to the xy-plane. From our work in Section 2, such planes have equations of the form $z = c$. If we substitute c for z in the given equation, we obtain $x^2 + y^2 = c$. It follows that if $c > 0$, then the trace in the plane having equation $z = c$ is a circle of radius $\sqrt{c}$. Three such circles are sketched in Fig. 8.10. If $c < 0$, then $x^2 + y^2 = c$ has no graph, and consequently none of the points beneath the xy-plane is on the surface. The trace in the xy-plane has equation $x^2 + y^2 = 0$, and therefore consists of only one point — the origin. Although traces in planes parallel to the xz- or yz-planes could be determined, those we have obtained are sufficient for an accurate description of the graph. The surface in this example may be regarded as having been generated by revolving about the z-axis the graph of the parabola in the yz-plane having equation $z = y^2$. This surface is called a *circular paraboloid* or *paraboloid of revolution.*

One of the simplest surfaces to sketch is a cylinder, which we now define.

(8.8) Definition of Cylinder

If C is a curve in a plane and l is a line not in the plane, then the set of points on all lines which intersect C and are parallel to l is called a *cylinder*.

The curve C in (8.8) is called a *directrix* for the cylinder, and each line through C and parallel to l is a *ruling* of the cylinder. The most familiar type of cylinder is a *right circular cylinder*, obtained by letting C be a circle in a plane and l a line perpendicular to the plane, as illustrated in (i) of Fig. 8.11. Although in the figure we have "cut off" the cylinder, it is to be understood that the rulings do not have finite length, but instead extend indefinitely. It is not required that the directrix C in (8.8) be a closed curve. This is illustrated in (ii) of Fig. 8.11, where C is a parabola.

In the discussion to follow we shall only consider the case in which the directrix C is on a coordinate plane and the line l is parallel to the coordinate axis which is not on the plane. Suppose, for example, that C is on the xy-plane and has equation $y = f(x)$, where f is a function, and that the rulings are parallel to the z-axis. As illustrated in Fig. 8.12, a point $P(x, y, z)$ is on the cylinder if and only if $Q(x, y, 0)$ is on C — that is, if and only if the first two coordinates x and y of P satisfy the equation $y = f(x)$. It follows that the equation of the cylinder is $y = f(x)$, and hence is the same as the equation of the directrix in the xy-plane.

Example 2. Sketch the graph of the equation

$$x^2/4 + y^2/9 = 1.$$

Solution: From our previous remarks, the graph is a cylinder with rulings parallel to the z-axis. We begin by sketching the graph of $x^2/4 + y^2/9 = 1$ in the xy-plane. This ellipse is a directrix for the cylinder. All traces in planes parallel to the xy-plane are ellipses congruent to this directrix. A

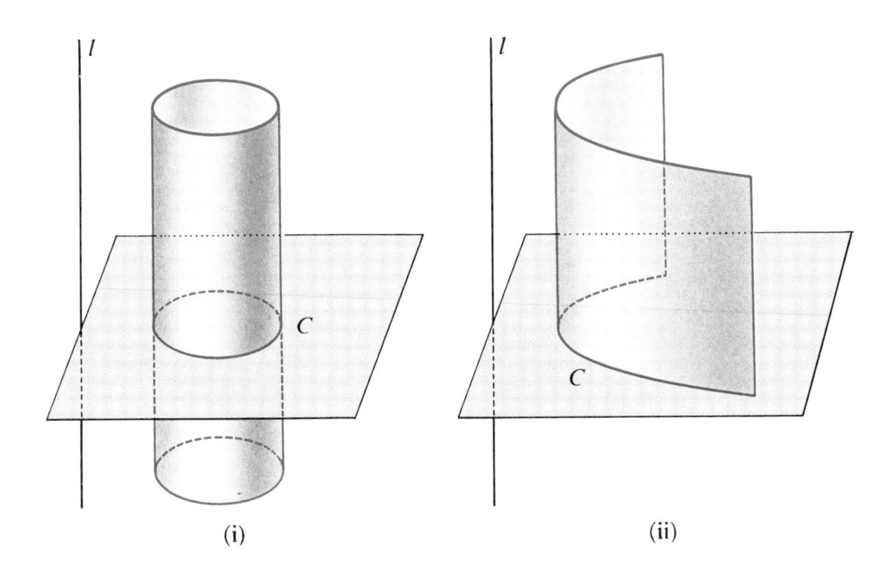

Figure 8.11

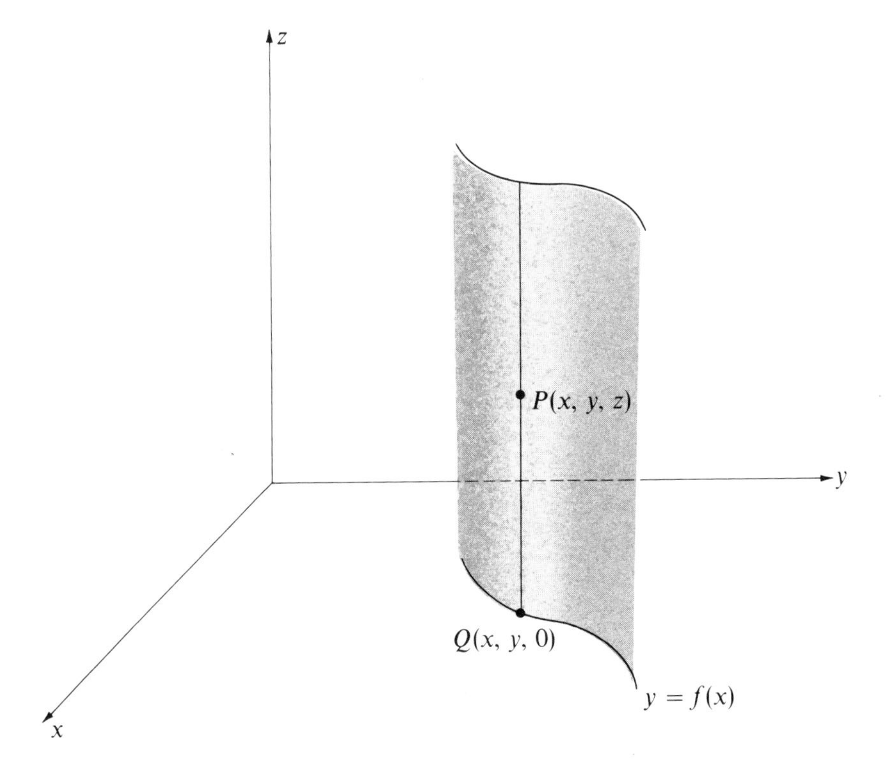

Figure 8.12

portion of the graph is shown in Fig. 8.13. This surface is called an *elliptic cylinder.*

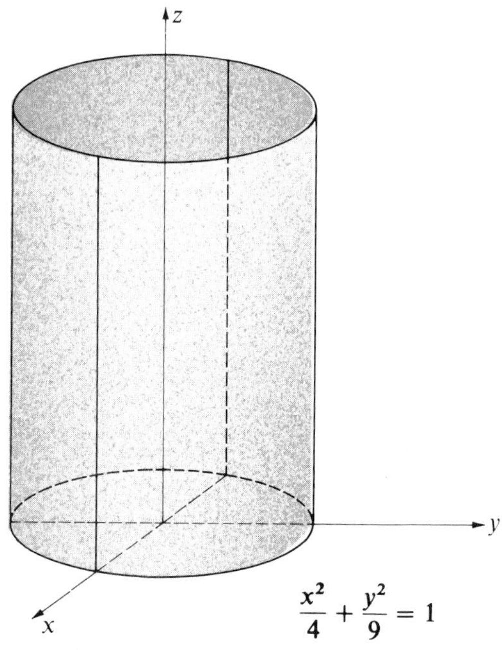

Figure 8.13

It can be shown that the graph of an equation which contains only the variables y and z is a cylinder with rulings parallel to the x-axis and whose trace (directrix) in the yz-plane is the graph of the given equation. Similarly, the graph of an equation which does not contain the variable y is a cylinder with rulings parallel to the y-axis and whose directrix is the graph of the given equation in the xz-plane.

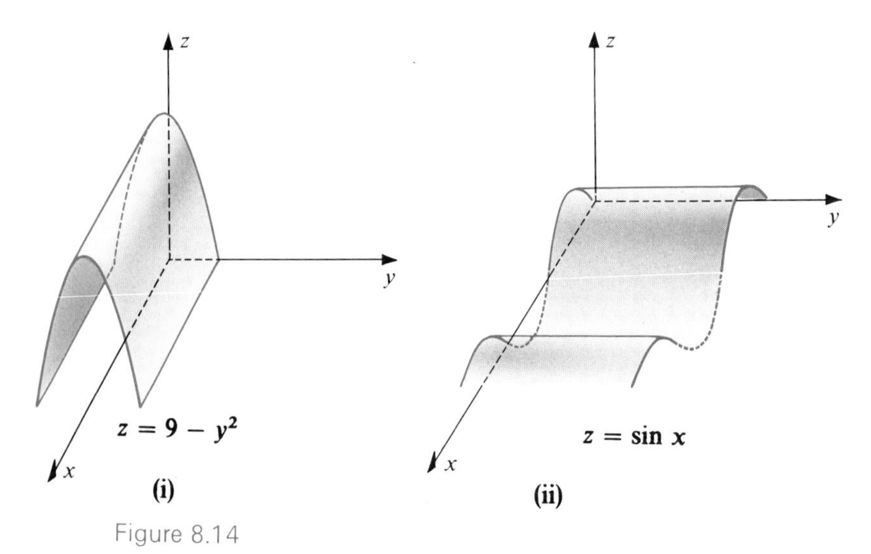

Figure 8.14

EXAMPLE 3. Sketch the graph of the equations

$$\text{(a)}\quad y^2 = 9 - z, \quad \text{(b)}\quad z = \sin x.$$

Solution: The graph of (a) is a cylinder with rulings parallel to the x-axis. A directrix in the yz-plane is the graph of the equation $y^2 = 9 - z$. Part of the graph is sketched in (i) of Fig. 8.14. This surface is called a *parabolic cylinder*.

The graph of (b) is a cylinder with rulings parallel to the y-axis and whose directrix in the xz-plane is the graph of the equation $z = \sin x$. A portion of the graph is sketched in (ii) of Fig. 8.14.

EXERCISES

In Exercises 1–4, after finding traces on various planes, sketch the graph of the given equation.

1. $y = x^2 + z^2$.
2. $x = y^2 + z^2$.
3. $z = 4x^2 + 9y^2$.
4. $y = 4x^2 + z^2 + 4$.

In Exercises 5–14 sketch the graph of the cylinder having the given equation.

5. $x^2 + y^2 = 9$.
6. $y^2 + z^2 = 16$.
7. $4y^2 + 9z^2 = 36$.
8. $x^2 + 5z^2 = 25$.
9. $x^2 = 9z$.
10. $x^2 - 4y = 0$.
11. $y^2 - x^2 = 16$.
12. $xz = 1$.
13. $z = e^y$.
14. $z = \log_{10} x$.

4 QUADRIC SURFACES

In the preceding chapter it was shown that, in two dimensions, the graph of a second degree equation in x and y is a conic section. In three-dimensional coordinate geometry, the graph of a second degree equation in x, y, and z is referred to as a *quadric surface*. In the present section we shall investigate standard equations for such surfaces.

The graph of an equation of the form

$$\frac{x^2}{a^2} + \frac{y^2}{b^2} + \frac{z^2}{c^2} = 1, \tag{8.9}$$

where a, b, and c are positive real numbers, is called an *ellipsoid*. Traces of this surface in planes parallel to coordinate planes are ellipses. For example, the trace in the xy-plane is the ellipse with equation $x^2/a^2 + y^2/b^2 = 1$. Similarly, the traces in the yz- and xz-planes are ellipses, as indicated in Fig. 8.15. Let us find the trace in an arbitrary plane parallel to the xy-plane — that is, in a plane whose equation is of the form $z = k$, where $k \in \mathbf{R}$. Substituting k for z in (8.9) leads to the equation

$$\frac{x^2}{a^2} + \frac{y^2}{b^2} = 1 - \frac{k^2}{c^2}. \tag{8.10}$$

If $|k| > c$, then $1 - k^2/c^2 < 0$ and there is no graph. Thus the graph lies between the planes having equations $z = -c$ and $z = c$. If $|k| < c$, then $1 - k^2/c^2 > 0$ and hence the trace in the plane whose equation is $z = k$ is an ellipse. Similarly, traces in planes parallel to the other two coordinate planes are ellipses, provided they do not intersect the x-axis outside of the closed interval $[-a, a]$ or the y-axis outside of $[-b, b]$.

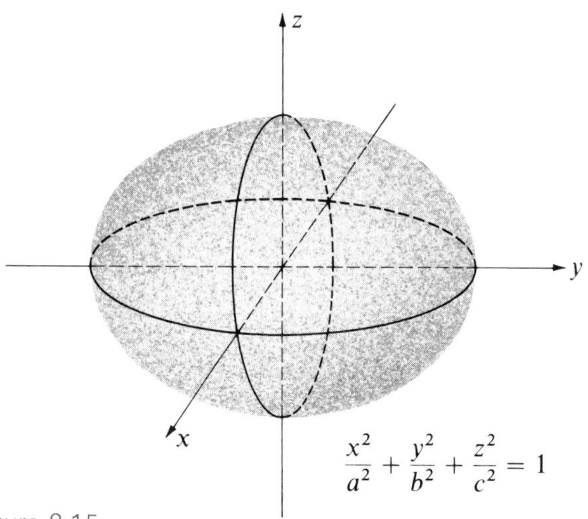

Figure 8.15

Note that if $a = b = c$ in (8.9), then the graph is a sphere of radius a with center at the origin.

The graph of an equation of the form

$$\frac{x^2}{a^2} + \frac{y^2}{b^2} - \frac{z^2}{c^2} = 1 \tag{8.11}$$

is called a *hyperboloid of one sheet.* The traces in the xz- and yz-planes are hyperbolas which have equations

$$\frac{x^2}{a^2} - \frac{z^2}{c^2} = 1 \quad \text{and} \quad \frac{y^2}{b^2} - \frac{z^2}{c^2} = 1,$$

respectively. Traces on planes parallel to the xy-plane have equations of the form

$$\frac{x^2}{a^2} + \frac{y^2}{b^2} = 1 + \frac{k^2}{c^2},$$

where $k \in \mathbf{R}$, and therefore are ellipses. The graph of (8.11) is sketched in Fig. 8.16. The z-axis is called the *axis* of the hyperboloid. The graph of the equation

$$\frac{x^2}{a^2} - \frac{y^2}{b^2} + \frac{z^2}{c^2} = 1$$

is also a hyperboloid of one sheet; however, in this case the axis of the hyperboloid is the y-axis. If the term involving x^2 is negative and the other terms are positive, then the axis of the hyperboloid coincides with the x-axis.

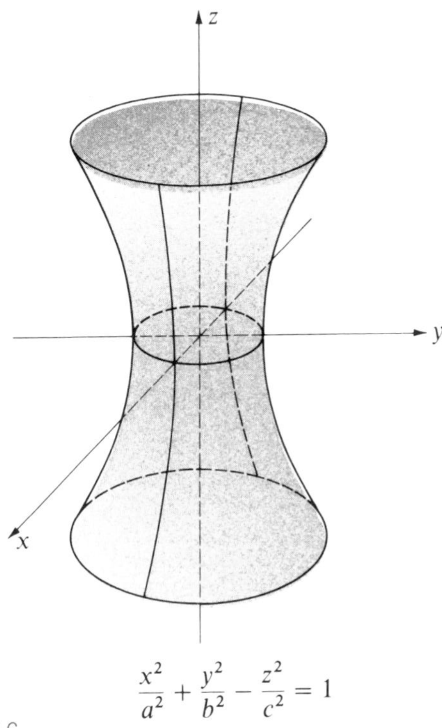

Figure 8.16

The graph of

$$\frac{x^2}{a^2} - \frac{y^2}{b^2} - \frac{z^2}{c^2} = 1 \tag{8.12}$$

is known as a *hyperboloid of two sheets.* Traces in the xy- and xz-planes are hyperbolas, whereas traces in planes with equations of the form $x = k$, $|k| > a$, are ellipses. We leave it to the reader to show that the graph of (8.12) has the appearance shown in Fig. 8.17. The x-axis is called the axis of the hyperboloid. By using minus signs on different terms, one can obtain a hyperboloid whose axis is the y- or the z-axis.

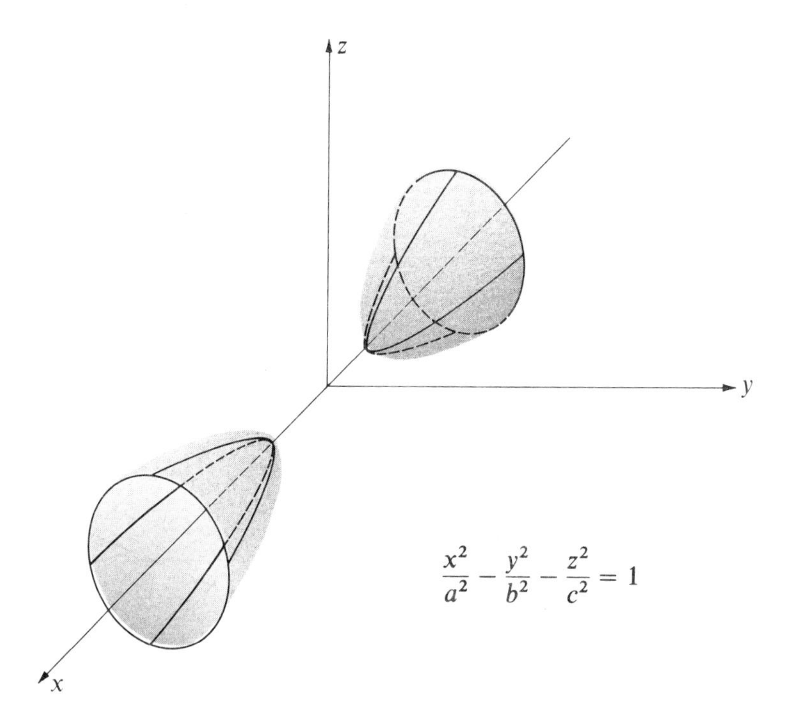

Figure 8.17

The graph of

$$\frac{x^2}{a^2} + \frac{y^2}{b^2} - \frac{z^2}{c^2} = 0 \tag{8.13}$$

is a *cone*, having, for its axis, the z-axis. The trace in the yz-plane has equation $y^2/b^2 - z^2/c^2 = 0$. Solving for y we obtain $y = \pm(b/c)z$, which gives us the equations of two straight lines through the origin in the yz-plane. Similarly, the trace in the xz-plane is a pair of straight lines which intersect at the origin. Traces in planes parallel to the xy-plane are ellipses (Why?). The graph is sketched in Fig. 8.18. By changing signs on the terms in (8.13) we obtain a cone whose axis is either the x- or the y-axis.

The graph of an equation of the form

$$\frac{x^2}{a^2} + \frac{y^2}{b^2} = cz \tag{8.14}$$

is called a *paraboloid*. Example 1 on p. 309 is the special case of (8.14) in which $a = b = c = 1$. If $c > 0$, then the graph of (8.14) is similar to that shown in Fig. 8.10, except that if $a \neq b$, then traces in planes parallel to the xy-plane are ellipses instead of circles. If $c < 0$, then the

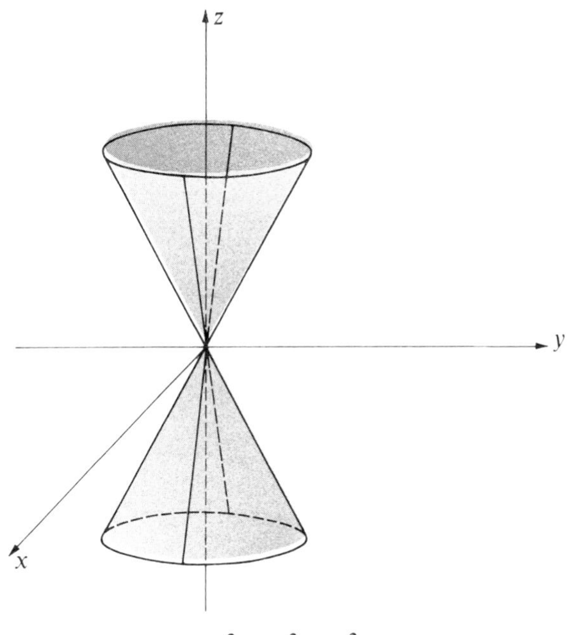

Figure 8.18 $\frac{x^2}{a^2} + \frac{y^2}{b^2} - \frac{z^2}{c^2} = 0$

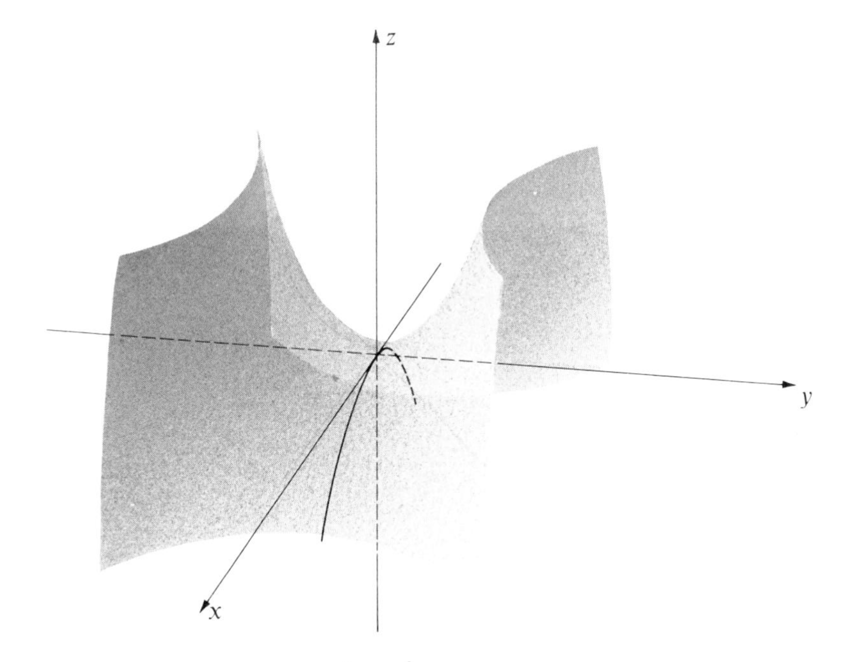

Figure 8.19 $\frac{y^2}{a^2} - \frac{x^2}{b^2} = cz$

paraboloid opens "downward." The z-axis is called the *axis* of the paraboloid. The graphs of the equations

$$\frac{x^2}{a^2} + \frac{z^2}{b^2} = cy \qquad \text{and} \qquad \frac{y^2}{a^2} + \frac{z^2}{b^2} = cx$$

are paraboloids whose axes are the y-axis and x-axis, respectively.

Finally, the graph of the equation

$$\frac{y^2}{a^2} - \frac{x^2}{b^2} = cz \tag{8.15}$$

is called a *hyperbolic paraboloid.* A typical sketch of this "saddle-shaped" surface for the case $c > 0$ is shown in Fig. 8.19. Variations are obtained by interchanging x, y, and z in (8.15).

It is possible to obtain formulas for translation or rotation of axes in three dimensions which are analogous to those in two dimensions. These, in turn, can be used to show that the graph of an equation of degree two in x, y, and z, when it exists, is one of the surfaces discussed in this chapter — except, as usual, for degenerate cases.

EXERCISES

Name and sketch the graph of each of the following equations.

1. $9x^2 + 4y^2 + z^2 = 36.$
2. $16x^2 - 4y^2 - z^2 + 1 = 0.$
3. $16x^2 - 25y^2 + 100z^2 = 200.$
4. $36x = 9y^2 + z^2.$
5. $y^2 - 9x^2 - z^2 - 9 = 0.$
6. $z^2 - x^2 - y^2 = 1.$
7. $16y = x^2 + 4z^2.$
8. $4y^2 + 9z^2 = 9x^2.$
9. $36x^2 - 16y^2 + 9z^2 = 0.$
10. $4y^2 + 25z^2 + 100x = 0.$
11. $4y = x^2 - z^2.$
12. $4x^2 + 16x = z^2.$

5 REVIEW EXERCISES

Oral

Define or discuss each of the following:

1. Ordered triple.
2. Rectangular coordinate system in three dimensions.
3. The distance formula.
4. The midpoint formula.
5. The trace of a graph on a plane.
6. Equations of planes.
7. Cylinder.
8. Quadric surfaces.
9. Ellipsoid.
10. Hyperboloid of one sheet.
11. Hyperboloid of two sheets.
12. Cone.
13. Paraboloid.
14. Hyperbolic paraboloid.

Written

1. Given the points $A(5, -3, 2)$ and $B(-1, -4, 3)$, find:
   (a) $d(A, B)$.
   (b) the coordinates of the midpoint of the line segment AB.
   (c) an equation of the sphere with center B and tangent to the xz-plane.
   (d) an equation of the plane through A which is parallel to the xy-plane.
   (e) an equation of the plane through B parallel to the xz-plane.

2. Find an equation of:
   (a) the plane through $A(0, 4, 9)$ and $B(0, -3, 7)$ which is perpendicular to the yz-plane.
   (b) the plane which has x-intercept 5, y-intercept -2, and z-intercept 6.
   (c) the cylinder which is perpendicular to the xy-plane and which has, for its directrix, the circle in the xy-plane with center $C(4, -3, 0)$ and radius 5.
   (d) an ellipsoid which has x-intercept 8, y-intercept 3, and z-intercept 1.

In Exercises 3–15 identify and sketch the graph of the given equation.

3. $x^2 + y^2 + z^2 - 14x + 6y - 8z + 10 = 0.$
4. $y + 4 = 0.$
5. $xyz = 0.$
6. $3x - 5y + 2z = 10.$
7. $4y - 3z - 15 = 0.$
8. $9x^2 + 4z^2 = 36.$
9. $y = z^2 + 1.$
10. $z^2 - 4x^2 = 9 - 4y^2.$
11. $x^2 + 4y + 9z^2 = 0.$
12. $z^2 - 4x^2 - y^2 = 4.$
13. $2x^2 + 4z^2 - y^2 = 0.$
14. $x^2 - 4y^2 = 4z.$
15. $x^2 + 2y^2 + 4z^2 = 16.$

Supplementary Questions

1. Derive formulas for translation of axes in three dimensions and study the effect of a translation on the equations of quadric surfaces. Use translations of axes to discuss the graph of an equation of the form

$$Ax^2 + By^2 + Cz^2 + Dx + Ey + Fz + G = 0.$$

2. In the system of *cylindrical coordinates*, a point P in three dimensions is represented by an ordered triple of the form (r, θ, z), where r and θ are polar coordinates for the projection of P onto the xy-plane. Derive formulas which give the relationship between rectangular and cylindrical coordinates. Find equations in r, θ, and z which correspond to equations of planes, cylinders, spheres and quadric surfaces. What are the graphs of the equations

(a) $r = k$? (b) $\theta = k$? (c) $z = k$?

3. In the system of *spherical coordinates* a point P is represented by an ordered triple of the form (ρ, θ, ϕ) where $\rho = d(O, P)$, θ is the angle described in question 2, and ϕ is the angle between the positive z-axis and the line segment OP. Derive formulas which give the relationship between rectangular and spherical coordinates. Find equations in ρ, θ, and ϕ which correspond to equations of planes, cylinders, spheres and quadric surfaces. What are the graphs of the equations

(a) $\rho = k$? (b) $\theta = k$? (c) $\phi = k$?

appendix I Mathematical Induction

The method of proof called *mathematical induction* is very important in all branches of mathematics. This method may be used to show that certain statements or formulas are true for all positive integers. For example, if n is a positive integer, let P_n denote the statement

$$(xy)^n = x^n y^n,$$

where x and y are real numbers. Thus P_1 represents the statement $(xy)^1 = x^1 y^1$, P_2 denotes $(xy)^2 = x^2 y^2$, P_3 is $(xy)^3 = x^3 y^3$, and so on. It is easy to show that P_1, P_2, and P_3 are *true* statements. However, since the set $\mathbf{N}$ of positive integers is infinite, it is impossible to check the validity of P_n for *every* positive integer n. In order to give a general proof, a method such as that provided by (A.2) is required. The latter method is based on the following fundamental axiom for the set $\mathbf{N}$.

(A.1) Axiom of Mathematical Induction

Let S be a subset of $\mathbf{N}$ with the following properties:

(i) $1 \in S$,
(ii) if $k \in S$, then $k + 1 \in S$.

Then $S = \mathbf{N}$.

There should be little reluctance about accepting (A.1). Thus if S is a set of natural numbers satisfying property (ii), then whenever S contains an arbitrary natural number k, it must also contain the "next" natural number $k + 1$. If S also satisfies property (i), then S contains 1 and hence, by (ii), S contains $1 + 1$, or 2. Applying (ii) again, we see that S contains $2 + 1$, or 3. If we continue in this manner, it can be argued that if n is any *specific* natural number n, then $n \in S$, since we can proceed a step at a time as above, eventually reaching n. Although this argument does not *prove* (A.1), it certainly makes it seem very plausible.

We shall use (A.1) to establish the following fundamental theorem.

(A.2) Principle of Mathematical Induction

Suppose that with each positive integer n there is associated a statement P_n. Then all the statements P_n are true provided:

(i) P_1 is true,
(ii) if k is an arbitrary positive integer such that P_k is true, then P_{k+1} is also true.

Proof: Assume that (i) and (ii) of (A.2) hold and let

$$S = \{n \in \mathbf{N} \mid P_n \text{ is true}\}.$$

By assumption P_1 is true and consequently $1 \in S$. Thus S satisfies property (i) of (A.1). Next suppose $k \in S$. Then P_k is true and hence, by assumption (ii) of (A.2) P_{k+1} is also true. Therefore $k + 1 \in S$. This shows that if $k \in S$, then $k + 1 \in S$ and therefore property (ii) of (A.1) is true. Consequently $S = \mathbf{N}$, which means that P_n is true for every positive integer n.

There are other variations of the principle of mathematical induction. A slight extension appears below in (A.9). In most of our work the statement P_n will usually be given in the form of an equation involving the arbitrary positive integer n, such as in our illustration $(xy)^n = x^n y^n$.

When applying (A.2), the following two steps should always be followed:

(A.3)

(i) Prove that P_1 is true.
(ii) *Assume* that P_k is true and *prove* that P_{k+1} is true.

Step (ii) is usually the most confusing for the beginning student. One does not *prove* that P_k is true (except for $k = 1$). Rather, one shows that *if* P_k is true, *then* the next statement P_{k+1} is true. This is all that is necessary according to (A.2). The assumption that P_k is true is sometimes referred to as the *induction hypothesis*.

Many interesting formulas concerning the positive integers can be established by mathematical induction. Two of these are illustrated in Examples 1 and 2. Others appear in the Exercises.

EXAMPLE 1. Prove that for any positive integer n, the sum of the first n positive integers is $n(n + 1)/2$.

Solution: If n is a positive integer, we let P_n denote the statement

(A.4) $$1 + 2 + 3 + \cdots + n = n(n + 1)/2,$$

where, by convention, when $n \leq 4$, the left side is adjusted so that there are precisely n terms in the sum. We wish to show that P_n is true for all n. Although it is instructive to check (A.4) for several values of n, it is unnecessary to do so. We need only follow the two steps (i) and (ii) of (A.3).

(i) If we substitute $n = 1$ in (A.4), then by convention, the left side collapses to 1 and the right side is $[1(1 + 1)]/2$, which also equals 1. This proves that P_1 is true.

(ii) We *assume* that P_k is true. Thus the induction hypothesis is

(A.5) $$1 + 2 + 3 + \cdots + k = \frac{k(k + 1)}{2}.$$

Our goal is to prove that P_{k+1} is true, that is,

(A.6) $$1 + 2 + 3 + \cdots + (k + 1) = \frac{(k + 1)[(k + 1) + 1]}{2}.$$

By the induction hypothesis we already have a formula for the sum of the first k positive integers. Hence a formula for the sum of the first $k + 1$ positive integers may be found simply by adding $(k + 1)$ to both sides of (A.5). Doing this and simplifying, we obtain

$$\begin{aligned} 1 + 2 + 3 + \cdots + k + (k + 1) &= \frac{k(k + 1)}{2} + (k + 1) \\ &= \frac{k(k + 1) + 2(k + 1)}{2} \\ &= \frac{k^2 + 3k + 2}{2} \\ &= \frac{(k + 1)(k + 2)}{2}, \end{aligned}$$

which can be written in the form (A.6). This shows that P_{k+1} is true, and the proof by mathematical induction is completed.

The laws of exponents can be proved by mathematical induction. In order to apply (A.3), we shall use the following definition of exponents.

(A.7) Definition of Exponents

Let x denote any real number. Then

(i) $x^1 = x$;

(ii) if k is a positive integer such that x^k is defined, let $x^{k+1} = x^k \cdot x$.

A definition such as (A.7) is called a *recursive definition.* In general, if a concept is defined for every positive integer n in such a way that the case corresponding to $n = 1$ is given and if, in addition, it is stated how any case after the first is obtained from the preceding one, then the definition is a recursive definition. Thus by (i) of (A.7) we have $x^1 = x$. Next applying (ii) of (A.7), we obtain

$$x^2 = x^{1+1} = x^1 \cdot x = x \cdot x.$$

Since x^2 is now defined we may employ (ii) again (with $k = 2$), obtaining

$$x^3 = x^{2+1} = x^2 \cdot x = (x \cdot x) \cdot x.$$

Hence x^3 is defined and (ii) of (A.7) is used again to obtain x^4. Thus

$$x^4 = x^{3+1} = x^3 \cdot x = [(x \cdot x) \cdot x] \cdot x.$$

Notice that this agrees with our previous formulation of x^n as a product of x with itself n times. It can be shown (by mathematical induction) that (A.7) defines x^n for all positive integers n.

EXAMPLE 2. If x is a real number, prove that $x^m \cdot x^n = x^{m+n}$ for all positive integers m and n.

Solution: Let m be an arbitrary positive integer. For each positive integer n, let P_n denote the statement

$$x^m \cdot x^n = x^{m+n}. \qquad \text{(A.8)}$$

We shall use (A.3) to prove that P_n is true for all positive integers n.

(i) To show that P_1 is true we write

$$\begin{aligned} x^m \cdot x^1 &= x^m \cdot x && \text{(i) of (A.7)} \\ &= x^{m+1} && \text{(ii) of (A.7)} \end{aligned}$$

which is the same as (A.8) when $n = 1$.

(ii) Assume that P_k is true for a positive integer k. Therefore the induction hypothesis is

$$x^m \cdot x^k = x^{m+k}.$$

We wish to prove that P_{k+1} is true, that is,

$$x^m \cdot x^{k+1} = x^{m+(k+1)}.$$

The proof may be arranged as follows:

$$\begin{aligned} x^m \cdot x^{k+1} &= x^m \cdot (x^k \cdot x) && \text{(ii) of (A.7)} \\ &= (x^m \cdot x^k) \cdot x && \text{(associative law in } \mathbf{R}) \\ &= x^{m+k} \cdot x && \text{(induction hypothesis)} \\ &= x^{(m+k)+1} && \text{(ii) of (A.7)} \\ &= x^{m+(k+1)} && \text{(associative law for integers).} \end{aligned}$$

Hence (A.8) is true for all positive integers n.

Consider a positive integer j and suppose that with each integer $n \geq j$ there is associated a statement P_n. For example, if $j = 6$, then the statements are numbered P_6, P_7, P_8, and so on. The principle of mathematical induction may be extended to cover this situation. Just as before, two steps are used. Specifically, to prove that all statements S_n are true for $n \geq j$:

(A.9)
(i′) prove that S_j is true,
(ii′) *Assume* that S_k is true for $k \geq j$ and *prove* that S_{k+1} is true.

EXAMPLE 3. Let a be a nonzero real number such that $a > -1$. Prove that $(1 + a)^n > 1 + na$ for all integers $n \geq 2$.

Solution: If n is a positive integer, we let P_n denote the inequality $(1 + a)^n > 1 + na$. Note that P_1 is *false*, since $(1 + a)^1 = 1 + 1 \cdot a$. However, we can show that P_n is true for $n \geq 2$ by using (A.9) with $j = 2$.

(i′) We write $(1 + a)^2 = 1 + 2a + a^2$. Since $a \neq 0$, we have $a^2 > 0$ and therefore $1 + 2a + a^2 > 1 + 2a$. This gives us $(1 + a)^2 > 1 + 2a$, and hence P_2 is true.
(ii′) Assume that P_k is true. Thus the induction hypothesis is

$$(1 + a)^k > 1 + ka. \tag{A.10}$$

We wish to show that P_{k+1} is true, that is, $(1 + a)^{k+1} > 1 + (k + 1)a$. Since $a > -1$, $a + 1 > 0$, and consequently multiplying both sides of (A.10) by $1 + a$ will not affect the sense of the inequality. Thus

$$(1 + a)^k(1 + a) > (1 + ka)(1 + a),$$

which may be rewritten

$$(1 + a)^{k+1} > 1 + ka + a + ka^2$$

or

$$(1 + a)^{k+1} > 1 + (k + 1)a + ka^2.$$

Since $ka^2 > 0$, we have

$$1 + (k + 1)a + ka^2 > 1 + (k + 1)a$$

and consequently

$$(1 + a)^{k+1} > 1 + (k + 1)a.$$

Thus P_{k+1} is true and the proof is completed.

EXERCISES

In Exercises 1–10 prove that the given statement is true for every positive integer n.

1. $2 + 4 + 6 + \cdots + 2n = n(n + 1)$.
2. $1 + 4 + 7 + \cdots + (3n - 2) = \frac{1}{2}n(3n - 1)$.
3. $1^2 + 2^2 + 3^2 + \cdots + n^2 = \dfrac{n(n + 1)(2n + 1)}{6}$.
4. $1^3 + 2^3 + 3^3 + \cdots + n^3 = \left[\dfrac{n(n + 1)}{2}\right]^2$.
5. $\dfrac{1}{1 \cdot 2} + \dfrac{1}{2 \cdot 3} + \dfrac{1}{3 \cdot 4} + \cdots + \dfrac{1}{n(n + 1)} = \dfrac{n}{n + 1}$.
6. $\dfrac{1}{1 \cdot 2 \cdot 3} + \dfrac{1}{2 \cdot 3 \cdot 4} + \dfrac{1}{3 \cdot 4 \cdot 5} + \cdots + \dfrac{1}{n(n + 1)(n + 2)} = \dfrac{n(n + 3)}{4(n + 1)(n + 2)}$.
7. $1^2 + 3^2 + 5^2 + \cdots + (2n - 1)^2 = \dfrac{n(2n - 1)(2n + 1)}{3}$.
8. $1^3 + 3^3 + 5^3 + \cdots + (2n - 1)^3 = n^2(2n^2 - 1)$.
9. $n < 2^n$.
10. $1 + 2n \leq 3^n$.
11. If a and b are real numbers, prove by mathematical induction that $(ab)^n = a^n b^n$ for every positive integer n.
12. If a is a real number, prove that $(a^m)^n = a^{mn}$ for all positive integers m and n.
13. Use mathematical induction to prove that if a is any real number greater than 1, then $a^n > 1$ for every positive integer n.
14. If $a \neq 1$, prove that

$$1 + a + a^2 + \cdots + a^{n-1} = \frac{a^n - 1}{a - 1},$$

for every positive integer n.

15. Use mathematical induction to prove that $a - b$ is a factor of $a^n - b^n$ for every positive integer n. [*Hint:* $a^{k+1} - b^{k+1} = a^k(a - b) + (a^k - b^k)b$.]
16. Prove that $a + b$ is a factor of $a^{2n-1} + b^{2n-1}$ for every positive integer n.

appendix II Tables

N	0	1	2	3	4	5	6	7	8	9
1.0	.0000	.0043	.0086	.0128	.0170	.0212	.0253	.0294	.0334	.0374
1.1	.0414	.0453	.0492	.0531	.0569	.0607	.0645	.0682	.0719	.0755
1.2	.0792	.0828	.0864	.0899	.0934	.0969	.1004	.1038	.1072	.1106
1.3	.1139	.1173	.1206	.1239	.1271	.1303	.1335	.1367	.1399	.1430
1.4	.1461	.1492	.1523	.1553	.1584	.1614	.1644	.1673	.1703	.1732
1.5	.1761	.1790	.1818	.1847	.1875	.1903	.1931	.1959	.1987	.2014
1.6	.2041	.2068	.2095	.2122	.2148	.2175	.2201	.2227	.2253	.2279
1.7	.2304	.2330	.2355	.2380	.2405	.2430	.2455	.2480	.2504	.2529
1.8	.2553	.2577	.2601	.2625	.2648	.2672	.2695	.2718	.2742	.2765
1.9	.2788	.2810	.2833	.2856	.2878	.2900	.2923	.2945	.2967	.2989
2.0	.3010	.3032	.3054	.3075	.3096	.3118	.3139	.3160	.3181	.3201
2.1	.3222	.3243	.3263	.3284	.3304	.3324	.3345	.3365	.3385	.3404
2.2	.3424	.3444	.3464	.3483	.3502	.3522	.3541	.3560	.3579	.3598
2.3	.3617	.3636	.3655	.3674	.3692	.3711	.3729	.3747	.3766	.3784
2.4	.3802	.3820	.3838	.3856	.3874	.3892	.3909	.3927	.3945	.3692
2.5	.3979	.3997	.4014	.4031	.4048	.4065	.4082	.4099	.4116	.4133
2.6	.4150	.4166	.4183	.4200	.4216	.4232	.4249	.4265	.4281	.4298
2.7	.4314	.4330	.4346	.4362	.4378	.4393	.4409	.4425	.4440	.4456
2.8	.4472	.4487	.4502	.4518	.4533	.4548	.4564	.4579	.4594	.4609
2.9	.4624	.4639	.4654	.4669	.4683	.4698	.4713	.4728	.4742	.4757
3.0	.4771	.4786	.4800	.4814	.4829	.4843	.4857	.4871	.4886	.4900
3.1	.4914	.4928	.4942	.4955	.4969	.4983	.4997	.5011	.5024	.5038
3.2	.5051	.5065	.5079	.5092	.5105	.5119	.5132	.5145	.5159	.5172
3.3	.5185	.5198	.5211	.5224	.5237	.5250	.5263	.5276	.5289	.5302
3.4	.5315	.5328	.5340	.5353	.5366	.5378	.5391	.5403	.5416	.5428
3.5	.5441	.5453	.5465	.5478	.5490	.5502	.5514	.5527	.5539	.5551
3.6	.5563	.5575	.5587	.5599	.5611	.5623	.5635	.5647	.5658	.5670
3.7	.5682	.5694	.5705	.5717	.5729	.5740	.5752	.5763	.5775	.5786
3.8	.5798	.5809	.5821	.5832	.5843	.5855	.5866	.5877	.5888	.5899
3.9	.5911	.5922	.5933	.5944	.5955	.5966	.5977	.5988	.5999	.6010
4.0	.6021	.6031	.6042	.6053	.6064	.6075	.6085	.6096	.6107	.6117
4.1	.6128	.6138	.6149	.6160	.6170	.6180	.6191	.6201	.6212	.6222
4.2	.6232	.6243	.6253	.6263	.6274	.6284	.6294	.6304	.6314	.6325
4.3	.6335	.6345	.6355	.6365	.6375	.6385	.6395	.6405	.6415	.6425
4.4	.6435	.6444	.6454	.6464	.6474	.6484	.6493	.6503	.6513	.6522
4.5	.6532	.6542	.6551	.6561	.6571	.6580	.6590	.6599	.6609	.6618
4.6	.6628	.6637	.6646	.6656	.6665	.6675	.6684	.6693	.6702	.6712
4.7	.6721	.6730	.6739	.6749	.6758	.6767	.6776	.6785	.6794	.6803
4.8	.6812	.6821	.6830	.6839	.6848	.6857	.6866	.6875	.6884	.6893
4.9	.6902	.6911	.6920	.6928	.6937	.6946	.6955	.6964	.6972	.6981
5.0	.6990	.6998	.7007	.7016	.7024	.7033	.7042	.7050	.7059	.7067
5.1	.7076	.7084	.7093	.7101	.7110	.7118	.7126	.7135	.7143	.7152
5.2	.7160	.7168	.7177	.7185	.7193	.7202	.7210	.7218	.7226	.7235
5.3	.7243	.7251	.7259	.7267	.7275	.7284	.7292	.7300	.7308	.7316
5.4	.7324	.7332	.7340	.7348	.7356	.7364	.7372	.7380	.7388	.7396

N	0	1	2	3	4	5	6	7	8	9
5.5	.7404	.7412	.7419	.7427	.7435	.7443	.7451	.7459	.7466	.7474
5.6	.7482	.7490	.7497	.7505	.7513	.7520	.7528	.7536	.7543	.7551
5.7	.7559	.7566	.7574	.7582	.7589	.7597	.7604	.7612	.7619	.7627
5.8	.7634	.7642	.7649	.7657	.7664	.7672	.7679	.7686	.7694	.7701
5.9	.7709	.7716	.7723	.7731	.7738	.7745	.7752	.7760	.7767	.7774
6.0	.7782	.7789	.7796	.7803	.7810	.7818	.7825	.7832	.7839	.7846
6.1	.7853	.7860	.7868	.7875	.7882	.7889	.7896	.7903	.7910	.7917
6.2	.7924	.7931	.7938	.7945	.7952	.7959	.7966	.7973	.7980	.7987
6.3	.7993	.8000	.8007	.8014	.8021	.8028	.8035	.8041	.8048	.8055
6.4	.8062	.8069	.8075	.8082	.8089	.8096	.8102	.8109	.8116	.8122
6.5	.8129	.8136	.8142	.8149	.8156	.8162	.8169	.8176	.8182	.8189
6.6	.8195	.8202	.8209	.8215	.8222	.8228	.8235	.8241	.8248	.8254
6.7	.8261	.8267	.8274	.8280	.8287	.8293	.8299	.8306	.8312	.8319
6.8	.8325	.8331	.8338	.8344	.8351	.8357	.8363	.8370	.8376	.8382
6.9	.8388	.8395	.8401	.8407	.8414	.8420	.8426	.8432	.8439	.8445
7.0	.8451	.8457	.8463	.8470	.8476	.8482	.8488	.8494	.8500	.8506
7.1	.8513	.8519	.8525	.8531	.8537	.8543	.8549	.8555	.8561	.8567
7.2	.8573	.8579	.8585	.8591	.8597	.8603	.8609	.8615	.8621	.8627
7.3	.8633	.8639	.8645	.8651	.8657	.8663	.8669	.8675	.8681	.8686
7.4	.8692	.8698	.8704	.8710	.8716	.8722	.8727	.8733	.8739	.8745
7.5	.8751	.8756	.8762	.8768	.8774	.8779	.8785	.8791	.8797	.8802
7.6	.8808	.8814	.8820	.8825	.8831	.8837	.8842	.8848	.8854	.8859
7.7	.8865	.8871	.8876	.8882	.8887	.8893	.8899	.8904	.8910	.8915
7.8	.8921	.8927	.8932	.8938	.8943	.8949	.8954	.8960	.8965	.8971
7.9	.8976	.8982	.8987	.8993	.8998	.9004	.9009	.9015	.9020	.9025
8.0	.9031	.9036	.9042	.9047	.9053	.9058	.9063	.9069	.9074	.9079
8.1	.9085	.9090	.9096	.9101	.9106	.9112	.9117	.9122	.9128	.9133
8.2	.9138	.9143	.9149	.9154	.9159	.9165	.9170	.9175	.9180	.9186
8.3	.9191	.9196	.9201	.9206	.9212	.9217	.9222	.9227	.9232	.9238
8.4	.9243	.9248	.9253	.9258	.9263	.9269	.9274	.9279	.9284	.9289
8.5	.9294	.9299	.9304	.9309	.9315	.9320	.9325	.9330	.9335	.9340
8.6	.9345	.9350	.9355	.9360	.9365	.9370	.9375	.9380	.9385	.9390
8.7	.9395	.9400	.9405	.9410	.9415	.9420	.9425	.9430	.9435	.9440
8.8	.9445	.9450	.9455	.9460	.9465	.9469	.9474	.9479	.9484	.9489
8.9	.9494	.9499	.9504	.9509	.9513	.9518	.9523	.9528	.9533	.9538
9.0	.9542	.9547	.9552	.9557	.9562	.9566	.9571	.9576	.9581	.9586
9.1	.9590	.9595	.9600	.9605	.9609	.9614	.9619	.9624	.9628	.9633
9.2	.9638	.9643	.9647	.9652	.9657	.9661	.9666	.9671	.9675	.9680
9.3	.9685	.9689	.9694	.9699	.9703	.9708	.9713	.9717	.9722	.9727
9.4	.9731	.9736	.9741	.9745	.9750	.9754	.9759	.9763	.9768	.9773
9.5	.9777	.9782	.9786	.9791	.9795	.9800	.9805	.9809	.9814	.9818
9.6	.9823	.9827	.9832	.9836	.9841	.9845	.9850	.9854	.9859	.9863
9.7	.9868	.9872	.9877	.9881	.9886	.9890	.9894	.9899	.9903	.9908
9.8	.9912	.9917	.9921	.9926	.9930	.9934	.9939	.9943	.9948	.9952
9.9	.9956	.9961	.9965	.9969	.9974	.9978	.9983	.9987	.9991	.9996

n	n^2	$\sqrt{n}$	n^3	$\sqrt[3]{n}$	n	n^2	$\sqrt{n}$	n^3	$\sqrt[3]{n}$
1	1	1.000	1	1.000	**51**	2,601	7.141	132,651	3.708
2	4	1.414	8	1.260	**52**	2,704	7.211	140,608	3.733
3	9	1.732	27	1.442	**53**	2,809	7.280	148,877	3.756
4	16	2.000	64	1.587	**54**	2,916	7.348	157,464	3.780
5	25	2.236	125	1.710	**55**	3,025	7.416	166,375	3.803
6	36	2.449	216	1.817	**56**	3,136	7.483	175,616	3.826
7	49	2.646	343	1.913	**57**	3,249	7.550	185,193	3.849
8	64	2.828	512	2.000	**58**	3,364	7.616	195,112	3.871
9	81	3.000	729	2.080	**59**	3,481	7.681	205,379	3.893
10	100	3.162	1,000	2.154	**60**	3,600	7.746	216,000	3.915
11	121	3.317	1,331	2.224	**61**	3,721	7.810	226,981	3.936
12	144	3.464	1,728	2.289	**62**	3,844	7.874	238,328	3.958
13	169	3.606	2,197	2.351	**63**	3,969	7.937	250,047	3.979
14	196	3.742	2,744	2.410	**64**	4,096	8.000	262,144	4.000
15	225	3.873	3,375	2.466	**65**	4,225	8.062	274,625	4.021
16	256	4.000	4,096	2.520	**66**	4,356	8.124	287,496	4.041
17	289	4.123	4,913	2.571	**67**	4,489	8.185	300,763	4.062
18	324	4.243	5,832	2.621	**68**	4,624	8.246	314,432	4.082
19	361	4.359	6,859	2.668	**69**	4,761	8.307	328,509	4.102
20	400	4.472	8,000	2.714	**70**	4,900	8.367	343,000	4.121
21	441	4.583	9,261	2.759	**71**	5,041	8.426	357,911	4.141
22	484	4.690	10,648	2.802	**72**	5,184	8.485	373,248	4.160
23	529	4.796	12,167	2.844	**73**	5,329	8.544	389,017	4.179
24	576	4.899	13,824	2.884	**74**	5,476	8.602	405,224	4.198
25	625	5.000	15,625	2.924	**75**	5,625	8.660	421,875	4.217
26	676	5.099	17,576	2.962	**76**	5,776	8.718	438,976	4.236
27	729	5.196	19,683	3.000	**77**	5,929	8.775	456,533	4.254
28	784	5.292	21,952	3.037	**78**	6,084	8.832	474,552	4.273
29	841	5.385	24,389	3.072	**79**	6,241	8.888	493,039	4.291
30	900	5.477	27,000	3.107	**80**	6,400	8.944	512,000	4.309
31	961	5.568	29,791	3.141	**81**	6,561	9.000	531,441	4.327
32	1,024	5.657	32,768	3.175	**82**	6,724	9.055	551,368	4.344
33	1,089	5.745	35,937	3.208	**83**	6,889	9.110	571,787	4.362
34	1,156	5.831	39,304	3.240	**84**	7,056	9.165	592,704	4.380
35	1,225	5.916	42,875	3.271	**85**	7,225	9.220	614,125	4.397
36	1,296	6.000	46,656	3.302	**86**	7,396	9.274	636,056	4.414
37	1,369	6.083	50,653	3.332	**87**	7,569	9.327	658,503	4.431
38	1,444	6.164	54,872	3.362	**88**	7,744	9.381	681,472	4.448
39	1,521	6.245	59,319	3.391	**89**	7,921	9.434	704,969	4.465
40	1,600	6.325	64,000	3.420	**90**	8,100	9.487	729,000	4.481
41	1,681	6.403	68,921	3.448	**91**	8,281	9.539	753,571	4.498
42	1,764	6.481	74,088	3.476	**92**	8,464	9.592	778,688	4.514
43	1,849	6.557	79,507	3.503	**93**	8,649	9.644	804,357	4.531
44	1,936	6.633	85,184	3.530	**94**	8,836	9.695	830,584	4.547
45	2,025	6.708	91,125	3.557	**95**	9,025	9.747	857,375	4.563
46	2,116	6.782	97,336	3.583	**96**	9,216	9.798	884,736	4.579
47	2,209	6.856	103,823	3.609	**97**	9,409	9.849	912,673	4.595
48	2,304	6.928	110,592	3.634	**98**	9,604	9.899	941,192	4.610
49	2,401	7.000	117,649	3.659	**99**	9,801	9.950	970,299	4.626
50	2,500	7.071	125,000	3.684	**100**	10,000	10.000	1,000,000	4.642

t	t degrees	sin t	tan t	cot t	cos t		
.0000	**0° 00′**	.0000	.0000	—	1.0000	**90° 00′**	1.5708
.0029	10	.0029	.0029	343.77	1.0000	50	1.5679
.0058	20	.0058	.0058	171.89	1.0000	40	1.5650
.0087	30	.0087	.0087	114.59	1.0000	30	1.5621
.0116	40	.0116	.0116	85.940	.9999	20	1.5592
.0145	50	.0145	.0145	68.750	.9999	10	1.5563
.0175	**1° 00′**	.0175	.0175	57.290	.9998	**89° 00′**	1.5533
.0204	10	.0204	.0204	49.104	.9998	50	1.5504
.0233	20	.0233	.0233	42.964	.9997	40	1.5475
.0262	30	.0262	.0262	38.188	.9997	30	1.5446
.0291	40	.0291	.0291	34.368	.9996	20	1.5417
.0320	50	.0320	.0320	31.242	.9995	10	1.5388
.0349	**2° 00′**	.0349	.0349	28.636	.9994	**88° 00′**	1.5359
.0378	10	.0378	.0378	26.432	.9993	50	1.5330
.0407	20	.0407	.0407	24.542	.9992	40	1.5301
.0436	30	.0436	.0437	22.904	.9990	30	1.5272
.0465	40	.0465	.0466	21.470	.9989	20	1.5243
.0495	50	.0494	.0495	20.206	.9988	10	1.5213
.0524	**3° 00′**	.0523	.0524	19.081	.9986	**87° 00′**	1.5184
.0553	10	.0552	.0553	18.075	.9985	50	1.5155
.0582	20	.0581	.0582	17.169	.9983	40	1.5126
.0611	30	.0610	.0612	16.350	.9981	30	1.5097
.0640	40	.0640	.0641	15.605	.9980	20	1.5068
.0669	50	.0669	.0670	14.924	.9978	10	1.5039
.0698	**4° 00′**	.0698	.0699	14.301	.9976	**86° 00′**	1.5010
.0727	10	.0727	.0729	13.727	.9974	50	1.4981
.0756	20	.0756	.0758	13.197	.9971	40	1.4952
.0785	30	.0785	.0787	12.706	.9969	30	1.4923
.0814	40	.0814	.0816	12.251	.9967	20	1.4893
.0844	50	.0843	.0846	11.826	.9964	10	1.4864
.0873	**5° 00′**	.0872	.0875	11.430	.9962	**85° 00′**	1.4835
.0902	10	.0901	.0904	11.059	.9959	50	1.4806
.0931	20	.0929	.0934	10.712	.9957	40	1.4777
.0960	30	.0958	.0963	10.385	.9954	30	1.4748
.0989	40	.0987	.0992	10.078	.9951	20	1.4719
.1018	50	.1016	.1022	9.7882	.9948	10	1.4690
.1047	**6° 00′**	.1045	.1051	9.5144	.9945	**84° 00′**	1.4661
.1076	10	.1074	.1080	9.2553	.9942	50	1.4632
.1105	20	.1103	.1110	9.0098	.9939	40	1.4603
.1134	30	.1132	.1139	8.7769	.9936	30	1.4573
.1164	40	.1161	.1169	8.5555	.9932	20	1.4544
.1193	50	.1190	.1198	8.3450	.9929	10	1.4515
.1222	**7° 00′**	.1219	.1228	8.1443	.9925	**83° 00′**	1.4486
		cos t	cot t	tan t	sin t	t degrees	t

t	t degrees	sin t	tan t	cot t	cos t		
.1222	**7° 00′**	.1219	.1228	8.1443	.9925	**83° 00′**	1.4486
.1251	10	.1248	.1257	7.9530	.9922	50	1.4457
.1280	20	.1276	.1287	7.7704	.9918	40	1.4428
.1309	30	.1305	.1317	7.5958	.9914	30	1.4399
.1338	40	.1334	.1346	7.4287	.9911	20	1.4370
.1367	50	.1363	.1376	7.2687	.9907	10	1.4341
.1396	**8° 00′**	.1392	.1405	7.1154	.9903	**82° 00′**	1.4312
.1425	10	.1421	.1435	6.9682	.9899	50	1.4283
.1454	20	.1449	.1465	6.8269	.9894	40	1.4254
.1484	30	.1478	.1495	6.6912	.9890	30	1.4224
.1513	40	.1507	.1524	6.5606	.9886	20	1.4195
.1542	50	.1536	.1554	6.4348	.9881	10	1.4166
.1571	**9° 00′**	.1564	.1584	6.3138	.9877	**81° 00′**	1.4137
.1600	10	.1593	.1614	6.1970	.9872	50	1.4108
.1629	20	.1622	.1644	6.0844	.9868	40	1.4079
.1658	30	.1650	.1673	5.9758	.9863	30	1.4050
.1687	40	.1679	.1703	5.8708	.9858	20	1.4021
.1716	50	.1708	.1733	5.7694	.9853	10	1.3992
.1745	**10° 00′**	.1736	.1763	5.6713	.9848	**80° 00′**	1.3963
.1774	10	.1765	.1793	5.5764	.9843	50	1.3934
.1804	20	.1794	.1823	5.4845	.9838	40	1.3904
.1833	30	.1822	.1853	5.3955	.9833	30	1.3875
.1862	40	.1851	.1883	5.3093	.9827	20	1.3846
.1891	50	.1880	.1914	5.2257	.9822	10	1.3817
.1920	**11° 00′**	.1908	.1944	5.1446	.9816	**79° 00′**	1.3788
.1949	10	.1937	.1974	5.0658	.9811	50	1.3759
.1978	20	.1965	.2004	4.9894	.9805	40	1.3730
.2007	30	.1994	.2035	4.9152	.9799	30	1.3701
.2036	40	.2022	.2065	4.8430	.9793	20	1.3672
.2065	50	.2051	.2095	4.7729	.9787	10	1.3643
.2094	**12° 00′**	.2079	.2126	4.7046	.9781	**78° 00′**	1.3614
.2123	10	.2108	.2156	4.6382	.9775	50	1.3584
.2153	20	.2136	.2186	4.5736	.9769	40	1.3555
.2182	30	.2164	.2217	4.5107	.9763	30	1.3526
.2211	40	.2193	.2247	4.4494	.9757	20	1.3497
.2240	50	.2221	.2278	4.3897	.9750	10	1.3468
.2269	**13° 00′**	.2250	.2309	4.3315	.9744	**77° 00′**	1.3439
.2298	10	.2278	.2339	4.2747	.9737	50	1.3410
.2327	20	.2306	.2370	4.2193	.9730	40	1.3381
.2356	30	.2334	.2401	4.1653	.9724	30	1.3352
.2385	40	.2363	.2432	4.1126	.9717	20	1.3323
.2414	50	.2391	.2462	4.0611	.9710	10	1.3294
.2443	**14° 00′**	.2419	.2493	4.0108	.9703	**76° 00′**	1.3265
		cos t	cot t	tan t	sin t	t degrees	t

t	t degrees	sin t	tan t	cot t	cos t		
.2443	**14° 00′**	.2419	.2493	4.0108	.9703	**76° 00′**	1.3265
.2473	10	.2447	.2524	3.9617	.9696	50	1.3235
.2502	20	.2476	.2555	3.9136	.9689	40	1.3206
.2531	30	.2504	.2586	3.8667	.9681	30	1.3177
.2560	40	.2532	.2617	3.8208	.9674	20	1.3148
.2589	50	.2560	.2648	3.7760	.9667	10	1.3119
.2618	**15° 00′**	.2588	.2679	3.7321	.9659	**75° 00′**	1.3090
.2647	10	.2616	.2711	3.6891	.9652	50	1.3061
.2676	20	2644	.2742	3.6470	.9644	40	1.3032
.2705	30	.2672	.2773	3.6059	.9636	30	1.3003
.2734	40	.2700	.2805	3.5656	.9628	20	1.2974
.2763	50	.2728	.2836	3.5261	.9621	10	1.2945
.2793	**16° 00′**	.2756	.2867	3.4874	.9613	**74° 00′**	1.2915
.2822	10	.2784	.2899	3.4495	.9605	50	1.2886
.2851	20	.2812	.2931	3.4124	.9596	40	1.2857
.2880	30	.2840	.2962	3.3759	.9588	30	1.2828
.2909	40	.2868	.2994	3.3402	.9580	20	1.2799
.2938	50	.2896	.3026	3.3052	.9572	10	1.2770
.2967	**17° 00′**	.2924	.3057	3.2709	.9563	**73° 00′**	1.2741
.2996	10	.2952	.3089	3.2371	.9555	50	1.2712
.3025	20	.2979	.3121	3.2041	.9546	40	1.2683
.3054	30	.3007	.3153	3.1716	.9537	30	1.2654
.3083	40	.3035	.3185	3.1397	.9528	20	1.2625
.3113	50	.3062	.3217	3.1084	.9520	10	1.2595
.3142	**18° 00′**	.3090	.3249	3.0777	.9511	**72° 00′**	1.2566
.3171	10	.3118	.3281	3.0475	.9502	50	1.2537
.3200	20	.3145	.3314	3.0178	.9492	40	1.2508
.3229	30	.3173	.3346	2.9887	.9483	30	1.2479
.3258	40	.3201	.3378	2.9600	.9474	20	1.2450
.3287	50	.3228	.3411	2.9319	.9465	10	1.2421
.3316	**19° 00′**	.3256	.3443	2.9042	.9455	**71° 00′**	1.2392
.3345	10	.3283	.3476	2.8770	.9446	50	1.2363
.3374	20	.3311	.3508	2.8502	.9436	40	1.2334
.3403	30	.3338	.3541	2.8239	.9426	30	1.2305
.3432	40	.3365	.3574	2.7980	.9417	20	1.2275
.3462	50	.3393	.3607	2.7725	.9407	10	1.2246
.3491	**20° 00′**	.3420	.3640	2.7475	.9397	**70° 00′**	1.2217
.3520	10	.3448	.3673	2.7228	.9387	50	1.2188
.3549	20	.3475	.3706	2.6985	.9377	40	1.2159
.3578	30	.3502	.3739	2.6746	.9367	30	1.2130
.3607	40	.3529	.3772	2.6511	.9356	20	1.2101
.3636	50	.3557	.3805	2.6279	.9346	10	1.2072
.3665	**21° 00′**	.3584	.3839	2.6051	.9336	**69° 00′**	1.2043
		cos t	cot t	tan t	sin t	t degrees	t

t	t degrees	sin t	tan t	cot t	cos t		
.3665	**21° 00′**	.3584	.3839	2.6051	.9336	**69° 00′**	1.2043
.3694	10	.3611	.3872	2.5826	.9325	50	1.2014
.3723	20	.3638	.3906	2.5605	.9315	40	1.1985
.3752	30	.3665	.3939	2.5386	.9304	30	1.1956
.3782	40	.3692	.3973	2.5172	.9293	20	1.1926
.3811	50	.3719	.4006	2.4960	.9283	10	1.1897
.3840	**22° 00′**	.3746	.4040	2.4751	.9272	**68° 00′**	1.1868
.3869	10	.3773	.4074	2.4545	.9261	50	1.1839
.3898	20	.3800	.4108	2.4342	.9250	40	1.1810
.3927	30	.3827	.4142	2.4142	.9239	30	1.1781
.3956	40	.3854	.4176	2.3945	.9228	20	1.1752
.3985	50	.3881	.4210	2.3750	.9216	10	1.1723
.4014	**23° 00′**	.3907	.4245	2.3559	.9205	**67° 00′**	1.1694
.4043	10	.3934	.4279	2.3369	.9194	50	1.1665
.4072	20	.3961	.4314	2.3183	.9182	40	1.1636
.4102	30	.3987	.4348	2.2998	.9171	30	1.1606
.4131	40	.4014	.4383	2.2817	.9159	20	1.1577
.4160	50	.4041	.4417	2.2637	.9147	10	1.1548
.4189	**24° 00′**	.4067	.4452	2.2460	.9135	**66° 00′**	1.1519
.4218	10	.4094	.4487	2.2286	.9124	50	1.1490
.4247	20	.4120	.4522	2.2113	.9112	40	1.1461
.4276	30	.4147	.4557	2.1943	.9100	30	1.1432
.4305	40	.4173	.4592	2.1775	.9088	20	1.1403
.4334	50	.4200	.4628	2.1609	.9075	10	1.1374
.4363	**25° 00′**	.4226	.4663	2.1445	.9063	**65° 00′**	1.1345
.4392	10	.4253	.4699	2.1283	.9051	50	1.1316
.4422	20	.4279	.4734	2.1123	.9038	40	1.1286
.4451	30	.4305	.4770	2.0965	.9026	30	1.1257
.4480	40	.4331	.4806	2.0809	.9013	20	1.1228
.4509	50	.4358	.4841	2.0655	.9001	10	1.1199
.4538	**26° 00′**	.4384	.4877	2.0503	.8988	**64° 00′**	1.1170
.4567	10	.4410	.4913	2.0353	.8975	50	1.1141
.4596	20	.4436	.4950	2.0204	.8962	40	1.1112
.4625	30	.4462	.4986	2.0057	.8949	30	1.1083
.4654	40	.4488	.5022	1.9912	.8936	20	1.1054
.4683	50	.4514	.5059	1.9768	.8923	10	1.1025
.4712	**27° 00′**	.4540	.5095	1.9626	.8910	**63° 00′**	1.0996
.4741	10	.4566	.5132	1.9486	.8897	50	1.0966
.4771	20	.4592	.5169	1.9347	.8884	40	1.0937
.4800	30	.4617	.5206	1.9210	.8870	30	1.0908
.4829	40	.4643	.5243	1.9074	.8857	20	1.0879
.4858	50	.4669	.5280	1.8940	.8843	10	1.0850
.4887	**28° 00′**	.4695	.5317	1.8807	.8829	**62° 00′**	1.0821
		cos t	cot t	tan t	sin t	t degrees	t

t	t degrees	sin t	tan t	cot t	cos t		
.4887	**28° 00′**	.4695	.5317	1.8807	.8829	**62° 00′**	1.0821
.4916	10	.4720	.5354	1.8676	.8816	50	1.0792
.4945	20	.4746	.5392	1.8546	.8802	40	1.0763
.4974	30	.4772	.5430	1.8418	.8788	30	1.0734
.5003	40	.4797	.5467	1.8291	.8774	20	1.0705
.5032	50	.4823	.5505	1.8165	.8760	10	1.0676
.5061	**29° 00′**	.4848	.5543	1.8040	.8746	**61° 00′**	1.0647
.5091	10	.4874	.5581	1.7917	.8732	50	1.0617
.5120	20	.4899	.5619	1.7796	.8718	40	1.0588
.5149	30	.4924	.5658	1.7675	.8704	30	1.0559
.5178	40	.4950	.5696	1.7556	.8689	20	1.0530
.5207	50	.4975	.5735	1.7437	.8675	10	1.0501
.5236	**30° 00′**	.5000	.5774	1.7321	.8660	**60° 00′**	1.0472
.5265	10	.5025	.5812	1.7205	.8646	50	1.0443
.5294	20	.5050	.5851	1.7090	.8631	40	1.0414
.5323	30	.5075	.5890	1.6977	.8616	30	1.0385
.5352	40	.5100	.5930	1.6864	.8601	20	1.0356
.5381	50	.5125	.5969	1.6753	.8587	10	1.0327
.5411	**31° 00′**	.5150	.6009	1.6643	.8572	**59° 00′**	1.0297
.5440	10	.5175	.6048	1.6534	.8557	50	1.0268
.5469	20	.5200	.6088	1.6426	.8542	40	1.0239
.5498	30	.5225	.6128	1.6319	.8526	30	1.0210
.5527	40	.5250	.6168	1.6212	.8511	20	1.0181
.5556	50	.5275	.6208	1.6107	.8496	10	1.0152
.5585	**32° 00′**	.5299	.6249	1.6003	.8480	**58° 00′**	1.0123
.5614	10	.5324	.6289	1.5900	.8465	50	1.0094
.5643	20	.5348	.6330	1.5798	.8450	40	1.0065
.5672	30	.5373	.6371	1.5697	.8434	30	1.0036
.5701	40	.5398	.6412	1.5597	.8418	20	1.0007
.5730	50	.5422	.6453	1.5497	.8403	10	.9977
.5760	**33° 00′**	.5446	.6494	1.5399	.8387	**57° 00′**	.9948
.5789	10	.5471	.6536	1.5301	.8371	50	.9919
.5818	20	.5495	.6577	1.5204	.8355	40	.9890
.5847	30	.5519	.6619	1.5108	.8339	30	.9861
.5876	40	.5544	.6661	1.5013	.8323	20	.9832
.5905	50	.5568	.6703	1.4919	.8307	10	.9803
.5934	**34° 00′**	.5592	.6745	1.4826	.8290	**56° 00′**	.9774
.5963	10	.5616	.6787	1.4733	.8274	50	.9745
.5992	20	.5640	.6830	1.4641	.8258	40	.9716
.6021	30	.5664	.6873	1.4550	.8241	30	.9687
.6050	40	.5688	.6916	1.4460	.8225	20	.9657
.6080	50	.5712	.6959	1.4370	.8208	10	.9628
.6109	**35° 00′**	.5736	.7002	1.4281	.8192	**55° 00′**	.9599
		cos t	cot t	tan t	sin t	t degrees	t

t	t degrees	sin t	tan t	cot t	cos t		
.6109	**35° 00′**	.5736	.7002	1.4281	.8192	**55° 00′**	.9599
.6138	10	.5760	.7046	1.4193	.8175	50	.9570
.6167	20	.5783	.7089	1.4106	.8158	40	.9541
.6196	30	.5807	.7133	1.4019	.8141	30	.9512
.6225	40	.5831	.7177	1.3934	.8124	20	.9483
.6254	50	.5854	.7221	1.3848	.8107	10	.9454
.6283	**36° 00′**	.5878	.7265	1.3764	.8090	**54° 00′**	.9425
.6312	10	.5901	.7310	1.3680	.8073	50	.9396
.6341	20	.5925	.7355	1.3597	.8056	40	.9367
.6370	30	.5948	.7400	1.3514	.8039	30	.9338
.6400	40	.5972	.7445	1.3432	.8021	20	.9308
.6429	50	.5995	.7490	1.3351	.8004	10	.9279
.6458	**37° 00′**	.6018	.7536	1.3270	.7986	**53° 00′**	.9250
.6487	10	.6041	.7581	1.3190	.7969	50	.9221
.6516	20	.6065	.7627	1.3111	.7951	40	.9192
.6545	30	.6088	.7673	1.3032	.7934	30	.9163
.6574	40	.6111	.7720	1.2954	.7916	20	.9134
.6603	50	.6134	.7766	1.2876	.7898	10	.9105
.6632	**38° 00′**	.6157	.7813	1.2799	.7880	**52° 00′**	.9076
.6661	10	.6180	.7860	1.2723	.7862	50	.9047
.6690	20	.6202	.7907	1.2647	.7844	40	.9018
.6720	30	.6225	.7954	1.2572	.7826	30	.8988
.6749	40	.6248	.8002	1.2497	.7808	20	.8959
.6778	50	.6271	.8050	1.2423	.7790	10	.8930
.6807	**39° 00′**	.6293	.8098	1.2349	.7771	**51° 00′**	.8901
.6836	10	.6316	.8146	1.2276	.7753	50	.8872
.6865	20	.6338	.8195	1.2203	.7735	40	.8843
.6894	30	.6361	.8243	1.2131	.7716	30	.8814
.6923	40	.6383	.8292	1.2059	.7698	20	.8785
.6952	50	.6406	.8342	1.1988	.7679	10	.8756
.6981	**40° 00′**	.6428	.8391	1.1918	.7660	**50° 00′**	.8727
.7010	10	.6450	.8441	1.1847	.7642	50	.8698
.7039	20	.6472	.8491	1.1778	.7623	40	.8668
.7069	30	.6494	.8541	1.1708	.7604	30	.8639
.7098	40	.6517	.8591	1.1640	.7585	20	.8610
.7127	50	.6539	.8642	1.1571	.7566	10	.8581
.7156	**41° 00′**	.6561	.8693	1.1504	.7547	**49° 00′**	.8552
.7185	10	.6583	.8744	1.1436	.7528	50	.8523
.7214	20	.6604	.8796	1.1369	.7509	40	.8494
.7243	30	.6626	.8847	1.1303	.7490	30	.8465
.7272	40	.6648	.8899	1.1237	.7470	20	.8436
.7301	50	.6670	.8952	1.1171	.7451	10	.8407
.7330	**42° 00′**	.6691	.9004	1.1106	.7431	**48° 00′**	.8378
		cos t	cot t	tan t	sin t	t degrees	t

t	t degrees	sin t	tan t	cot t	cos t		
.7330	**42° 00′**	.6691	.9004	1.1106	.7431	**48° 00′**	.8378
.7359	10	.6713	.9057	1.1041	.7412	50	.8348
.7389	20	.6734	.9110	1.0977	.7392	40	.8319
.7418	30	.6756	.9163	1.0913	.7373	30	.8290
.7447	40	.6777	.9217	1.0850	.7353	20	.8261
.7476	50	.6799	.9271	1.0786	.7333	10	.8232
.7505	**43° 00′**	.6820	.9325	1.0724	.7314	**47° 00′**	.8203
.7534	10	.6841	.9380	1.0661	.7294	50	.8174
.7563	20	.6862	.9435	1.0599	.7274	40	.8145
.7592	30	.6884	.9490	1.0538	.7254	30	.8116
.7621	40	.6905	.9545	1.0477	.7234	20	.8087
.7650	50	.6926	.9601	1.0416	.7214	10	.8058
.7679	**44° 00′**	.6947	.9657	1.0355	.7193	**46° 00′**	.8029
.7709	10	.6967	.9713	1.0295	.7173	50	.7999
.7738	20	.6988	.9770	1.0235	.7153	40	.7970
.7767	30	.7009	.9827	1.0176	.7133	30	.7941
.7796	40	.7030	.9884	1.0117	.7112	20	.7912
.7825	50	.7050	.9942	1.0058	.7092	10	.7883
.7854	**45° 00′**	.7071	1.0000	1.0000	.7071	**45° 00′**	.7854
		cos t	cot t	tan t	sin t	t degrees	t

angles	log sin	log cos	log tan	log cot	
0° 00′		10.0000			90° 00′
10	7.4637	.0000	7.4637	12.5363	50
20	.7648	.0000	.7648	.2352	40
30	7.9408	.0000	7.9409	12.0591	30
40	8.0658	.0000	8.0658	11.9342	20
50	.1627	10.0000	.1627	.8373	10
1° 00′	8.2419	9.9999	8.2419	11.7581	89° 00′
10	.3088	.9999	.3089	.6911	50
20	.3668	.9999	.3669	.6331	40
30	.4179	.9999	.4181	.5819	30
40	.4637	.9998	.4638	.5362	20
50	.5050	.9998	.5053	.4947	10
2° 00′	8.5428	9.9997	8.5431	11.4569	88° 00′
10	.5776	.9997	.5779	.4221	50
20	.6097	.9996	.6101	.3899	40
30	.6397	.9996	.6401	.3599	30
40	.6677	.9995	.6682	.3318	20
50	.6940	.9995	.6945	.3055	10
3° 00′	8.7188	9.9994	8.7194	11.2806	87° 00′
10	.7423	.9993	.7429	.2571	50
20	.7645	.9993	.7652	.2348	40
30	.7857	.9992	.7865	.2135	30
40	.8059	.9991	.8067	.1933	20
50	.8251	.9990	.8261	.1739	10
4° 00′	8.8436	9.9989	8.8446	11.1554	86° 00′
10	.8613	.9989	.8624	.1376	50
20	.8783	.9988	.8795	.1205	40
30	.8946	.9987	.8960	.1040	30
40	.9104	.9986	.9118	.0882	20
50	.9256	.9985	.9272	.0728	10
5° 00′	8.9403	9.9983	8.9420	11.0580	85° 00′
10	.9545	.9982	.9563	.0437	50
20	.9682	.9981	.9701	.0299	40
30	.9816	.9980	.9836	.0164	30
40	8.9945	.9979	8.9966	11.0034	20
50	9.0070	.9977	9.0093	10.9907	10
6° 00′	9.0192	9.9976	9.0216	10.9784	84° 00′
10	.0311	.9975	.0336	.9664	50
20	.0426	.9973	.0453	.9547	40
30	.0539	.9972	.0567	.9433	30
40	.0648	.9971	.0678	.9322	20
50	.0755	.9969	.0786	.9214	10
7° 00′	9.0859	9.9968	9.0891	10.9109	83° 00′
	log cos	log sin	log cot	log tan	angles

* Add −10 to all logarithms.

angles	log sin	log cos	log tan	log cot	
7° 00′	9.0859	9.9968	9.0891	10.9109	83° 00′
10	.0961	.9966	.0995	.9005	50
20	.1060	.9964	.1096	.8904	40
30	.1157	.9963	.1194	.8806	30
40	.1252	.9961	.1291	.8709	20
50	.1345	.9959	.1385	.8615	10
8° 00′	9.1436	9.9958	9.1478	10.8522	82° 00′
10	.1525	.9956	.1569	.8431	50
20	.1612	.9954	.1658	.8342	40
30	.1697	.9952	.1745	.8255	30
40	.1781	.9950	.1831	.8169	20
50	.1863	.9948	.1915	.8085	10
9° 00′	9.1943	9.9946	9.1997	10.8003	81° 00′
10	.2022	.9944	.2078	.7922	50
20	.2100	.9942	.2158	.7842	40
30	.2176	.9940	.2236	.7764	30
40	.2251	.9938	.2313	.7687	20
50	.2324	.9936	.2389	.7611	10
10° 00′	9.2397	9.9934	9.2463	10.7537	80° 00′
10	.2468	.9931	.2536	.7464	50
20	.2538	.9929	.2609	.7391	40
30	.2606	.9927	.2680	.7320	30
40	.2674	.9924	.2750	.7250	20
50	.2740	.9922	.2819	.7181	10
11° 00′	9.2806	9.9919	9.2887	10.7113	79° 00′
10	.2870	.9917	.2953	.7047	50
20	.2934	.9914	.3020	.6980	40
30	.2997	.9912	.3085	.6915	30
40	.3058	.9909	.3149	.6851	20
50	.3119	.9907	.3212	.6788	10
12° 00′	9.3179	9.9904	9.3275	10.6725	78° 00′
10	.3238	.9901	.3336	.6664	50
20	.3296	.9899	.3397	.6603	40
30	.3353	.9896	.3458	.6542	30
40	.3410	9893	.3517	.6483	20
50	.3466	.9890	.3576	.6424	10
13° 00′	9.3521	9.9887	9.3634	10.6366	77° 00′
10	.3575	.9884	.3691	.6309	50
20	.3629	.9881	.3748	.6252	40
30	.3682	.9878	.3804	.6196	30
40	.3734	.9875	.3859	.6141	20
50	.3786	.9872	.3914	.6086	10
14° 00′	9.3837	9.9869	9.3968	10.6032	76° 00′
	log cos	log sin	log cot	log tan	angles

angles	log sin	log cos	log tan	log cot	
14° 00′	9.3837	9.9869	9.3968	10.6032	76° 00′
10	.3887	.9866	.4021	.5979	50
20	.3937	.9863	.4074	.5926	40
30	.3986	.9859	.4127	.5873	30
40	.4035	.9856	.4178	.5822	20
50	.4083	.9853	.4230	.5770	10
15° 00′	9.4130	9.9849	9.4281	10.5719	75° 00′
10	.4177	.9846	.4331	.5669	50
20	.4223	.9843	.4381	.5619	40
30	.4269	.9839	.4430	.5570	30
40	.4314	.9836	.4479	.5521	20
50	.4359	.9832	.4527	.5473	10
16° 00′	9.4403	9.9828	9.4575	10.5425	74° 00′
10	.4447	.9825	.4622	.5378	50
20	.4491	.9821	.4669	.5331	40
30	.4533	.9817	.4716	.5284	30
40	.4576	.9814	.4762	.5238	20
50	.4618	.9810	.4808	.5192	10
17° 00′	9.4659	9.9806	9.4853	10.5147	73° 00′
10	.4700	.9802	.4898	.5102	50
20	.4741	.9798	.4943	.5057	40
30	.4781	.9794	.4987	.5013	30
40	.4821	.9790	.5031	.4969	20
50	.4861	.9786	.5075	.4925	10
18° 00′	9.4900	9.9782	9.5118	10.4882	72° 00′
10	.4939	.9778	.5161	.4839	50
20	.4977	.9774	.5203	.4797	40
30	.5015	.9770	.5245	.4755	30
40	.5052	.9765	.5287	.4713	20
50	.5090	.9761	.5329	.4671	10
19° 00′	9.5126	9.9757	9.5370	10.4630	71° 00′
10	.5163	.9752	.5411	.4589	50
20	.5199	.9748	.5451	.4549	40
30	.5235	.9743	.5491	.4509	30
40	.5270	.9739	.5531	.4469	20
50	.5306	.9734	.5571	.4429	10
20° 00′	9.5341	9.9730	9.5611	10.4389	70° 00′
10	.5375	.9725	.5650	.4350	50
20	.5409	.9721	.5689	.4311	40
30	.5443	.9716	.5727	.4273	30
40	.5477	.9711	.5766	.4234	20
50	.5510	.9706	.5804	.4196	10
21° 00′	9.5543	9.9702	9.5842	10.4158	69° 00′
	log cos	log sin	log cot	log tan	angles

angles	log sin	log cos	log tan	log cot	
21° 00′	9.5543	9.9702	9.5842	10.4158	69° 00′
10	.5576	.9697	.5879	.4121	50
20	.5609	.9692	.5917	.4083	40
30	.5641	.9687	.5954	.4046	30
40	.5673	.9682	.5991	.4009	20
50	.5704	.9677	.6028	.3972	10
22° 00′	9.5736	9.9672	9.6064	10.3936	68° 00′
10	.5767	.9667	.6100	.3900	50
20	.5798	.9661	.6136	.3864	40
30	.5828	.9656	.6172	.3828	30
40	.5859	.9651	.6208	.3792	20
50	.5889	.9646	.6243	.3757	10
23° 00′	9.5919	9.9640	9.6279	10.3721	67° 00′
10	.5948	.9635	.6314	.3686	50
20	.5978	.9629	.6348	.3652	40
30	.6007	.9624	.6383	.3617	30
40	.6036	.9618	.6417	.3583	20
50	.6065	.9613	.6452	.3548	10
24° 00′	9.6093	9.9607	9.6486	10.3514	66° 00′
10	.6121	.9602	.6520	.3480	50
20	.6149	.9596	.6553	.3447	40
30	.6177	.9590	.6587	.3413	30
40	.6205	.9584	.6620	.3380	20
50	.6232	.9579	.6654	.3346	10
25° 00′	9.6259	9.9573	9.6687	10.3313	65° 00′
10	.6286	.9567	.6720	.3280	50
20	.6313	.9561	.6752	.3248	40
30	.6340	.9555	.6785	.3215	30
40	.6366	.9549	.6817	.3183	20
50	.6392	.9543	.6850	.3150	10
26° 00′	9.6418	9.9537	9.6882	10.3118	64° 00′
10	.6444	.9530	.6914	.3086	50
20	.6470	.9524	.6946	.3054	40
30	.6495	.9518	.6977	.3023	30
40	.6521	.9512	.7009	.2991	20
50	.6546	.9505	.7040	.2960	10
27° 00′	9.6570	9.9499	9.7072	10.2928	63° 00′
10	.6595	.9492	.7103	.2897	50
20	.6620	.9486	.7134	.2866	40
30	.6644	.9479	.7165	.2835	30
40	.6668	.9473	.7196	.2804	20
50	.6692	.9466	.7226	.2774	10
28° 00′	9.6716	9.9459	9.7257	10.2743	62° 00′
	log cos	log sin	log cot	log tan	angles

angles	log sin	log cos	log tan	log cot	
28° 00′	9.6716	9.9459	9.7257	10.2743	62° 00′
10	.6740	.9453	.7287	.2713	50
20	.6763	.9446	.7317	.2683	40
30	.6787	.9439	.7348	.2652	30
40	.6810	.9432	.7378	.2622	20
50	.6833	.9425	.7408	.2592	10
29° 00′	9.6856	9.9418	9.7438	10.2562	61° 00′
10	.6878	.9411	.7467	.2533	50
20	.6901	.9404	.7497	.2503	40
30	.6923	.9397	.7526	.2474	30
40	.6946	.9390	.7556	.2444	20
50	.6968	.9383	.7585	.2415	10
30° 00′	9.6990	9.9375	9.7614	10.2386	60° 00′
10	.7012	.9368	.7644	.2356	50
20	.7033	.9361	.7673	.2327	40
30	.7055	.9353	.7701	.2299	30
40	.7076	.9346	.7730	.2270	20
50	.7097	.9338	.7759	.2241	10
31° 00′	9.7118	9.9331	9.7788	10.2212	59° 00′
10	.7139	.9323	.7816	.2184	50
20	.7160	.9315	.7845	.2155	40
30	.7181	.9308	.7873	.2127	30
40	.7201	.9300	.7902	.2098	20
50	.7222	.9292	.7930	.2070	10
32° 00′	9.7242	9.9284	9.7958	10.2042	58° 00′
10	.7262	.9276	.7986	.2014	50
20	.7282	.9268	.8014	.1986	40
30	.7302	.9260	.8042	.1958	30
40	.7322	.9252	.8070	.1930	20
50	.7342	.9244	.8097	.1903	10
33° 00′	9.7361	9.9236	9.8125	10.1875	57° 00′
10	.7380	.9228	.8153	.1847	50
20	.7400	.9219	.8180	.1820	40
30	.7419	.9211	.8208	.1792	30
40	.7438	.9203	.8235	.1765	20
50	.7457	.9194	.8263	.1737	10
34° 00′	9.7476	9.9186	9.8290	10.1710	56° 00′
10	.7494	.9177	.8317	.1683	50
20	.7513	.9169	.8344	.1656	40
30	.7531	.9160	.8371	.1629	30
40	.7550	.9151	.8398	.1602	20
50	.7568	.9142	.8425	.1575	10
35° 00′	9.7586	9.9134	9.8452	10.1548	55° 00′
	log cos	log sin	log cot	log tan	angles

angles	log sin	log cos	log tan	log cot	
35° 00′	9.7586	9.9134	9.8452	10.1548	55° 00′
10	.7604	.9125	.8479	.1521	50
20	.7622	.9116	.8506	.1494	40
30	.7640	.9107	.8533	.1467	30
40	.7657	.9098	.8559	.1441	20
50	.7675	.9089	.8586	.1414	10
36° 00′	9.7692	9.9080	9.8613	10.1387	54° 00′
10	.7710	.9070	.8639	.1361	50
20	.7727	.9061	.8666	.1334	40
30	.7744	.9052	.8692	.1308	30
40	.7761	.9042	.8718	.1282	20
50	.7778	.9033	.8745	.1255	10
37° 00′	9.7795	9.9023	9.8771	10.1229	53° 00′
10	.7811	.9014	.8797	.1203	50
20	.7828	.9004	.8824	.1176	40
30	.7844	.8995	.8850	.1150	30
40	.7861	.8985	.8876	.1124	20
50	.7877	.8975	.8902	.1098	10
38° 00′	9.7893	9.8965	9.8928	10.1072	52° 00′
10	.7910	.8955	.8954	.1046	50
20	.7926	.8945	.8980	.1020	40
30	.7941	.8935	.9006	.0994	30
40	.7957	.8925	.9032	.0968	20
50	.7973	.8915	.9058	.0942	10
39° 00′	9.7989	9.8905	9.9084	10.0916	51° 00′
10	.8004	.8895	.9110	.0890	50
20	.8020	.8884	.9135	.0865	40
30	.8035	.8874	.9161	.0839	30
40	.8050	.8864	.9187	.0813	20
50	.8066	.8853	.9212	.0788	10
40° 00′	9.8081	9.8843	9.9238	10.0762	50° 00′
10	.8096	.8832	.9264	.0736	50
20	.8111	.8821	.9289	.0711	40
30	.8125	.8810	.9315	.0685	30
40	.8140	.8800	.9341	.0659	20
50	.8155	.8789	.9366	.0634	10
41° 00′	9.8169	9.8778	9.9392	10.0608	49° 00′
10	.8184	.8767	.9417	.0583	50
20	.8198	.8756	.9443	.0557	40
30	.8213	.8745	.9468	.0532	30
40	.8227	.8733	.9494	.0506	20
50	.8241	.8722	.9519	.0481	10
42° 00′	9.8255	9.8711	9.9544	10.0456	48° 00′
	log cos	log sin	log cot	log tan	angles

angles	log sin	log cos	log tan	log cot	
42° 00′	9.8255	9.8711	9.9544	10.0456	48° 00′
10	.8269	.8699	.9570	.0430	50
20	.8283	.8688	.9595	.0405	40
30	.8297	.8676	.9621	.0379	30
40	.8311	.8665	.9646	.0354	20
50	.8324	.8653	.9671	.0329	10
43° 00′	9.8338	9.8641	9.9697	10.0303	47° 00′
10	.8351	.8629	.9722	.0278	50
20	.8365	.8618	.9747	.0253	40
30	.8378	.8606	.9772	.0228	30
40	.8391	.8594	.9798	.0202	20
50	.8405	.8582	.9823	.0177	10
44° 00′	9.8418	9.8569	9.9848	10.0152	46° 00′
10	.8431	.8557	.9874	.0126	50
20	.8444	.8545	.9899	.0101	40
30	.8457	.8532	.9924	.0076	30
40	.8469	.8520	.9949	.0051	20
50	.8482	.8507	.9975	.0025	10
45° 00′	9.8495	9.8495	10.0000	10.0000	45° 00′
	log cos	log sin	log cot	log tan	angles

Answers to Odd-Numbered Exercises

CHAPTER ONE

Section 1 (page 5)

1. (a) True. (b) False, $2 \notin \{1, 3\}$. (c) True.
(d) True. (e) False, $\{2\} \subseteq \{1, 2\}$. (f) True.
(g) False, $2 \in \{1, 2\}$. (h) False, $\{2\} \subseteq \{2\}$.

3. $\{a, b, c, d\}$, $\{a, b, c\}$, $\{a, b, d\}$, $\{b, c, d\}$, $\{a, c, d\}$, $\{a, b\}$, $\{a, c\}$, $\{a, d\}$, $\{b, c\}$, $\{b, d\}$, $\{c, d\}$, $\{a\}$, $\{b\}$, $\{c\}$, $\{d\}$, $\varnothing$.

5. (a) $T \subseteq S$. (b) $S \subseteq T$. (c) $S = \varnothing$. (d) Any S.
(e) S and T disjoint. (f) $S = T = \varnothing$.
(g) Any S and T. (h) Any S and T.

7.

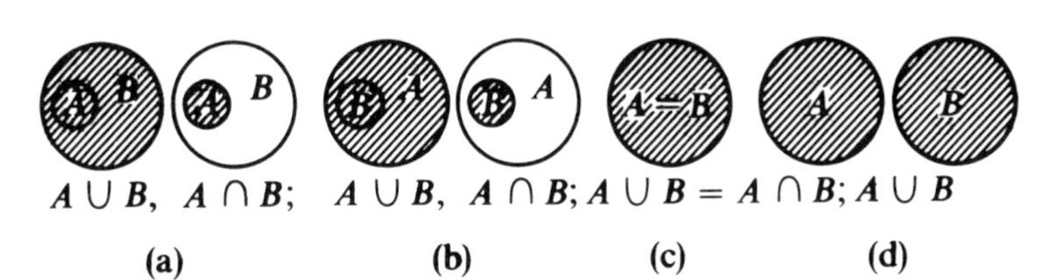

Section 2 (page 11)

1. (a) $>$. (b) $<$. (c) $>$.
(d) $>$. (e) $=$. (f) $<$.

3. (a) 3. (b) 7. (c) 7.
(d) 3. (e) $2 - \sqrt{3}$. (f) -1.

5. If $a < b$ and $b < c$, then $a < c$. If $a < b$, then $a + c < b + c$. If $a < b$ and $c > 0$, then $ac < bc$. If $a < b$ and $c < 0$, then $ac > bc$.

7. $a > 8/3$. **9.** $a > -2$. **11.** $a > -7/3$.

13. $-13 < a < 5$. **15.** $-2 < a < 3$. **17.** $-1 < a < 2$.

19. Multiply both sides by -1.

21. $\{-1/42\}$. **23.** $\{5, -3/2\}$. **25.** $\{23/3\}$.

27. $\{4 + 3i, 4 - 3i\}$. **29.** $\{x \mid x > -2/5\}$. **31.** $\{x \mid 4/3 < x < 2\}$.

33. $\{x \mid x > 2\} \cup \{x \mid x < -1\}$. **35.** $\{x \mid -2 < x < 5\}$.

Section 3 (page 19)

1. (a) 7. (b) 4. (c) 4. (d) 3.

3. (a) 7. (b) -4. (c) 4. (d) 3.

5.

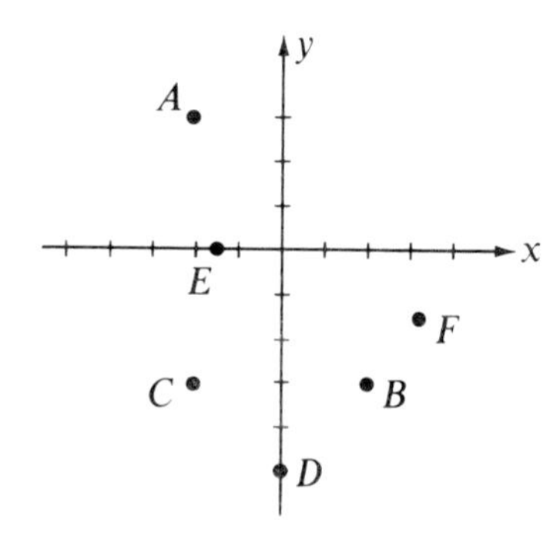

7. (a) The points on the coordinate axes.
(b) The points in quadrants I and III (not on axes).
(c) The points in quadrants II or IV (not on axes).

9. 5. **11.** $\sqrt{26}$.

17. $P(1, -1)$ is on the perpendicular bisector since $d(A, P) = \sqrt{13} = d(B, P)$.

19. Let a, b, and c be the coordinates of J, K, and L, respectively. Then $\overline{JK} + \overline{KL} = (b - a) + (c - b) = c - a = \overline{JL}$.

Section 4 (page 26)

1. (a) $S \times T = \{(1, x), (1, y), (2, x), (2, y), (3, x), (3, y)\}$.
(b) $T \times S = \{(x, 1), (x, 2), (x, 3), (y, 1), (y, 2), (y, 3)\}$.
(c) $S \times S = \{(1, 1), (1, 2), (1, 3), (2, 1), (2, 2), (2, 3), (3, 1), (3, 2), (3, 3)\}$.
(d) $T \times T = \{(x, x), (x, y), (y, x), (y, y)\}$.

3.

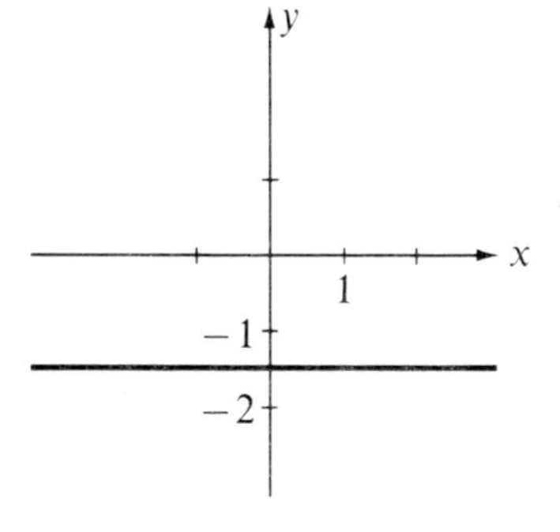

5.

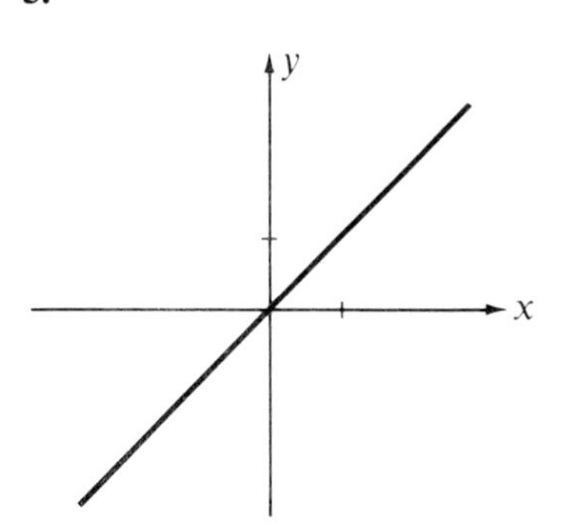

7.

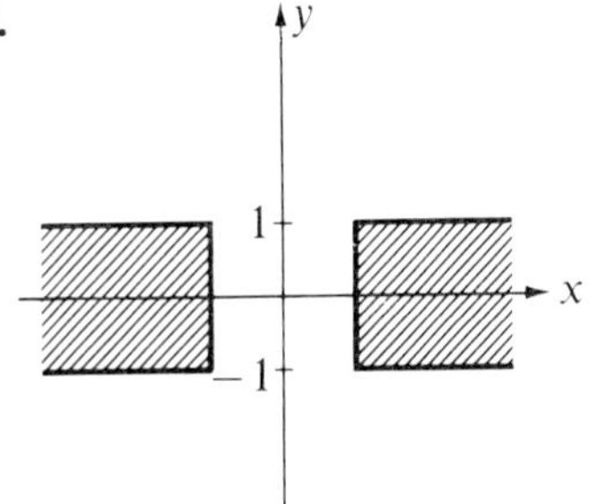

9.

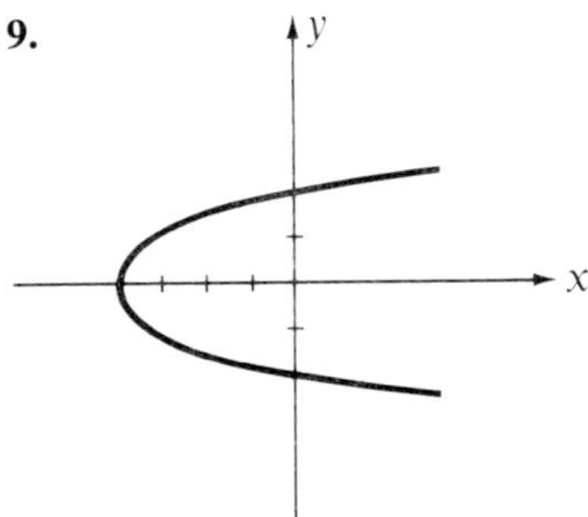

11.

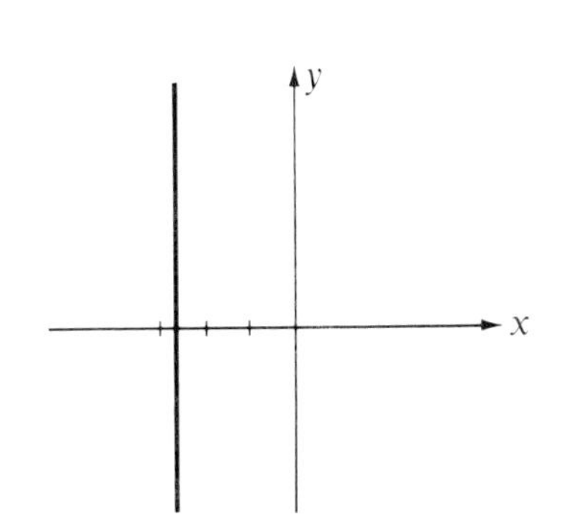

13.

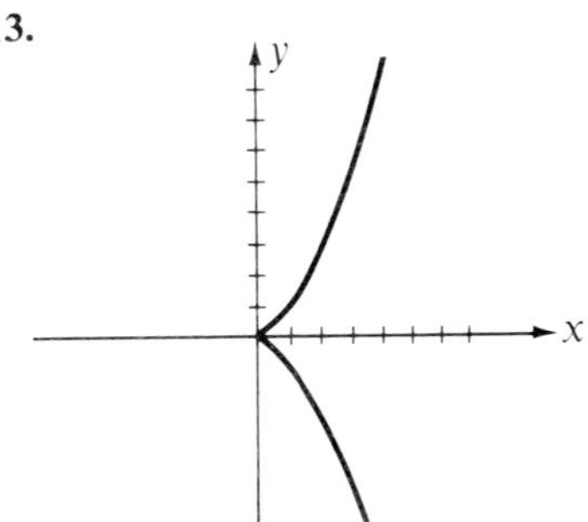

15. $x^2 + y^2 + 4x - 8y + 11 = 0$.

17. $(x - 3)^2 + (y + 5)^2 = 4$.

19. $(x - 4)^2 + (y + 7)^2 = 65$.

Section 5 (page 32)

1. (a) 3. (b) 5. (c) 5. (d) $11 - 5\sqrt{2}$.
(e) $a^2 - 3a + 5$.
(f) $a^2 + 3a + 5$.
(g) $-a^2 + 3a - 5$.
(h) $a^2 + 2ah + h^2 - 3a - 3h + 5$.
(i) $a^2 + h^2 - 3(a + h) + 10$.
(j) $2a - 3 + h$.
(k) $2x - 3 + h$.

3. (a) $1/2$. (b) $1/4$. (c) 1. (d) $\sqrt{2}/2$.
(e) $1/(a + 1)$.
(f) $1/(1 - a)$.
(g) $-1/(a + 1)$.
(h) $1/(a + h + 1)$.
(i) $1/(a + 1) + 1/(h + 1)$.
(j) $-1/(a + h + 1)(a + 1)$.
(k) $-1/(x + h + 1)(x + 1)$.

5. (a) $a^2/(1 + a^2)$.
(b) $a^2 + 1$.
(c) $1/(a^4 + 1)$.
(d) $1/(a^4 + 2a^2 + 1)$.
(e) $1/(x + 1)$.
(f) $\sqrt{x^2 + 1}/(x^2 + 1)$.

7. $17/2$; $(a^2 + 1)/2$; $\{x \mid x \geq 0\}$.

9. $\{x \mid x \geq 1\}$.

11. **R**.

13. One-to-one.

15. Not one-to-one.

17. Even.

19. Odd.

21. Odd.

23. Yes.

25. No.

27. No.

29. Yes.

Section 6 (page 39)

1. $\{x \in \mathbf{R} \mid x \neq \pm 2\}$.

3. $\{x \in \mathbf{R} \mid x > -2, \quad x < 3\}$.

5.

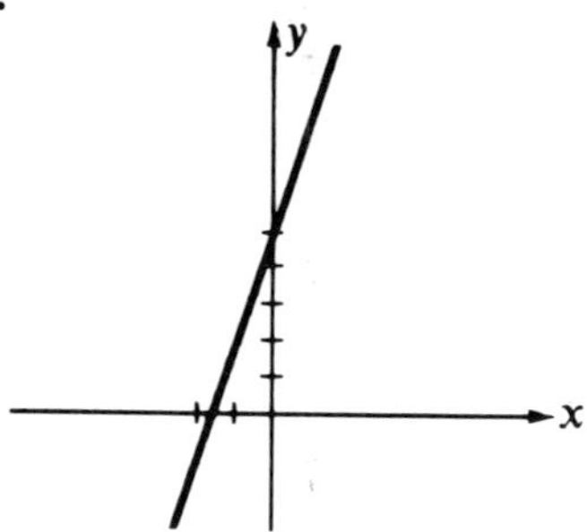

7.

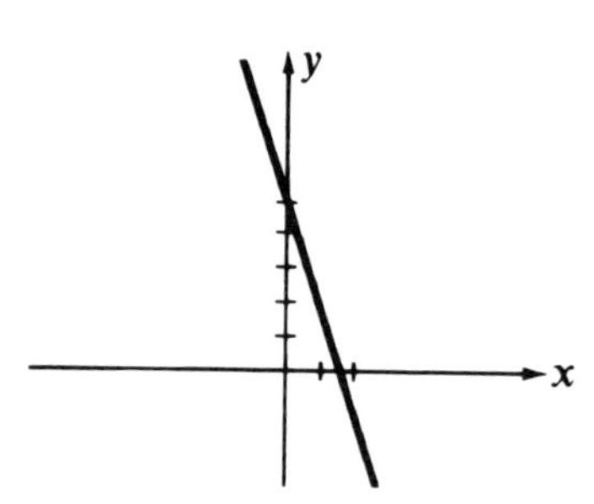

9.

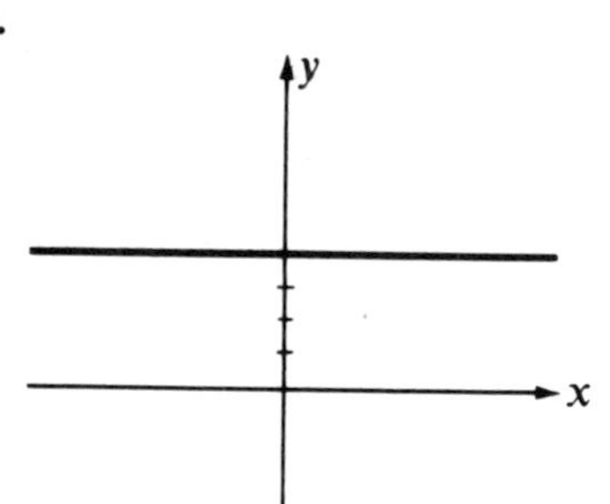

11.

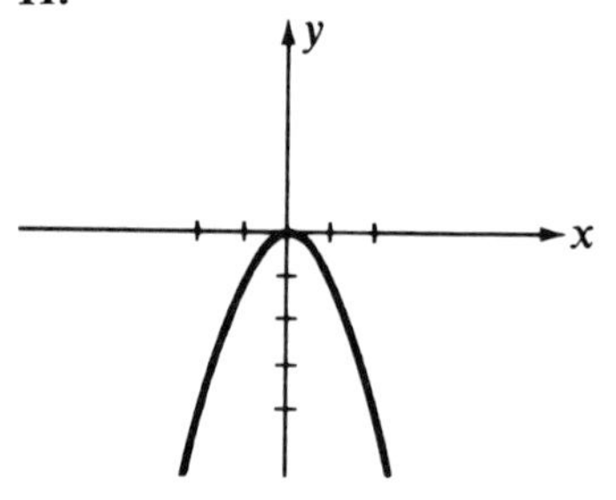

13.

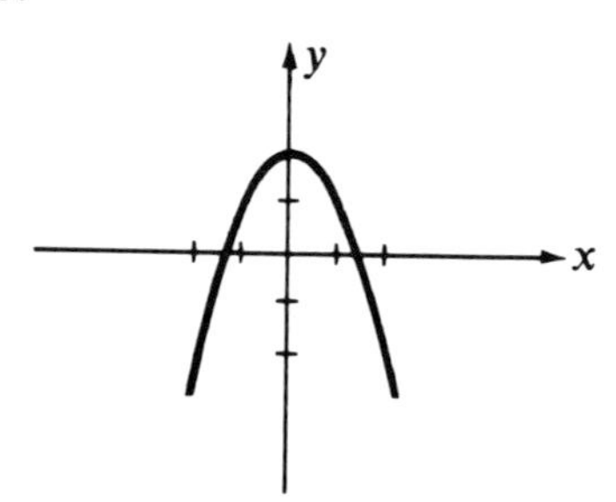

15.

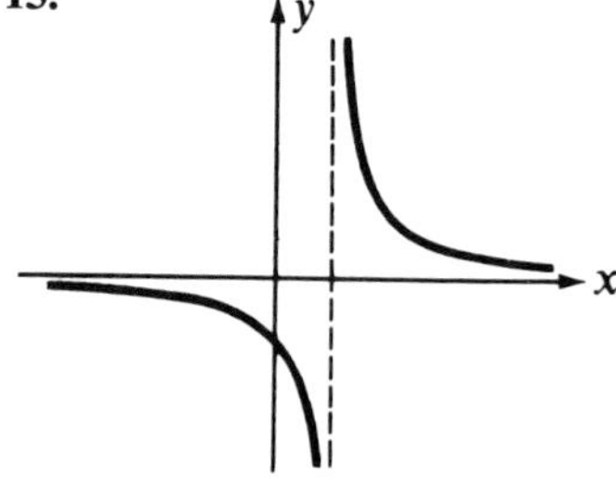

17.

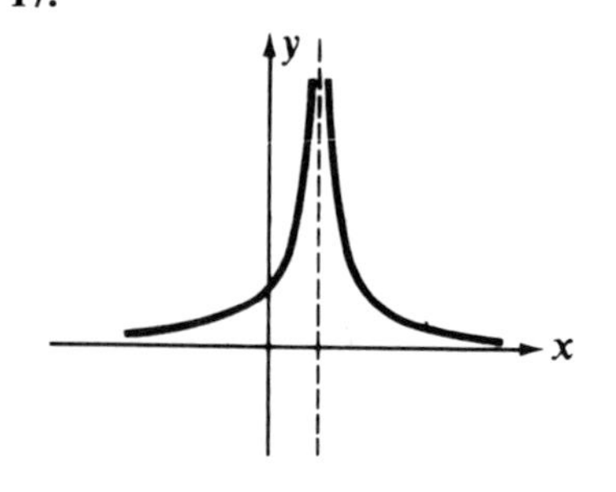

19.

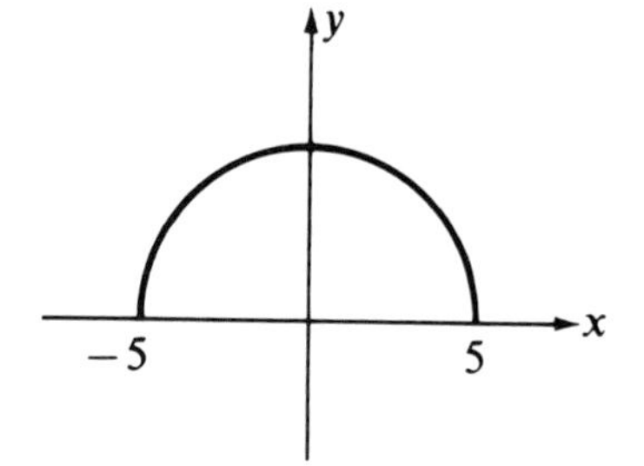

21.

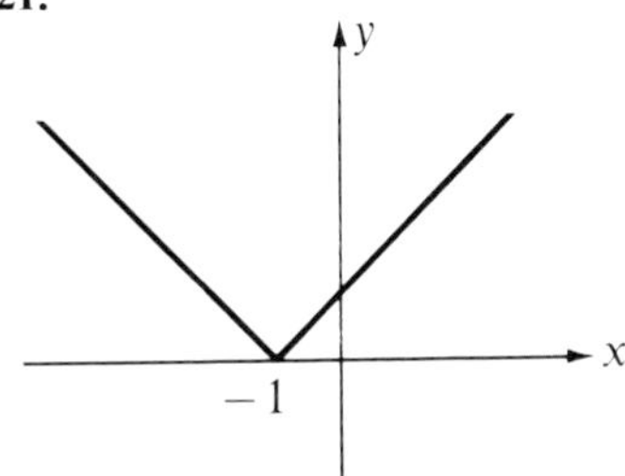

23.

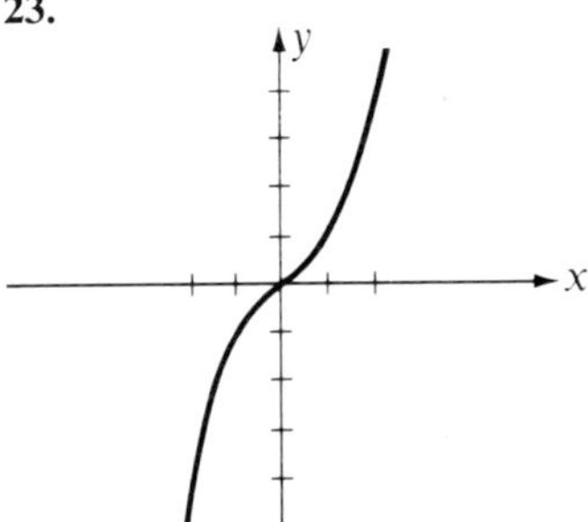

25.

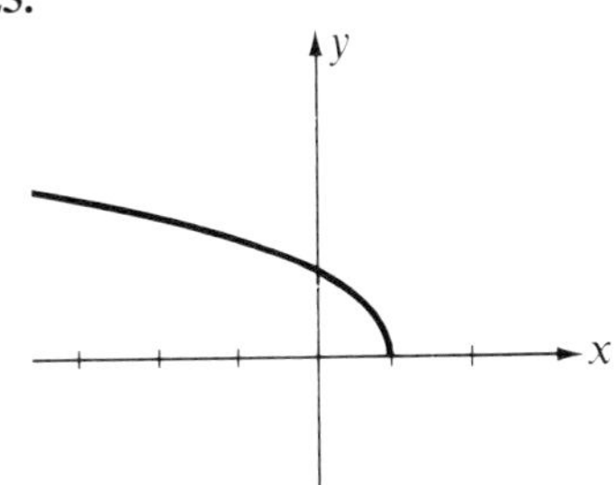

27.

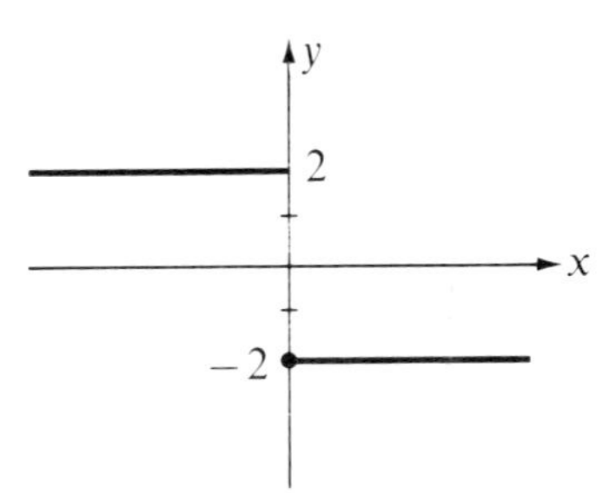

29.

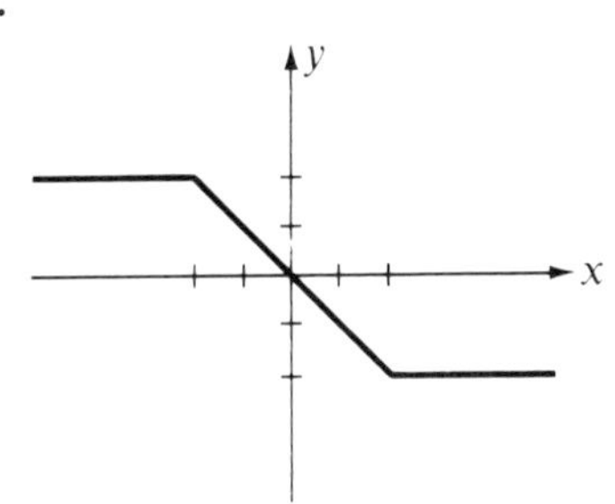

Section 7 (page 45)

1. $3x^2 + 1/(2x - 3)$; $3x^2 - 1/(2x - 3)$; $3x^2/(2x - 3)$; $3x^2(2x - 3)$.

3. $2x$; $2/x$; $(x^4 - 1)/x^2$; $(x^2 + 1)/(x^2 - 1)$.

7. $(f \circ g)(x) = 98x^2 - 112x + 37$; $(g \circ f)(x) = -14x^2 - 31$.

9. $(f \circ g)(x) = (x + 1)^3$; $(g \circ f)(x) = x^3 + 1$.

11. $(f \circ g)(x) = \dfrac{3}{(3x^2 + 2)^2} + 2$; $(g \circ f)(x) = \dfrac{1}{27x^4 + 36x^2 + 14}$.

13. $(f \circ g)(x) = \sqrt{2x^2 + 7}$; $(g \circ f)(x) = 2x + 4$.

Section 8 (page 50)

5. $f^{-1}(x) = (x - 8)/11$.

7. $f^{-1}(x) = \sqrt{6 - x}$, $z \le x \le 6$.

9. $f^{-1}(x) = (x^2 + 2)/7$, $x \ge 0$.

Review Exercises (page 51)

1. (a) $\{6\}$. (b) $\{6, 2\}$. (c) $\{1, 2, 5, 6\}$. (d) $\{1, 2, 5, 6\}$.

3. (a) $22/7 - \pi$. (b) 0. (c) -3.

5. (a) $\{x \mid x > 7/2\}$. (b) $\{x \mid 9.7 < x < 10.3\}$.
(c) $\{x \mid x > -1/3\} \cup \{x \mid x < -1\}$.
(d) $\{x \mid 0 < x < 4\}$.

7. 70 square units.

9. (a) $x^2 + y^2 - 8x + 10y - 33 = 0$.
(b) $x^2 + y^2 + 2x - 2y + 1 = 0$.

11. (a) $x^2 + 5x$, $x^2 + x + 2$, $2x^3 + 5x^2 - x - 1$, $(x^2 + 3x + 1)/(2x - 1)$, $4x^2 + 2x - 1$, $2x^2 + 6x + 1$.
(b) $x^2 + 4 + \sqrt{2x + 5}$, $x^2 + 4 - \sqrt{2x + 5}$, $(x^2 + 4)\sqrt{2x + 5}$, $(x^2 + 4)/\sqrt{2x + 5}$, $2x + 9$, $\sqrt{2x^2 + 13}$.
(c) $5x + 2 + (1/x^2)$, $5x + 2 - (1/x^2)$, $(5x + 2)/x^2$, $5x^3 + 2x^2$, $2 + (5/x^2)$, $1/(5x + 2)^2$.

CHAPTER TWO

Section 1 (page 60)

1. $x^4 - 2x^3 - 2x^2 + 8x + 3$; 4. **3.** $5x^2 - 10x$; 2.

5. $6y^5 + 4y^4 - 11y^3 + 13y^2 + 14y - 20$; 5.

7. The degree of the product is 9, and the sum of the degrees of the factors is 9.

13. Multiplicative inverses do not exist. All other field properties are true.

Section 3 (page 68)

1. $(2x + 1)(x - 5)$. **3.** $(x^4 + 1)(x^2 + 1)(x + 1)(x - 1)$.

5. $(2x - 1)(x + 1)(x - 1)$.

7. $q(x) = 2x^2 - 5x + 20$, $r(x) = -8x + 103$.

9. $q(x) = (3/2)x$, $r(x) = 18x$.

11. $q(x) = 0$, $r(x) = 2x^3 - x^2 + 1$.

17. 14/3. **19.** $f(3) = 0$.

21. $f(c) > 0$ for all $c \in \mathbf{R}$.

23. If $f(x) = x^n - y^n$, then $f(y) = 0$. If n is even, then $f(-y) = 0$.

Section 4 (page 72)

1. $2x^2 + 5x + 7$; 19. **3.** $x^2 - 5x + 17$; -78.

5. $x^4 - 2x^3 + 4x^2 - 7x + 14$; -27.

7. $2x^3 + x^2 - \frac{1}{2}x - \frac{1}{4}$; $\frac{7}{8}$.

9. $x^3 + 3ix^2 - 3x - 4i$; 9.

11. $f(2) = -19$; $f(-2) = 57$.

13. $f(1 + 2i) = -3 - 6i$; $f(1 - 2i) = 6i - 3$.

Section 5 (page 76)

1. $x^3 + 3x^2 + (i - 3)x + 4 + 4i$.

3. $x^3 - 7x^2 + 17x - 15$.

5. $x^4 - 2x^3 - 11x^2 + 12x + 36$.

7. -2 (multiplicity two), $\sqrt{2}i$ and $-\sqrt{2}i$ (multiplicity one).

9. 0 (multiplicity two), 5 and -1 (multiplicity one).

11. 0 (multiplicity three), 2 and -2 (multiplicity two).

13. -3 (multiplicity two), 1 and -2 (multiplicity one).

15. $f(x) = (x - 3)^2(x - 2i)(x + 2i)$.

17. $f(x) = (x - 1)^3(x + 4)$; -4.

19. If $f(c) = g(c)$, then $h(c) = 0$.

21. $A = -1, B = -1, C = 0$.

23. $A = 15/4, B = 2, C = 3/4$.

Section 6 (page 83)

1. $x^2 - 6x + 13$.

3. $x^4 - 6x^3 + 18x^2 - 24x + 16$.

5. $2, \; -3, \; 5/2$.

7. $2/3, \; -2, -1/2$.

9. $4/3, \; -2 \pm i$.

11. $-2, \; 3, \; -1 \pm \sqrt{2}i$.

13. $1/3$.

15. $4, \; -7, \; \pm\sqrt{2}$.

21. By (2.29) complex zeros occur in conjugate pairs.

Review Exercises (page 84)

1. (a) $2x^3 + x^2 + 7$, $2x^3 - x^2 - 2x + 3$, $2x^5 + 2x^4 + 3x^3 + 4x^2 + 3x + 10$.
(b) $14x^4 - x^3 + x^2 + 4x - 1$, $x^3 + x^2 - 4x - 1$, $49x^8 - 7x^7 + 7x^6 + 27x^5 - 7x^4 + 5x^3 - 4x$.
(c) $x + 3$, $3 - x$, $3x$.
(d) $2x^2 - 7x + 4$, $-2x^2 + 7x - 4$, 0.

3. (a) $(x^2 + 4)(x + 2)(x - 2)$. (b) $(2x - 3)(4x^2 + 6x + 9)$.
(c) $x(x - 1)(x + 1)(2x - 1)$.

5. (a) -65. (b) $f(-3) = 0$.

7. (a) $x^3 - 6x^2 + 11x - 6$. (b) $x^3 - 9x^2 + 31x - 39$.
(c) $x^3 - (1 + i)x^2 + (2 + i)x - 2$.

9. $(x + 2)^3(x - 1)(x + 3)$.

CHAPTER THREE

Section 1 (page 90)

1. (a) $\{x \mid -3 < x < 5\}$. (b) $\{x \mid -2 \le x \le 3\}$.
(c) $\{x \mid 0 < x < 5\}$. (d) $\{x \mid x > -3\}$.
(e) $\{x \mid x < -1\}$.

3. (a) If $x_1 < x_2$ and $m > 0$, then $mx_1 < mx_2$. Therefore $mx_1 + b < mx_2 + b$; that is, $f(x_1) < f(x_2)$.
(b) If $x_1 < x_2$ and $m < 0$, then $mx_1 > mx_2$. Therefore $mx_1 + b > mx_2 + b$; that is, $f(x_1) > f(x_2)$.

5. Increasing in $(-\infty, \infty)$. **7.** Decreasing in $(-\infty, \infty)$.

9. Neither increasing nor decreasing.

11. Increasing in $(-\infty, 0]$, decreasing in $[0, \infty)$.

13. Increasing in $(-\infty, 0]$, decreasing in $[0, \infty)$.

15. Decreasing in $(-\infty, 1)$ and in $(1, \infty)$.

17. Increasing in $(-\infty, 1)$, decreasing in $(1, \infty)$.

19. Increasing in $[-5, 0]$, decreasing in $[0, 5]$.

21. Decreasing in $(-\infty, -1)$, increasing in $(-1, \infty)$.

23. Increasing in $(-\infty, \infty)$. **25.** Decreasing in $(-\infty, 1)$.

27. Neither increasing nor decreasing.

29. Decreasing in $[-2, 2]$.

Section 2 (page 94)

1. $f(-1) = 1$ and $f(0) = -1$.
(a) Between -0.4 and -0.3. (b) Between -0.35 and -0.34.

3. Between -7 and -6; -1 and 0; 1 and 2.

5. If $a \in \mathbf{R}$, let $z = (a - b)/m$. Then $f(z) = a$.

Section 3 (page 97)

1.

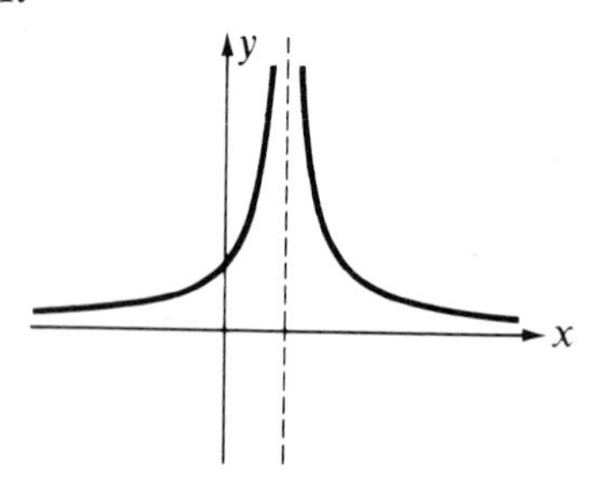

3.

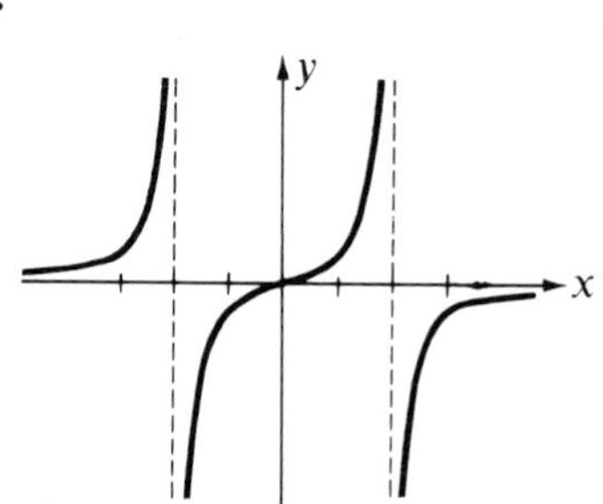

5.

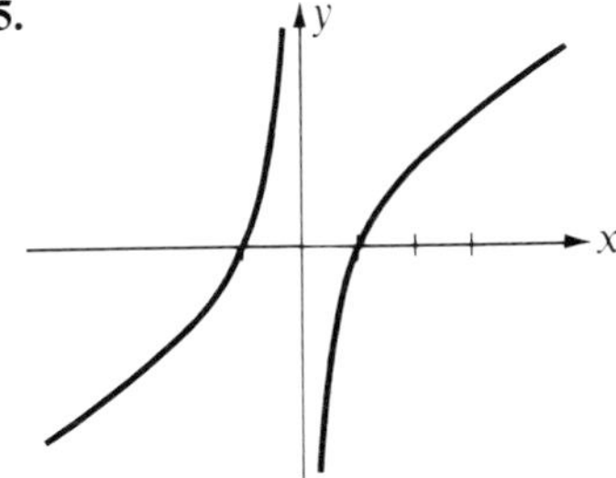

Section 4 (page 100)

1. $f(x) = (4 + 10x - 2x^2)/5$; **R**.
3. $f(x) = \sqrt{1 - x^2}$; $\{x \mid -1 \le x \le 1\}$.
5. $f(x) = (x^3 - 1)/(x - 4)$; $\{x \mid x \ne 4\}$.
7. $f(x) = x + \sqrt{x}$; $\{x \mid x \ge 0\}$.
9. $f(x) = x - 2\sqrt{x} + 1$; $\{x \mid x \ge 0\}$.

Review Exercises (page 101)

1. $f(x) \ge f(c)$ for all x in $[a, b]$.
3. $f(a)$ is a minimum value in $[a, b]$ since if $a < x$, then $a^3 < x^3$; that is, $f(a) < f(x)$. Similarly $f(b)$ is a maximum value since if $x < b$, then $f(x) < f(b)$.
5. Between 1.4 and 1.5.
7. There are many examples. One is the function f defined by $f(x) = x$ if $0 < x \le 1$ and $f(0) = 1$.

CHAPTER FOUR

Section 1 (page 108)

1. $-1/108$.
3. $9/2$.
5. $20y/x^3$.
7. $9x^{10}y^{14}$.
9. $2a^5/5b^6c^{11}$.
11. a^4b^8.
13. 0.
15. 0.04.
17. $24x^{3/2}$.
19. 1.
21.

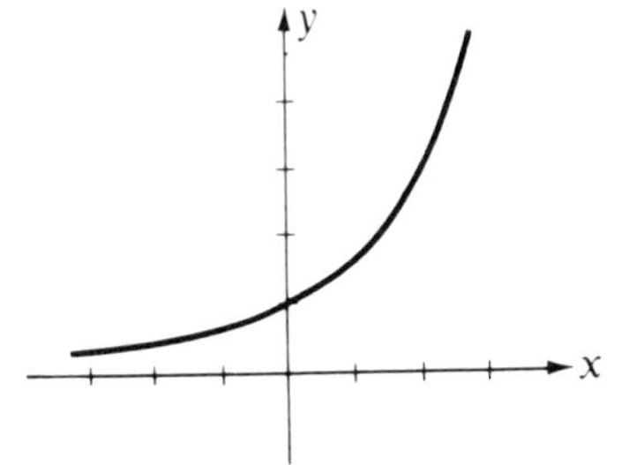

23.

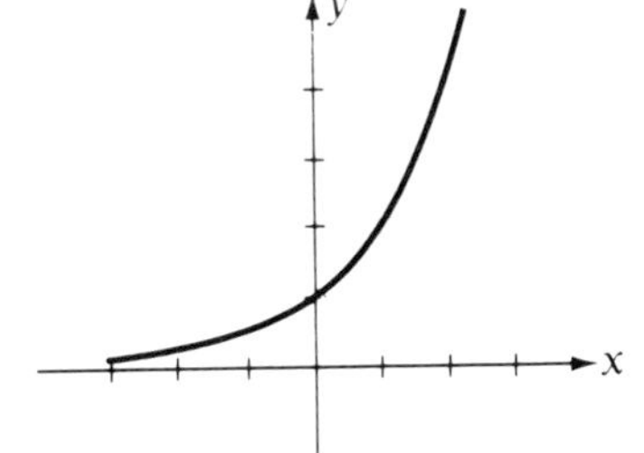

25.

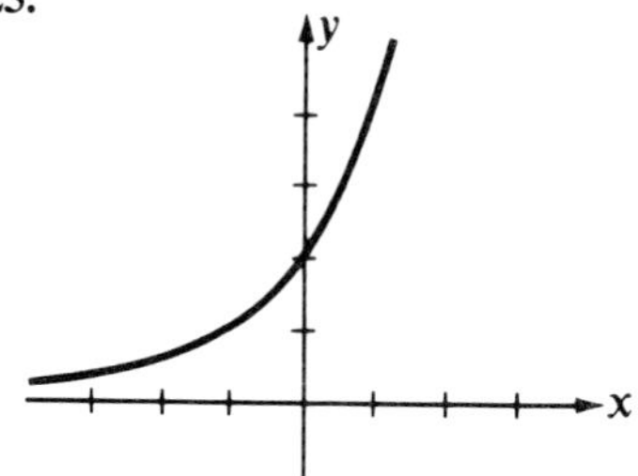

27.

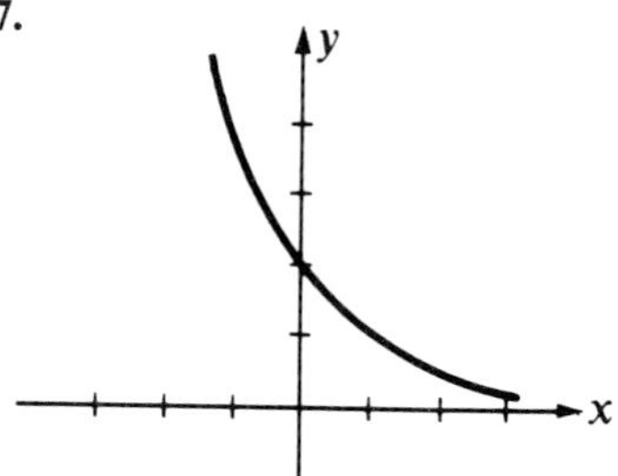

29.

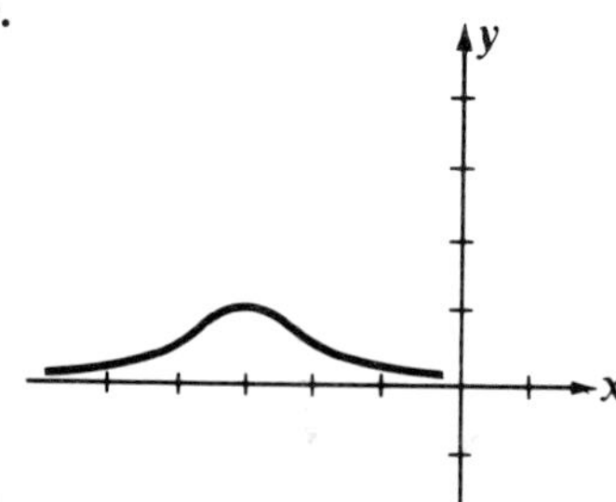

31. $P(c, d)$ is on the graph of $y = a^x$ if and only if $P(-c, d)$ is on the graph of $y = a^{-x}$.

33. If $0 < a < 1$, then $1/a > 1$. Hence if $r < s$, then $(1/a)^r < (1/a)^s$. This implies that $a^s < a^r$.

35. $k = -1/1600$.

Section 2 (page 114)

1. $\log_3 9 = 2$.
3. $\log_{10} 0.001 = -3$.
5. $4^3 = 64$.
7. $a^0 = 1$.
9. -3.
11. 4.
13. 5.
15. 1/3.
17. $\{-14/3\}$.
19. $\{\sqrt{10}/10^3\}$.
21. $\{\sqrt{5}\}$.
23. $\{x \mid -1 < x < 15\}$.
25. $\{x \mid 100 \leq x \leq 1000\}$.
27. $\{15/2\}$.
29. $\{4/3\}$.
31. $\{9/8\}$.
33. $\{70\}$.
35. $\{4\}$.
37. $\log_a x + 3 \log_a y - 2 \log_a z$.
39. $(1/2) \log_a x + (1/4) \log_a y$.
41. $\log [x^2\sqrt[5]{x + 1}/(x - 1)^3]$.
43. $\log_{1/a} x = -3 = \log_a (1/x)$.
45. Multiply the expression in parentheses by $(x - \sqrt{x^2 - 1})/(x - \sqrt{x^2 - 1})$ and simplify.

Section 3 (page 118)

1.

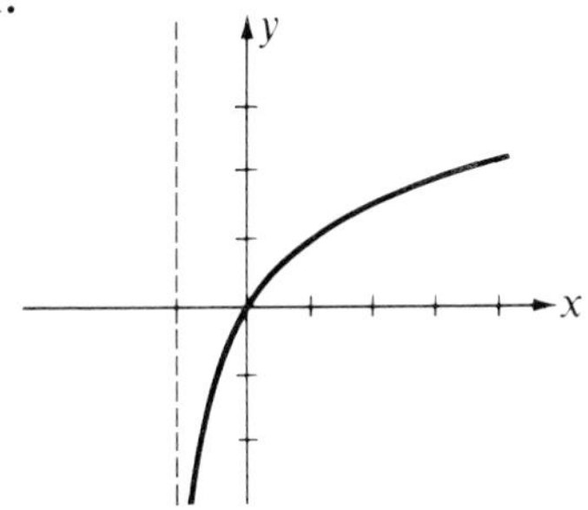

3.

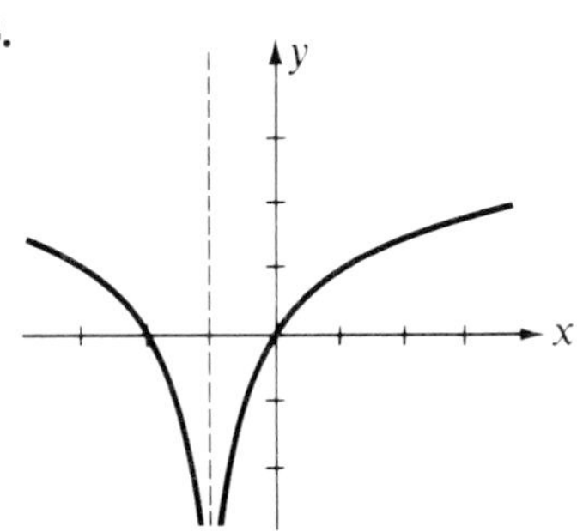

5.

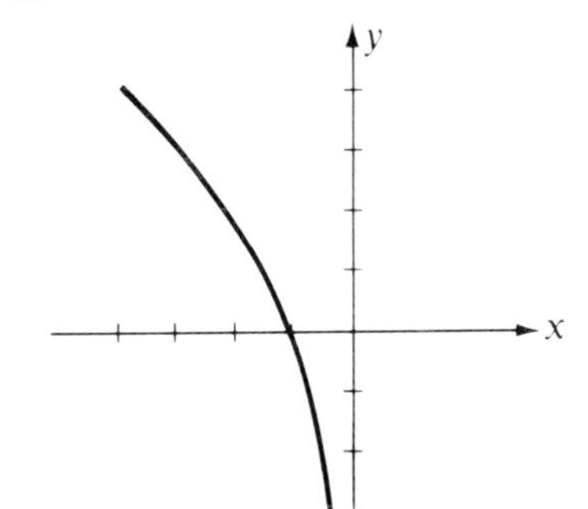

7.

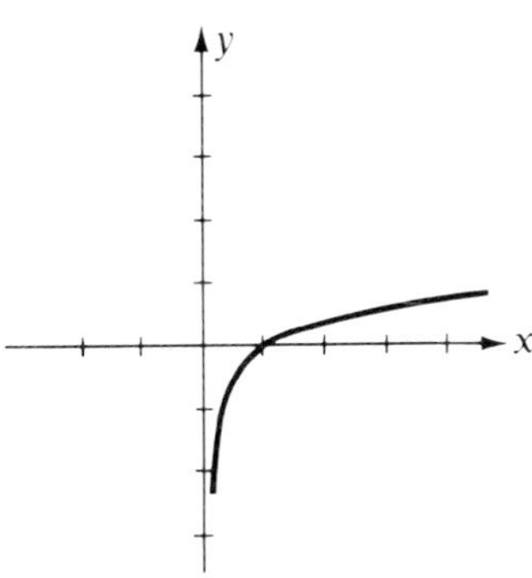

9.

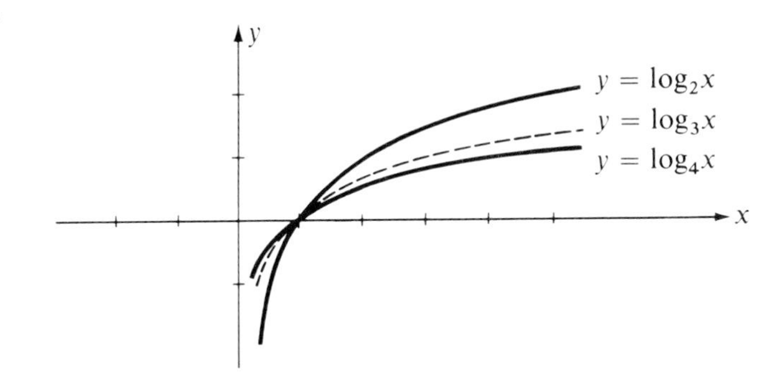

Lies between the graphs of the first two equations.

Section 4 (page 122)

1. 0.8267; 2.8267; 0.8267 − 3. **3.** 1.9868; 5.9868; 0.9868 − 1.

5. 0.8494 − 5; 1.8494; 2.8494. **7.** 3.8938; 0.9734; 0.1062 − 4.

9. 0.1568 − 2; 1.8432; 0.7952 − 1.

11. 1.2239. **13.** 8.3490. **15.** 6460. **17.** 1.05.

19. 0.0757. **21.** 0.0725. **23.** 0.957.

Section 5 (page 126)

1. 1.5532. **3.** 3.8729. **5.** 0.2287 − 3.
7. 5.3025. **9.** 275.3 **11.** 2.793.
13. 0.04278. **15.** 0.000000002338.

Section 6 (page 128)

1. 15.5 **3.** 8.08. **5.** 715.
7. 0.719. **9.** 23.1. **11.** 1.97.
13. −0.129. **15.** 0.0000238. **17.** 1.30.
19. 3.33. **21.** 89.7 square units. **23.** 1.83 seconds.

Section 7 (page 131)

1. $\{1/\log 5\}$.
3. $\{\log(3/32)/\log 18\}$.
5. $\{\pm\sqrt{\log 6/\log 2}\}$.
7. $\{10\}$.
9. $\{\sqrt{11}/3\}$.
11. $(-\infty, -1)$.
13. $\{1, 10^{\sqrt{3}}, 10^{-\sqrt{3}}\}$.
15. $\{10{,}000\}$.
17. $\{10^{100}\}$.
19. $x = \log_a(y \pm \sqrt{y^2 - 1})$.
21. $t = -(L/R)\log_a[1 - (Ri/E)]$.

Review Exercises (page 133)

1. (a) 6. (b) 7. (c) 1/5. (d) 8.
3. (a) $\{19\}$. (b) $\{1/10, -1/10\}$. (c) $\{x \mid -1 < x < 24\}$.
(d) $\{x \mid 1/10 < x < 10\}$. (e) $\{5\}$.
5. (a) $2\log_a z + (1/2)\log_a x - 3\log_a y$. (b) $\log_a(y^3/x^4)$.
7. (a) 42,780. (b) 0.1044. (c) 0.02338.
9. (a) $\{8/3\}$. (b) $\{\log 2/\log 6\}$. (c) $\{(\log 2 - 5\log 3)/(\log 2 + \log 3)\}$.
(d) $\{20\}$. (e) $\{x \mid x > -1\}$. (f) $\{1{,}100\}$.

CHAPTER FIVE

Section 1 (page 140)

3. (a) II. (b) III. (c) III. (d) IV. (e) II.
5. (a) $(-1, 0)$. (b) $(0, -1)$. (c) $(1, 0)$. (d) $(0, -1)$.
7. $(-1/2, -\sqrt{3}/2)$.
9. $(-\sqrt{3}/2, -1/2)$.
11. Any number of the form $\pi/6 + 2n\pi$ where $n \in \mathbf{Z}$.
13. (a) $(-4/5, -3/5)$. (b) $(4/5, -3/5)$.
(c) $(-4/5, -3/5)$. (d) $(-4/5, 3/5)$.
15. (a) $(-a, -b)$. (b) $(a, -b)$.
(c) $(-a, -b)$. (d) $(-a, b)$.

Section 2 (page 146)

5. IV. **7.** III.

The following are arranged in the order sin t, cos t, tan t, csc t, sec t, cot t:

9. $-2/3$, $\sqrt{5}/3$, $-2\sqrt{5}/5$, $-3/2$, $3\sqrt{5}/5$, $-\sqrt{5}/2$.

11. $-3/5$, $-4/5$, $3/4$, $-5/3$, $-5/4$, $4/3$.

13. $-12/13$, $-5/13$, $12/5$, $-13/12$, $-13/5$, $5/12$.

15. (a) 0, -1, 0, $-$, -1, $-$.
(b) -1, 0, $-$, -1, $-$, 0.

17. (a) $1/2$, $-\sqrt{3}/2$, $-\sqrt{3}/3$, 2, $-2\sqrt{3}/3$, $-\sqrt{3}$.
(b) $-\sqrt{2}/2$, $-\sqrt{2}/2$, 1, $-\sqrt{2}$, $-\sqrt{2}$, 1.

Section 3 (page 154)

3. All $\pi/6$. **5.** All $\pi/4$. **7.** $\pi - 1.9$. **9.** $2\pi - 5$.

11. (a) $\sin 7\pi/6 = -1/2$, $\cos 7\pi/6 = -\sqrt{3}/2$, $\tan 7\pi/6 = \sqrt{3}/3$.
(b) $\sin 5\pi/6 = 1/2$, $\cos 5\pi/6 = -\sqrt{3}/2$, $\tan 5\pi/6 = -\sqrt{3}/3$.
(c) $\sin(-\pi/6) = -1/2$, $\cos(-\pi/6) = \sqrt{3}/2$, $\tan(-\pi/6) = -\sqrt{3}/3$.

13. (a) $\sin 3\pi/4 = \sqrt{2}/2$, $\cos 3\pi/4 = -\sqrt{2}/2$, $\tan 3\pi/4 = -1$.
(b) $\sin(-3\pi/4) = -\sqrt{2}/2$, $\cos(-3\pi/4) = -\sqrt{2}/2$, $\tan(-3\pi/4) = 1$.
(c) $\sin 9\pi/4 = \sqrt{2}/2$, $\cos 9\pi/4 = \sqrt{2}/2$, $\tan 9\pi/4 = 1$.

15. 0.5592. **17.** 0.3665. **19.** -3.2041. **21.** 0.8744.

23. 0.5055. **25.** -0.6537. **27.** 0.5751. **29.** 0.3759.

Section 5 (page 165)

1. $585°$, $945°$, $-135°$, $-495°$.

3. $300°$, $660°$, $-420°$, $-780°$.

5. $270°$, $990°$, $-90°$, $-450°$.

7. $11\pi/4$, $19\pi/4$, $-5\pi/4$, $-13\pi/4$.

9. $\pi/2$, $5\pi/2$, $-3\pi/2$, $-7\pi/2$.

11. (a) $\pi/6$. (b) $-5\pi/2$. (c) $4\pi/3$.

13. (a) $-17\pi/4$. (b) $5\pi/12$. (c) $5\pi/9$.

15. (a) $150°$. (b) $-540°$. (c) $630°$.

17. (a) $480°$. (b) $-135°$. (c) $36°$.

19. $229°\ 11'\ 2''$.

21. (a) 5/3 radians, $300/\pi$ degrees. (b) 7.5 square units.

23. (a) 4.19 inches. (b) 12.57 square inches.

Section 6 (page 171)

1. (a) $50°$. (b) $15°$. (c) $63°$. (d) $62°\ 38'$.
(e) $1°$. (f) $60°$. (g) $17°\ 2'\ 47''$.

3. (a) 0.6111. (b) 0.2186. (c) 0.3062.
5. (a) -0.7002. (b) -0.5640. (c) 0.1593.
7. 0.4006.
9. 0.6173.
11. -0.1859.
13. $28° 25'$, $151° 35'$.
15. $49° 46'$, $229° 46'$.
17. $155° 56'$, $335° 56'$.
19. $136° 54'$, $223° 6'$.
21. $\sin\theta = \sqrt{5}/5$, $\cos\theta = -2\sqrt{5}/5$, $\tan\theta = -1/2$, $\csc\theta = \sqrt{5}$, $\sec\theta = -\sqrt{5}/2$, $\cot\theta = -2$.
23. $\sin\theta = -3\sqrt{10}/10$, $\cos\theta = \sqrt{10}/10$, $\tan\theta = -3$, $\csc\theta = -\sqrt{10}/3$, $\sec\theta = \sqrt{10}$, $\cot\theta = -1/3$.
25. 1, 0, —, 1, —, 0.
27. $\sqrt{2}/2$, $-\sqrt{2}/2$, -1, $\sqrt{2}$, $-\sqrt{2}$, -1.

Section 7 (page 179)

1. $\alpha \doteq 48°$, $b \doteq 17$, $c \doteq 26$.
3. $\beta \doteq 70° 40'$, $b \doteq 20.8$, $c \doteq 22.0$.
5. $\alpha \doteq 21° 36'$, $a \doteq 523$, $c \doteq 142$.
7. $\alpha \doteq 20°$, $\beta \doteq 70°$, $c \doteq 41$.
9. $\alpha \doteq 59°$, $\beta \doteq 31°$, $c \doteq 663$.
11. 31°.
13. 82.9 feet.
15. 163 feet.
17. 71.5 feet.
19. 447 feet.
21. 55 miles.
23. 3 miles.

Review Exercises (page 181)

1. $(-1, 0)$, $(0, -1)$, $(0, 1)$, $(1/2, \sqrt{3}/2)$, $(-\sqrt{2}/2, -\sqrt{2}/2)$.
3. (a) II. (b) II. (c) IV.
7. (a) $\pi/4$; $\pi/6$; $\pi/3$. (b) 40°; $32° 48'$; 31°.
9. (a) $-\sqrt{2}/2$. (b) $\sqrt{3}$. (c) $\sqrt{3}/2$. (d) -1.
   (e) $-\sqrt{3}$. (f) 2.
11. (a) $50° 49'$, $309° 11'$. (b) $237° 16'$, $302° 44'$.
    (c) $43° 56'$, $223° 56'$.
13. (a) $\beta = 30°$, $c = 20$, $a = \sqrt{300}$.
    (b) $\alpha \doteq 37° 50'$, $b \doteq 41.2$, $c \doteq 52.2$.
    (c) $\alpha \doteq 57°$, $\beta \doteq 33°$, $c \doteq 76$.

CHAPTER SIX

Section 2 (page 192)

1. All $\pi/6 + 2n\pi$ and $5\pi/6 + 2n\pi$, where $n \in \mathbf{Z}$.
3. All $2\pi/3 + 2n\pi$, $4\pi/3 + 2n\pi$, $5\pi/4 + 2n\pi$, and $7\pi/4 + 2n\pi$, where $n \in \mathbf{Z}$.
5. All $\pi/3 + n\pi$ and $2\pi/3 + n\pi$, where $n \in \mathbf{Z}$.

7. All $n \cdot (\pi/2)$, $\pi/6 + n\pi$, and $5\pi/6 + n\pi$, where $n \in \mathbf{Z}$.

9. $\{0°, 45°, 135°, 180°, 225°, 315°\}$.

11. $\{30°, 150°, 270°\}$.

13. $\{90°, 270°\}$.

15. $\{0°, 45°, 135°, 180°, 225°, 315°\}$.

17. $\varnothing$.

19. $\{0°, 120°\}$.

21. $\{0°, 90°\}$.

23. $\{0°\}$.

25. $\{u \in \mathbf{R} \mid u \neq \pi/2 + n\pi, \quad n \in \mathbf{Z}\}$.

27. $\{90°, 210°, 270°, 330°\}$.

29. $\{135°, 315°\}$.

31. $\{41°\ 50', 138°\ 10', 194°\ 30', 345°\ 30'\}$.

33. $\{15°\ 30', 164°\ 30'\}$.

Section 3 (page 198)

1. (a) $\cos 62°\ 46'$. (b) $\sin \pi/3$. (c) $\cot 22°\ 50'$.

3. (a) $(\sqrt{2} + 1)/2$. (b) $(\sqrt{6} + \sqrt{2})/4$.

5. (a) $-\sqrt{3}/3 - 1$. (b) $-2 - \sqrt{3}$.

7. $\cos(\alpha + \beta) = -36/85$, $\sin(\alpha + \beta) = 77/85$; II.

9. $44/125$, $117/125$, $44/117$, $-4/5$, $3/5$, $-4/3$.

23. $\sin u \cos v \cos w + \cos u \sin v \cos w + \cos u \cos v \sin w - \sin u \sin v \sin w$.

Section 4 (page 203)

1. $24/25$, $-7/25$, $-24/7$.

3. $-4\sqrt{21}/25$, $-17/25$, $4\sqrt{21}/17$.

5. $\sqrt{\dfrac{3 - \sqrt{5}}{6}}$, $\sqrt{\dfrac{3 + \sqrt{5}}{6}}$, $\dfrac{3 - \sqrt{5}}{2}$.

7. $-\dfrac{\sqrt{2 + \sqrt{2}}}{2}$, $\dfrac{\sqrt{2 - \sqrt{2}}}{2}$, $-1 - \sqrt{2}$.

9. $\dfrac{\sqrt{2 - \sqrt{2}}}{2}$, $\dfrac{\sqrt{2 + \sqrt{2}}}{2}$.

23. $\{0°, 120°, 180°, 240°\}$.

25. $\{60°, 180°, 300°\}$.

27. $\{0°, 180°\}$.

29. $\{0°, 60°, 300°\}$.

Section 5 (page 207)

1. $\cos 3x - \cos 7x$.

3. $\frac{1}{2}(\sin 5t - \sin t)$.

5. $\frac{3}{2}(\cos 14u + \cos 4u)$.

7. $2 \cos u \sin 5u$.

9. $2 \sin 6t \sin t$.

11. $-2 \cos 3t \sin t$.

19. $\{n \cdot 45° \mid n \in \mathbf{Z}\}$.

Section 7 (page 213)

1. (a) Amplitude $a = 3$, period $p = 2\pi$.
   (b) $a = 1$, $p = 2\pi/3$.
   (c) $a = 1/3$, $p = 2\pi$.
   (d) $a = 1$, $p = 6\pi$.
   (e) $a = 2$, $p = 6\pi$.
   (f) $a = 1/3$, $p = \pi$.

Section 8 (page 218)

13. Amplitude is 8, period is $\frac{1}{2}$ second, frequency is 2 cycles per second.

Section 9 (page 224)

1. $5\pi/6$. **3.** $-\pi/3$. **5.** $-\pi/2$.

7. 0.9687. **9.** 1.1316. **11.** $\sqrt{3}/2$.

13. 3/5. **15.** $\pi/4$. **17.** $(12\sqrt{5} - \sqrt{10})/30$.

19. 7/24. **21.** $-24/25$. **23.** $4\sqrt{17}/17$.

25. $u\sqrt{1 + u^2}/(1 + u^2)$. **27.** $\sqrt{2 + 2u}/2$.

Section 10 (page 231)

3. $\beta \doteq 62°$, $b \doteq 14.1$, $c \doteq 15.6$.

5. $\gamma \doteq 100°\ 10'$, $b \doteq 55.1$, $c \doteq 68.7$.

7. $\alpha \doteq 58°\ 40'$, $a \doteq 487$, $b \doteq 442$.

9. $\beta \doteq 53°\ 40'$, $\gamma \doteq 61°\ 10'$, $c \doteq 20.6$.

11. $\alpha \doteq 77°\ 30'$, $\beta \doteq 49°\ 10'$, $b \doteq 108$; $\alpha \doteq 102°\ 30'$, $\beta \doteq 24°\ 10'$, $b \doteq 58.7$.

13. $\alpha \doteq 20°\ 30'$, $\gamma \doteq 46°\ 20'$, $a \doteq 94.5$.

15. 219 yards. **17.** 50 feet. **19.** 2.7 miles.

21. Approximately 3.7 miles from A and 5.4 miles from B.

Section 11 (page 234)

1. $a \doteq 26$, $\beta \doteq 41°$, $\gamma \doteq 79°$.

3. $b \doteq 177$, $\alpha \doteq 25°\ 10'$, $\gamma \doteq 4°\ 50'$.

5. $c \doteq 2.8$, $\alpha \doteq 21°\ 10'$, $\beta = 43°\ 40'$.

7. $\alpha \doteq 29°$, $\beta \doteq 46°\ 30'$, $\gamma \doteq 104°\ 30'$.

9. $\alpha \doteq 12°\ 30'$, $\beta \doteq 136°\ 30'$, $\gamma \doteq 31°$.

11. 63, 87 inches. **13.** 92 feet. **15.** 24 miles. **17.** 39 miles.

Review Exercises (page 236)

3. (a) $(\sqrt{2}/4)(\sqrt{3} + 1)$. (b) $-2 + \sqrt{3}$.
(c) $(-\sqrt{2}/4)(\sqrt{3} + 1)$. (d) $\sqrt{2 + \sqrt{2}}/4$.

5. (a) $(1/2) \cos 2t - (1/2) \cos 8t$; $(1/2) \cos (u/12) + (1/2) \sin (7u/12)$; $2 \sin 3x + 2 \sin x$.
(b) $2 \cos 2\theta \sin 6\theta$; $2 \sin 6u \sin u$; $2 \cos (5t/12) \sin (t/12)$.

7. 10; 1/3; 3.

9. (a) $-\pi/3$. (b) $\pi/4$. (c) π. (d) $-\pi/6$.
(e) 1/2. (f) 0. (g) 161/189. (h) 100.
(i) $\pi/4$.

CHAPTER SEVEN

Section 1 (page 243)

1. $x^2 + y^2 + 4x - 10y + 20 = 0.$

3. $(x + 4)^2 + (y + 3)^2 = 81.$

5. $(x + 3/2)^2 + (y - 1/2)^2 = 25/2.$

7. $(x + 4)^2 + (y - 1)^2 = r^2.$

9. $(-2, 3)$; 3.

11. $(-3, 0)$; 3.

13. $(1/4, -1/4)$; $\sqrt{26}/4$.

15. $(1/3, -2/3)$; 2.

17. $x^2 + y^2 - 3x + 9y - 20 = 0.$

Section 2 (page 251)

1. 4.

3. 3/8.

5. No slope.

7. Slopes of AB and CD equal 1/4; slopes of AD and BC equal 3/2.

11. (a) 135°. (b) 30°. (c) $\text{Tan}^{-1}(2)$. (d) 20°.

13. (a) 0. (b) $-\sqrt{3}$. (c) 1. (d) 4/3.

15. Slope of AB = slope of BC.

17. $-1/4$.

19. $-8/3$.

21. $(-12, 0)$.

Section 3 (page 257)

1. $x - 2y = 14.$

3. $3x - 8y = 41.$

5. $x - 8y = 24.$

7. (a) $x = 10.$ (b) $y = -6.$

9. $5x + 2y = 29.$

11. $5x - 7y + 15 = 0.$

13. $x + 6y - 9 = 0$; $4x + y - 4 = 0$; $3x - 5y + 5 = 0$; $(15/23, 32/23)$.

15. $m = 3/4$, $b = 2$.

17. $m = -1/2$, $b = 0$.

19. $m = 0$, $b = 4$.

21. $m = -5/4$, $b = 5$.

23. $y - 7 = m(x + 5).$

25. $y = mx + 3/5.$

27. Slopes are $-4/5$.

29. Slopes are -2.

Section 5 (page 266)

1. $F(2, 0)$; $x = -2$.

3. $F(0, -3)$; $y = 3$.

5. $F(-3/8, 0)$; $x = 3/8$.

7. $F(0, 1/32)$; $y = -1/32$.

9. $F(0, 4)$; $x = -2$.

11. $F(2, -7/4)$; $y = -9/4$.

13. $y^2 = 8x.$

15. $(x - 6)^2 = 12(y - 1).$

17. $3x^2 = -4y.$

19. $(y + 11)^2 = 4p(x - 6).$

21. 9/16 feet from the vertex.

Section 6 (page 272)

1. $V(\pm 3, 0)$; $F(\pm\sqrt{5}, 0)$.

3. $V(0, \pm 4)$; $F(0, \pm 2\sqrt{3})$.

5. $V(0, \pm\sqrt{5})$; $F(0, \pm\sqrt{3})$.

7. $V(\pm 1/2, 0)$; $F(\pm\sqrt{21}/10, 0)$.

9. $\dfrac{x^2}{64} + \dfrac{y^2}{39} = 1.$

11. $\dfrac{4x^2}{9} + \dfrac{y^2}{25} = 1.$

13. $\dfrac{8x^2}{81} + \dfrac{y^2}{36} = 1.$

15. $\{(2, 2), (4, 1)\}.$

17. $\dfrac{x^2}{100} + \dfrac{y^2}{k} = 1, \quad 0 < |k| < 10; \qquad \dfrac{x^2}{100} + \dfrac{3y^2}{16} = 1.$

19. $2\sqrt{21}$ feet.

21. If the eccentricity is close to 1, the ellipse is very flat. If the eccentricity is close to zero, the ellipse is almost circular.

Section 7 (page 278)

1. $V(\pm 3, 0), \quad F(\pm\sqrt{13}, 0), \quad y = \pm 2x/3.$

3. $V(0, \pm 3), \quad F(0, \pm\sqrt{13}), \quad y = \pm 3x/2.$

5. $V(0, \pm 4), \quad F(0, \pm 2\sqrt{5}), \quad y = \pm 2x.$

7. $V(\pm 1, 0), \quad F(\pm\sqrt{2}, 0), \quad y = \pm x.$

9. $\dfrac{x^2}{9} - \dfrac{y^2}{16} = 1.$

11. $\dfrac{y^2}{21} - \dfrac{x^2}{4} = 1.$

13. $\dfrac{x^2}{9} - \dfrac{y^2}{36} = 1.$

15. $\{(0, 4), (8/3, -20/3)\}.$

17. $x^2 - y^2/25 = k, \quad k > 0.$

19. Conjugate hyperbolas have the same asymptotes.

Section 8 (page 284)

1. Parabola, vertex $(-1, 5)$, focus $(1, 5)$.

3. Ellipse, center $(4, 2)$, vertices $(1, 2)$ and $(7, 2)$, endpoints of minor axis $(4, 4)$ and $(4, 0)$.

5. Hyperbola, center $(-5, 1)$, vertices $(-9, 1)$ and $(-1, 1)$, endpoints of conjugate axis $(-5, 6)$ and $(-5, -4)$.

7. Ellipse, center $(-5, 3)$, vertices $(-5, 2)$ and $(-5, 4)$, endpoints of minor axis $(-11/2, 3)$ and $(-9/2, 3)$.

9. Hyperbola, center $(5, 0)$, vertices $(5, 4)$ and $(5, -4)$, endpoints of conjugate axis $(-5, 0)$ and $(15, 0)$.

11. Ellipse, center $(-3, 1)$, vertices $(-7, 1)$ and $(1, 1)$, endpoints of minor axis $(-3, 4)$ and $(-3, -2)$.

13. Ellipse, center $(5, 2)$, vertices $(5, 7)$ and $(5, -3)$, endpoints of minor axis $(3, 2)$ and $(7, 2)$.

15. Parabola, vertex $(-1/2, -25/16)$, focus $(-1/2, 7/16)$.

17. Hyperbola, center $(-2, -5)$, vertices $(-2, -2)$ and $(-2, -8)$, endpoints of conjugate axis $(4, -5)$ and $(-8, -5)$.

19. Hyperbola, center $(6, 2)$, vertices $(6, 4)$ and $(6, 0)$, endpoints of conjugate axis $(0, 2)$ and $(12, 2)$.

Section 9 (page 288)

The following answers contain equations in x' and y' resulting from a rotation of axes.

1. Ellipse, $(x')^2 + 16(y')^2 = 16.$

3. Hyperbola, $4(x')^2 - (y')^2 = 1.$

5. Ellipse, $(x')^2 + 9(y')^2 = 9.$

7. Parabola, $(y')^2 = 4(x' - 1).$

Section 10 (page 294)

1.

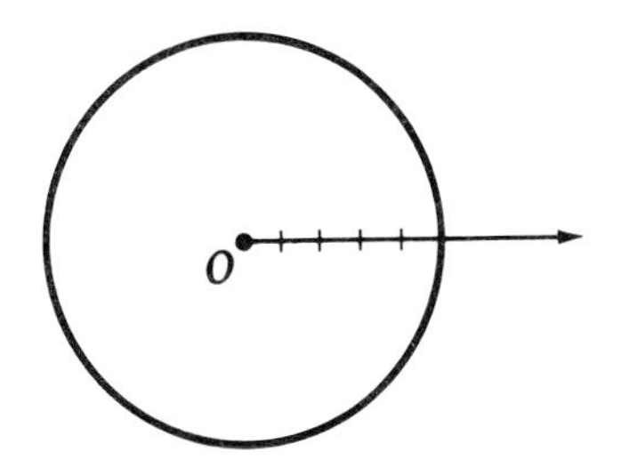

3.

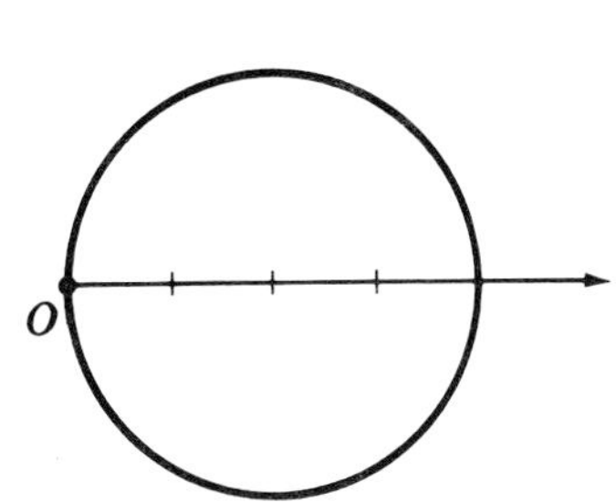

5.

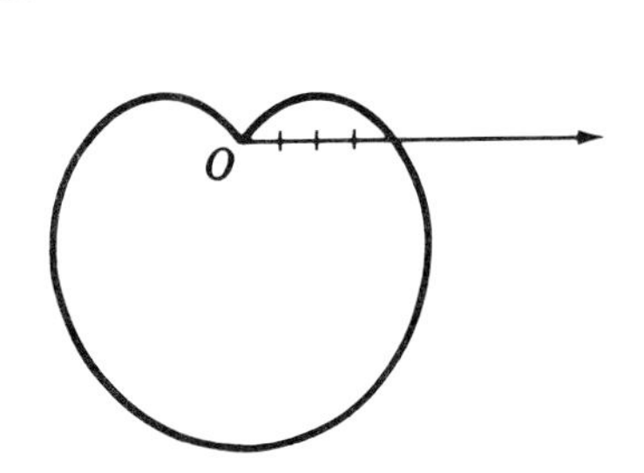

7.

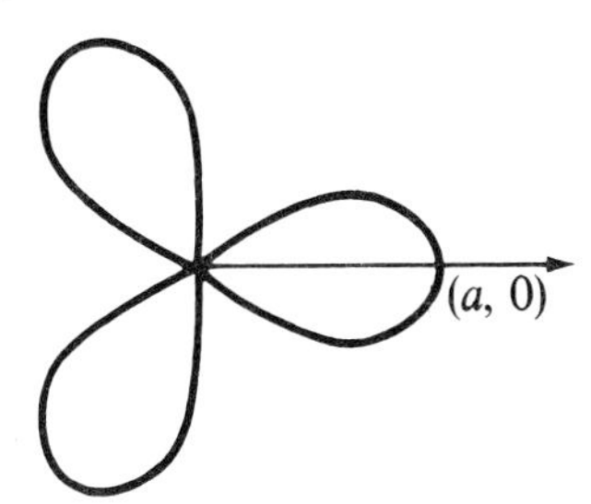

9.

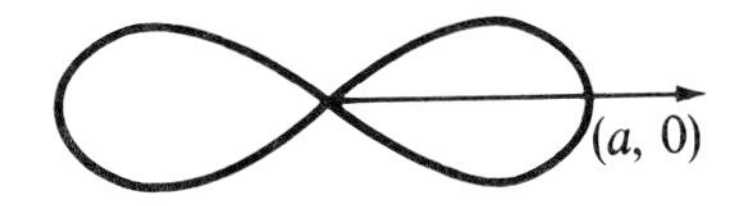

11.

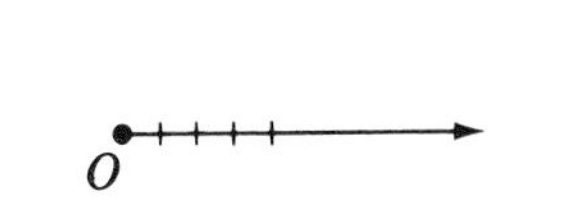

13.

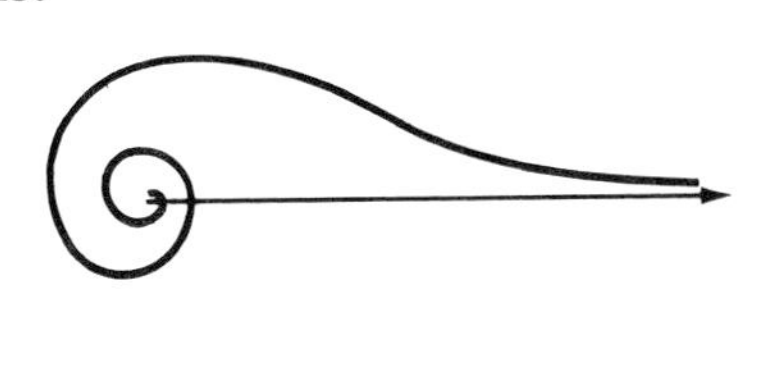

15.

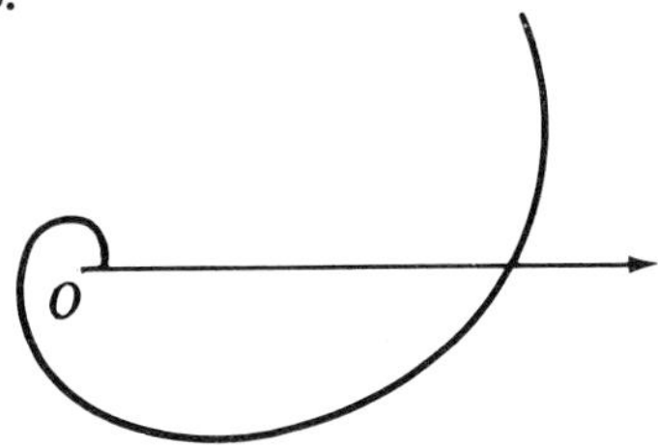

17. $r = 4$. **19.** $r = 6 \csc \theta$. **21.** $r^2 = 16 \sec 2\theta$.

23. $x^2 + y^2 - 6y = 0$. **25.** $x^2 + y^2 = a^2$. **27.** $y^2 = x^2(x^2 + y^2)$.

Review Exercises (page 296)

1. (a) $C(5, -7)$, $r = 9$. (b) $x^2 + y^2 - 10x + 14y = 0$.
(c) $x^2 + y^2 - 8x + 8y + 16 = 0$.

3. (a) Parabola; $F(16, 0)$; $V(0, 0)$.
(b) Parabola; $F(-2, 3)$; $V(-2, 1)$.
(c) Ellipse; $F(0, \pm\sqrt{7})$; $V(0, \pm 4)$.

(d) Hyperbola; $F(0, \pm 5)$; $V(0, \pm 4)$.
(e) Hyperbola; $F(\pm \sqrt{2}, 0)$; $V(\pm 1, 0)$.

5. (a) Ellipse; center $(-3, 2)$, vertices $(-6, 2)$ and $(0, 2)$, endpoints of minor axis $(-3, 0)$ and $(-3, 4)$.
(b) Hyperbola; center $(5, -4)$, vertices $(5, -2)$ and $(5, -6)$, endpoints of conjugate axis $(3, -4)$ and $(7, -4)$.
(c) Parabola; vertex $(2, -4)$, focus $(4, -4)$.

7. (a) $r \sin^2 \theta = 4 \cos \theta$. (b) $r = 3 \cos \theta - 4 \sin \theta$.
(c) $r(2 \cos \theta - 3 \sin \theta) = 8$.

CHAPTER EIGHT

Section 1 (page 304)

1. (a) $\sqrt{104}$. (b) $(3, 1, -1)$.

3. (a) $\sqrt{53}$. (b) $(-1/2, -1, 1)$.

5. $(2, 5, 1)$, $(-4, 2, 1)$, $(-4, 5, 1)$, $(2, 2, -3)$, $(-4, 5, -3)$, $(2, 2, 1)$.

7. $d(A, B)^2 + d(A, C)^2 = d(B, C)^2$; $3\sqrt{2}/2$.

9. $x^2 + y^2 + z^2 - 6x + 2y - 4z + 5 = 0$.

11. $4x^2 + 4y^2 + 4z^2 + 40x - 8z + 103 = 0$.

13. $C(-2, 1, -1)$, 2.

15. $C(4, 0, -4)$, 4.

17. $6x - 12y + 4z + 13 = 0$, a plane.

19. (a) A plane parallel to the xy-plane.
(b) A plane parallel to the xz-plane.
(c) The yz-plane.

Section 2 (page 308)

11. (a) $z = 4$. (b) $x = 6$. (c) $y = -7$.

13. $x - 3y + 11 = 0$.

15. $20x - 5y + 2z = 0$.

Section 4 (page 318)

1. Ellipsoid.

3. Hyperboloid of one sheet.

5. Hyperboloid of two sheets.

7. Paraboloid.

9. Cone.

11. Hyperbolic paraboloid.

Review Exercises (page 319)

1. (a) $\sqrt{38}$. (b) $(2, -7/2, 5/2)$.
(c) $x^2 + y^2 + z^2 + 2x + 8y - 6z + 10 = 0$.
(d) $z = 2$. (e) $y = -4$.

3. Sphere; center $C(7, -3, 4)$, radius 8.

5. The three coordinate planes.

7. A plane perpendicular to the yz-plane with y-intercept 15/4 and z-intercept -5.

9. A cylinder with rulings parallel to the x-axis.

11. Paraboloid.

13. Cone.

15. Ellipsoid.

Index